Biotransformation of Agricultural Waste and By-Products

Biotransformation of Agricultural Waste and By-Products

The Food, Feed, Fibre, Fuel (4F) Economy

Edited by

Palmiro Poltronieri

CNR-ISPA, Lecce, Italy

Oscar Fernando D'Urso

Bioesplora srl, San Michele Salentino, Brindisi, Italy

ELSEVIER

AMSTERDAM • BOSTON • HEIDELBERG • LONDON • NEW YORK • OXFORD
PARIS • SAN DIEGO • SAN FRANCISCO • SINGAPORE • SYDNEY • TOKYO

Elsevier
Radarweg 29, PO Box 211, 1000 AE Amsterdam, Netherlands
The Boulevard, Langford Lane, Kidlington, Oxford OX5 1GB, UK
50 Hampshire Street, 5th Floor, Cambridge, MA 02139, USA

British Library Cataloguing-in-Publication Data
A catalogue record for this book is available from the British Library

Library of Congress Cataloging-in-Publication Data
A catalog record for this book is available from the Library of Congress

ISBN: 978-0-12-803622-8

Publisher: John Fedor
Acquisition Editor: Kostas Marinakis
Editorial Project Manager: Sarah Jane Watson
Production Project Manager: Paul Prasad Chandramohan
Designer: Greg Harris

Typeset by TNQ Books and Journals

Contents

List of Figures .... *xiii*
List of Tables .... *xvii*
List of Abbreviations .... *xix*
About the Editors .... *xxiii*
About the Contributors .... *xxv*
Introduction .... *xxxv*

Chapter 1: Crop and Plant Biomass as Valuable Material for BBB. Alternatives for Valorization of Green Wastes **1**

T. Dietrich, M.C. Villaran Velasco, P.J. Echeverría, B. Pop and A. Rusu

Abbreviations 1
1.1 Introduction 1
1.2 Plant Biomass as Feedstock for Industrial Applications: General Considerations 3
1.2.1 Temporal Availability and Compositions of Biomass 5
1.2.2 Logistic Requirements of Biomass 6
1.3 Plant Biomass: A Sustainable Feedstock for Industrial Applications 8
1.3.1 Current Applications 8
1.3.2 Resource for Value-Added Compounds 9
1.3.3 Feedstock for the Fermentation Industry 14
1.4 Conclusion 15
Acknowledgment 16
References 16

Chapter 2: Fruit Processing in Central America and Mexico **21**

C. Velázquez, A.M. Pérez Carvajal and L.A. Prado Barragán

2.1 Introduction 21
2.2 Fruit Production in the Region 22
2.2.1 Costa Rica 22
2.2.2 Other Central American Countries 25
2.2.3 Mexico 27
2.3 Fruit Industrialization in the Region 27
2.3.1 Costa Rica 27
2.3.2 Central America 28

2.4 Physicochemical Composition and Biological Activity of Fruits of the Region.....30
2.4.1 Physicochemical Composition of Fruits of Costa Rica.....30
2.4.2 Physicochemical Composition of Fruits Produced in Mexico.....35
2.5 Biological Activities of Fruits From Costa Rica and the Central America Region.....39
2.5.1 Banana.....39
2.5.2 Mango.....39
2.5.3 Pineapple.....40
2.5.4 Watermelon.....41
2.6 Agro–Industrial Fruit Wastes and Its Potential as Biomass.....41
2.7 Conclusions.....42
References.....43

Chapter 3: Plants By-Products and Fibers' Industrial Exploitation 49

S. Narasimhan, B.S. Srikanth and P. Poltronieri

3.1 Introduction.....49
3.2 India Food Industry Wastes: Mango Case Study.....50
3.2.1 Processed Mango.....50
3.2.2 Barriers.....50
3.2.3 Opportunities.....53
3.3 Market Making in the Bioeconomy of Plants' By-Products in Europe and in the World.....53
3.3.1 Feedstock Biomasses.....55
3.3.2 Cellulose.....55
3.3.3 EU Textile Industries and Vegetal Fibers.....57
3.3.4 Innovative Companies Involved in Recycling of Plant Residues, Transforming Them in Textiles, Biobased Biochemicals, and Fiber Applications.....57
3.3.5 EU Business Analysis of Case Studies.....60
3.3.6 Cargill Starches.....63
3.3.7 Tomatoes Plant Waste Recycling Industrial Group.....63
3.3.8 Novamont Case Study on Bioplastics.....64
3.3.9 Wheatoleo.....65
3.4 Conclusion.....65
References.....66

Chapter 4: Organic Recycling of Agricultural Waste Today: Composting and Anaerobic Digestion 69

N. Mortier, F. Velghe and S. Verstichel

Abbreviations.....69
4.1 Introduction.....69
4.2 EU & American Policy on Organic Waste.....69
4.2.1 Europe.....70
4.2.2 South and Central America.....73
4.2.3 North America.....74

4.3 Industrial Composting 78
4.3.1 Principle 78
4.3.2 Input 79
4.3.3 Pretreatment Steps 80
4.3.4 Composting Process 80
4.3.5 Composting Methods 83
4.3.6 Posttreatment Steps 88
4.3.7 SWOT Analysis 88
4.3.8 Composting Within the TRANSBIO Project 89
4.4 Anaerobic Digestion 90
4.4.1 Principle 90
4.4.2 Input 90
4.4.3 Pretreatment Steps 91
4.4.4 Anaerobic Digestion Process 92
4.4.5 Digestion Methods 95
4.4.6 Posttreatment Steps 100
4.4.7 Biogas Treatment 100
4.4.8 Biogas Valorization 105
4.4.9 SWOT Analysis 106
4.4.10 Anaerobic Digestion Within the TRANSBIO Project 116
4.5 Conclusions 121
References 122

Chapter 5: Biomass Pretreatment Strategies (Technologies, Environmental Performance, Economic Considerations, Industrial Implementation)....125

C. Peral

5.1 Introduction 125
5.2 Physical and Chemical Characteristic of Lignocellulosic Biomass 126
5.2.1 Cellulose 127
5.2.2 Hemicellulose 128
5.2.3 Lignin 130
5.3 Pretreatments 131
5.4 Physical Pretreatments 131
5.4.1 Mechanical Comminution 131
5.4.2 Extrusion Pretreatment 133
5.4.3 Liquid Hot Water Pretreatment 133
5.4.4 Pyrolysis and Torrefaction 134
5.4.5 Freeze Pretreatment 135
5.4.6 Irradiation 135
5.4.7 Pulsed Electric Field Pretreatment 136
5.4.8 Ultrasonic Pretreatment 136
5.5 Chemical Pretreatments 137
5.5.1 Acid Hydrolysis 137
5.5.2 Alkaline Hydrolysis 138

5.5.3 Organosolv Pretreatment.. 139
5.5.4 Peroxide Pretreatment 140
5.5.5 Ozonolysis.. 140
5.5.6 Peroxyformic Acid 141
5.6 Physicochemical Pretreatments 141
5.6.1 Steam Explosion 141
5.6.2 Ammonia Fiber Explosion and Ammonia Pretreatment 142
5.6.3 Carbon Dioxide Explosion Pretreatment 143
5.6.4 Wet Oxidation Pretreatment.. 144
5.6.5 Ionic Liquid Pretreatment 145
5.7 Biological Pretreatment 145
5.7.1 Pretreatment With Fungal Species 146
5.7.2 Enzymatic Pretreatment 147
5.8 Combined Methods.. 147
5.8.1 Alkaline Pretreatment and Gamma Irradiation 147
5.8.2 Alkaline and Mechanical Pretreatment 147
5.8.3 Alkaline Pretreatment and Wet Oxidative Pretreatment 148
5.8.4 Combinatorial Biological Pretreatment 148
5.9 Industrial Implementation of Lignocellulosic Biomass Pretreatments.............. 148
5.10 Conclusions.. 155
References.. 155

Chapter 6: Tobacco Seed Oil for Biofuels 161

P. Poltronieri

6.1 Introduction.. 161
6.1.1 Feedstock Biomasses 163
6.1.2 Tobacco Mutagenesis and Genetic Studies 164
6.1.3 Selection of Tobacco Varieties for Biotic Stress, Pathogen, and Nematode Resistance 166
6.1.4 Bioeconomy of Biofuels 167
6.1.5 Biokerosene Deployment and Sustainability Issues 173
6.1.6 Bioeconomics of Tobacco Seeds Oil for Aircraft Fuel: Energy Tobacco Variety Solaris.. 178
6.1.7 Sustainability of Production and Exploitation of Solaris Tobacco; Protein Cake as Feed.. 181
6.2 Conclusion 185
References.. 185

Chapter 7: Fermentative Production Methods.. 189

L.A. Prado Barragán, J.J.B. Figueroa, L.V. Rodríguez Durán, C.N. Aguilar González and C. Hennigs

7.1 Introduction.. 189
7.2 Submerged Fermentation.. 189
7.2.1 Biotechnological Potential.. 190
7.2.2 Types of Bioreactors 191
7.2.3 Process Control 196
7.2.4 Microorganisms.. 200

7.2.5 Substrates 200
7.2.6 Biomolecules Production 201
7.3 Solid-State Fermentation 202
7.3.1 Biotechnological Potential 202
7.3.2 Types of Bioreactors 202
7.3.3 Process Control 206
7.3.4 Microorganisms 210
7.3.5 Substrates 211
7.3.6 Biomolecules Produced by Solid-State Fermentation 213
7.4 Solid-State Fermentation Versus Submerged Fermentation: General Considerations 213
7.5 Perspectives and Conclusions 215
References 215

Chapter 8: Composition of Plant Biomass Biotech Engineering of Cell Wall to Optimize Biofuel Production 219

F. Francocci and I.B. Reca

8.1 Introduction 219
8.2 The Plant Cell Wall: An Overview 219
8.3 Organization and Biosynthesis of Wall Components 220
8.3.1 Cellulose 220
8.3.2 Hemicellulose 221
8.3.3 Pectins 222
8.3.4 Lignin 222
8.4 Enzymes and Their Role in Cell Wall Modification (and Deconstruction) 223
8.4.1 Cell Wall Modification/Deconstruction: Learning From the Plant 223
8.4.2 Cell Wall Deconstruction/Modification: Learning From the Microbes 225
8.5 Manipulating Polysaccharides and Lignin Content or Composition to Optimize Biofuel Production 226
8.5.1 Plants Cell Wall as Source of Energy 226
8.5.2 Cell Wall Recalcitrance, the Major Obstacle to Bioconversion 227
8.5.3 Modification of the Cell Wall Structure to Improve the Saccharification Process 228
8.5.4 Lignin Modification to Improve Cellulose Accessibility 229
8.5.5 Hemicellulose and Pectin Modification to Overcome Cell Wall Recalcitrance 229
8.6 Conclusions and Outlook 230
References 231

Chapter 9: Production of Dicarboxylic Acid Platform Chemicals Using Yeasts: Focus on Succinic Acid 237

C. Pais, R. Franco-Duarte, P. Sampaio, J. Wildner, A. Carolas, D. Figueira and B.S. Ferreira

9.1 Markets and Applications 237
9.1.1 Drivers for Biobased Chemicals 237
9.1.2 Current Uses and Production of Dicarboxylic Acids 239

9.2 Selection and Improvement of Yeast Strains 243
9.2.1 Yeasts as Platform Organisms for Dicarboxylic Acid Production 244
9.2.2 Microorganisms Producing Succinic Acid: Natural Producers Versus Metabolic Engineered Strains 245
9.2.3 Succinate Formation Pathways 245
9.2.4 Metabolic Engineering Strategies for Biotechnological Production of Succinic Acid 248
9.3 Production of Biobased Dicarboxylic Acids 251
9.3.1 From Strain Isolation to Initial Physiological Studies 251
9.3.2 Linking Strain Physiology to Bioreaction Conditions 252
9.3.3 Requirements for Scale-Up Toward Commercial Operation 254
9.4 Downstream Processing of Biobased Dicarboxylic Acids 255
9.4.1 General Steps in Downstream Processing 255
9.4.2 Factors Influencing Downstream Processing 257
9.4.3 State of the Art DSP for Dicarboxylic Acids and New Trends 258
9.4.4 Designing an Effective Downstream Process 263
9.5 Challenges for Sustainable Microbial Production of Dicarboxylic Acids Platform Chemicals 264
9.6 Conclusion 265
References 265

Chapter 10: Production, Stabilization, and Uses of Enzymes From Fruit and Vegetable Byproducts 271

L.A. Prado Barragán, J.J. Buenrostro-Figueroa, C.N. Aguilar González and I. Marañon

10.1 Introduction 271
10.2 Markets for Microbial Enzymes 272
10.3 Enzymes in Technical Applications 272
10.4 Encapsulation of Enzymes and Use in the Industry 273
10.4.1 Immobilization 274
10.4.2 Encapsulation 276
10.5 Uses of Immobilized Enzymes 282
10.5.1 Laundry and Dish Applications 283
10.6 Conclusion and Perspectives 284
References 284

Chapter 11: Sustainability Considerations on the Valorization of Organic Waste 287

L. Six, F. Velghe, S. Verstichel and S. De Meester

Abbreviations 287
11.1 Introduction 287
11.2 Valorization Pathways for Organic Waste and the Waste Hierarchy 288
11.3 Sustainability 290
11.3.1 The Concept of Sustainability 290
11.3.2 Sustainability Assessment 291

11.4 Case Studies: Environmental Sustainability of Organic Waste Valorization Pathways ... 296
11.4.1 Reuse as Feed ... 296
11.4.2 Recycling by Conversion to Higher Value Chemicals ... 297
11.4.3 Recycling by Composting ... 298
11.4.4 Energy Recovery and Recycling by Anaerobic Digestion With Composting ... 299
11.4.5 Energy Recovery by Conversion to Biodiesel ... 302
11.4.6 Energy Recovery by Incineration ... 303
11.5 Conclusions ... 304
References ... 305

Chapter 12: GMO Acceptance in the World and Issues for the Overcoming of Restrictions: Cisgenesis, RNA Transfer, Rootstock to Shoot Delivery, Novel Methods of Transformation ... 309

P. Poltronieri

12.1 Introduction ... 309
12.1.1 Genetically Modified Plants and Their Regulation Process in Different Countries ... 309
12.1.2 Novel Plant Products ... 317
12.1.3 Gene Modification Using TALEN and Zinc-finger Nucleases, CRISPR/Cas9 Nuclease System ... 322
12.1.4 Cisgenesis and Intragenesis ... 324
12.1.5 Genetically Modified Rootstock Grafting ... 325
12.1.6 Other Plant Improvement Techniques ... 327
12.1.7 Accelerated Transformation of Woody Plants and Early Flowering ... 328
12.2 Non-GMO Methods for Plants Improvement: Mutagenesis ... 329
12.2.1 Insertional Mutagenesis ... 330
12.2.2 Chemical Mutagenesis ... 331
12.2.3 Mutagenesis by Physical Agents ... 333
12.3 Regulation Harmonization ... 333
12.4 Conclusion ... 338
References ... 338

Index ... 343

List of Figures

Figure 1.1	World production and world price indices (value 2000=100) (Database FAO/OECD Agricultural Outlook 2014–2023).	2
Figure 1.2	Asparagus waste.	3
Figure 1.3	Garlic residues.	3
Figure 1.4	Green bean by-products.	4
Figure 1.5	By-product calendar of subproducts available in the Ebro valley (Spain) (Personal communication with TRASA S.L.).	5
Figure 1.6	Transport costs of waste and by-products versus transport distance (AWARENET).	7
Figure 2.1	Central American major crops produced in the period 2009–2013. Sugar cane data reported against secondary axis.	26
Figure 2.2	Flow diagram of the preparation of bananas to be used as carbon source in lactic acid fermentation process.	42
Figure 4.1	Organic matter transformation into humus, biomass, and minerals.	78
Figure 4.2	Example of composting process (AP Business Technology Consultancy).	81
Figure 4.3	Typical temperature and acidity profile during composting.	82
Figure 4.4	Open air windrow system.	85
Figure 4.5	A fully automated turning machine.	86
Figure 4.6	Schematic overview of turning of table composting piles.	86
Figure 4.7	Drum composting reactor (DTEnvironmental).	87
Figure 4.8	The four steps in the anaerobic digestion process.	92
Figure 4.9	Evolution of biogas production on byproducts before bioproduct extraction.	118
Figure 4.10	Evolution of the biogas production of sweet corn before and after bioproduct extraction.	119
Figure 4.11	Evolution of the biogas production of potato waste before and after bioproduct extraction.	120
Figure 5.1	Structure of cellulose molecule.	127
Figure 5.2	Hydrogen bonds between cellulose polymer chains.	128
Figure 5.3	Structure of hemicellulose.	129
Figure 5.4	*p*-coumaryl-, coniferyl-, and sinapyl alcohol, building blocks of lignin.	130
Figure 6.1	Feedstock biomass options for production of ethanol, fuels, and valuable chemicals.	164
Figure 6.2	Prospects of transformation of municipal solid wastes in Europe into ethanol and fuels.	171

Figure 6.3	Scheme of production of biofuels and chemicals starting from municipal solid wastes.	172
Figure 6.4	Diagram of CO_2 emissions between 2005 and 2050. The graphic shows the evolution of GHG emission under two contrasting economic measures and policies: no action, and carbon neutral growth: by 2050 a net aviation carbon emission trajectory will reach a −50% reduction.	174
Figure 6.5	Comparison of Solaris variety (left) inflorescence size with that of control tobacco (right). The differences in size of inflorescences and seed content establish Solaris as a variety producing a high quantity of tobacco seed oil.	182
Figure 6.6	Solaris plants from genetics to market (G2M) flow. From research to agronomy studies, to process management, added value services, to market.	183
Figure 6.7	Tobacco seed oil from Solaris plants. Supply chain overview and flow chart of processes to obtain biofuel.	184
Figure 7.1	Principle of a batch stirred tank reactor.	191
Figure 7.2	Principle of a continuous stirred tank reactor.	192
Figure 7.3	Principle of a fed-batch stirred tank reactor.	193
Figure 7.4	Principle of a bubble column reactor.	194
Figure 7.5	Principle of an airlift bioreactor.	195
Figure 7.6	Principle of a fluidized bed bioreactor.	196
Figure 7.7	Principle of a trickle bed bioreactor.	197
Figure 7.8	Microbial growth dynamics.	198
Figure 7.9	Column reactor.	204
Figure 7.10	Tray reactor.	205
Figure 7.11	Rotary drum reactor.	206
Figure 7.12	Continuous screw reactor.	207
Figure 9.1	Main succinate production metabolic pathways in *Saccharomyces cerevisiae:* (A) oxidative branch of TCA cycle; (B) reductive branch of TCA cycle; (C) glyoxylate pathway. Not all enzymatic steps are shown.	246
Figure 9.2	Schematic principle and processing steps during (A) reactive extraction and (B) back extraction system with displacer as back extraction compound.	260
Figure 9.3	Schematic principle of electrodialysis.	262
Figure 10.1	Main methods of enzyme immobilization: (A) electrostatic adsorption, (B) covalent binding, (C) cross-linking, (D) entrapment, and (E) encapsulation.	275
Figure 10.2	Schematic diagram of a spray drying encapsulation process.	278
Figure 10.3	Schematic diagram of a bottom spray fluid bed coating process (Wurster process) to obtain enzymes with a double layer coating, changing the shell formulation atomized at the appropriated time.	279
Figure 10.4	Schema of a process combining spray drying and a fluid bed.	280
Figure 10.5	Schematic diagram of a top spray fluid bed agglomerating process.	280
Figure 10.6	Classification of methods to make and solidify drops.	281
Figure 11.1	The waste hierarchy.	289
Figure 11.2	Life cycle phases of a product (*dotted line* = recycling).	291
Figure 11.3	Life cycle assessment phases and their interactions.	292

Figure 11.4	Relationship between life cycle inventory, midpoint indicators, endpoint indicators, and single endpoint score in ReCiPe 2008 (from left to right).	295
Figure 11.5	Climate change impact of food waste valorization routes: dry feed production, wet feed production, composting, and landfilling.	297
Figure 11.6	Environmental impact of different composting technologies.	299
Figure 11.7	Landfilling everything or separating for 65% composting and 35% landfilling.	299
Figure 11.8	Climate change impact of composting, anaerobic digestion, and incineration technologies (over 100 years).	300
Figure 11.9	Endpoint impact of composting, anaerobic digestion, and incineration technologies (Eco-indicator 95+).	301
Figure 11.10	Environmental impact of municipal waste treatment with increasing diversion of food waste from incineration to anaerobic digestion with composting and small-scale composting.	301
Figure 11.11	Life cycle climate change impact of fuels derived from food waste, agricultural products, and fossil resources (per MJ energy).	303
Figure 11.12	Climate change impact ranges of electricity generation by renewable and nonrenewable sources.	303
Figure 11.13	Climate change impact ranges of heat production by renewable and nonrenewable sources.	304

List of Tables

Table 1.1	Vegetable Waste Amount in 2012.	4
Table 1.2	Valuable Compounds From Fruit and Vegetable Biomass.	12
Table 1.3	Currently Used Carbohydrate Carbon Sources in Industrial Fermentation.	15
Table 2.1	Main Seven Exported Fruits Form Costa Rica for the Year 2014, Expressed in Thousands of US Dollars and in Tons.	23
Table 2.2	Agricultural Production in Tons for Central American Countries in the Year 2013.	26
Table 2.3	Physicochemical Composition of Cultivated and Exported Fruits From Costa Rica and Central America.	31
Table 2.4	Vitamin C Content (mg/100 g) of the Main Cultivated Fruits in Costa Rica and Central America.	32
Table 2.5	Phenolic Compounds Content (mg GAE/100 g) of the Main Cultivated Fruits.	34
Table 2.6	Antioxidant Capacity Assessed by ORAC Method (TE Micromol/g) of the Traditional Tropical Fruits.	35
Table 2.7	Composition of Green Coffee of Varieties Arabica and Robusta, Produced in Mexico.	36
Table 2.8	Physicochemical Characterization of Pomegranate Seeds and Husk in g/100 g of Fruit Produced in Mexico.	38
Table 3.1	Potential of Mango Waste Components.	51
Table 4.1	Recycling Targets for Biodegradable Waste as Prescribed by the Landfill Directive.	71
Table 4.2	Main Characteristics of Some Biodegradable Materials.	79
Table 4.3	Duration of Some Industrial Composting Processes (Large-scale Compost; Deyerling & Fuchs; Productie Van Compost; Offaly).	82
Table 4.4	Levels of Rottegrad Based on the Potential Heating Ability of Compost.	83
Table 4.5	Overview Baumuster Composting Categories.	84
Table 4.6	Theoretical Amounts of Methane in Biogas.	93
Table 4.7	Analysis Results of Biogas Potential Tests on Byproducts Before Bioproduct Extraction.	117
Table 4.8	Analysis Results of Sweet Corn Products.	118
Table 4.9	Analysis Results of Potato Products.	119
Table 5.1	Effect of Various Pretreatments Methods on Composition of Lignocellulosic Biomass.	132
Table 5.2	List of Main Biorefinery Facilities, Status and Pretreatment Strategies.	150
Table 6.1	Content in Fatty Acids of Some Tobacco Varieties Selected.	180

Table 6.2	Characteristics of the Tobacco Oil Obtained by Pressing the Seed and Filtering.	181
Table 7.1	Main Characteristics of SSF Bioreactors.	203
Table 7.2	Differences Between Solid-State (SSF) and Submerged Liquid Fermentation (SmF).	214
Table 8.1	Worldwide Production of Bioethanol.	226
Table 9.1	Demonstration-Scale Facilities for the Production of Biobased Succinic Acid.	240
Table 9.2	Main Metabolic Engineering Strategies to Improve Succinate Production in Different Yeast Species.	249
Table 9.3	General Steps in Downstream Processing, Highlighting Common Principles and Unit Operations as well as the Respective Desired Outcome of Each Step.	256
Table 11.1	Impact Assessment Methods Used in the International Life Cycle Data System Midpoint Method.	293
Table 12.1	List of Genetically Modified Crops Cultivated Around the World.	311
Table 12.2	Genetically Modified Crops Approved in European Union.	312

List of Abbreviations

A4A	Airlines for America
AFEX	Ammonia fiber explosion
AG	Agronomic trait
AHAS	Acetohydroxy acid synthetase
ALH	Acetolactate synthetase
AM	Association mapping
AMIMCl	1-allyl-3-methylimidazonium chloride
APHIS	Animal and Plant Health Inspection Service
ASTM	American Society for Testing and Materials
ATP	Adenine trinucleotide phosphate
BDO	1,4-butanediol
BMIMCl	1-butyl-3-methylimidazonium chloride
Bt	Bacillus thuringensis
BTG	Biomass Technology Group
C4H	Cinnamate 4-hydroxylase
CAGR	Compound annual growth rate
CCoAOMT	CoA O-methyltransferase
CCR	Cinnamoyl CoA reductase
CDE	Carbon dioxide explosion
CesA	Cellulose synthase A
CESA	Cellulose synthase catalytic subunit
cgRNA	Chimeric guide RNAs
CHP	Combined heat and power
CRISPR	Clustered Regularly Interspaced Short Palindromic Repeats
DM	Dry Matter
DME	Dimethyl ether
DP	Degree of polymerization
DSBs	DNA double-strand breaks
DSP	Downstream processing
EBFPI	European Biofuels FlightPath Initiative
EBTP	European Biofuel Technology Platform
ECHA	European Chemical Agency
EDTA	Ethylene diamine tetracetic acid
EFSA	European Food Safety Agency
E-LCA	Environmental life cycle assessment
ELISA	Enzyme-linked immunosorbent assays
EMS	Ethyl Methane Sulphonate
ENI	Ente Nazionale Idrocarburi

EPA	Environmental Protection Agency
EPSPS	5-enolpyruvylshikimate-3-phosphate synthase
ETI	Effector triggered immunity
FAA	Federal Aviation Administration
FDA	Food and Drug Administration
FLT	FLOWERING LOCUS T
FR	Fungus resistance
FSK	Fully Synthetic Kerosene
FT	Fischer Tropsch
FVW	Fruit and vegetable waste
GHG	Greenhouse-Gas-Emissions
GMO	Genetically modified organisms
HDR	Homology-directed repair
HEFA	Hydrotreated Esters and Fatty Acids
HFTL	Heavy Fraction FT Liquids
HMF	5-hydroxymethylfurfural
HPLC	High performance liquid chromatography
HR	Homologous recombination
HR	Hypersensitive response
HRJ/HEFA	Hydroprocessed oils and fats
HT	Herbicide tolerance
HT-IR	Herbicide tolerance and insect resistance
IEA	International Energy Agency
IL	Ionic liquids
ILUC	Indirect land use change
IR	Insect resistance
KGD	α-ketoglutarate dehydrogenase
LCA	Life Cycle Analysis
LCC	Life cycle costing
LFTL	Light Fraction FT Liquids
LM	Linkage mapping
LTR	Long terminal repeat
MASBI	Midwest Aviation Sustainable Biofuels Initiative
MCC	Microcrystalline cellulose
MFC	Microfibrillar cellulose
MFTL	Medium Fraction FT Liquids
MiR	MicroRNA
MLG	Mixed linkage glucans
MS	Male sterility
MSW	Municipal solid waste
NADH	Nicotinamide adenine dinucleotide hydrate
NBGT	Nuclease-Based Gene Targeting
nDart1	Nonautonomous DNA-based active rice transposon1
NGS	Next-Generation Sequencing
NHEJ	Non-homologous end joining
NPP	Novel Plant Products
ODM	Oligonucleotide-directed mutagenesis
OMT	O-methyl transferase
OR	Origins of replication
PAL	Phenylalanine ammonia lyase
PAMP	Pathogen Associated Molecular Pattern

PBS	Polybutylene succinate
PEF	Pulsed-electric-field pre-treatment
PEG	Poly ethylene glycol
PGPR	Plant Growth Promoting
PH	Plant height
PHB	Polyhydroxybutirate
PLA	Poly lactic acid
PLL	Poly-L-Lysine
PMEIs	Pectin methylesterase inhibitors
PPO	Polyphenol oxidase
PQ	Product quality
PVP	Poly vinyl pyrrolidone
QTL	Quantitative trait loci
RB	Reverse breeding
RED	Renewable Energy Directives
REDD	Reducing Emissions from Deforestation and Forest Degradation
RG-II	Rhamnogalacturonan II
RNAi	RNA interference
RSB	Roundtable and Sustainable Materials
SA	Succinic acid
SAA	South african airlines
SAR	Systemic acquired resistance
SFE	Supercritical fluid extraction
sgRNA	Single-guide RNA
SIP	Synthesized iso-paraffine
SiRNAs	Small interfering RNAs
SKA	Synthetic paraffinic kerosene with aromatics
SkyNRG	Sky Energy
S-LCA	Social life cycle assessment
SmF	Submerged fermantation
SOD	Superoxide dismutase
SPK	Synthetic paraffinic kerosene
SPT	Seed Production Technology
SSF	Solid state fermentation
SWOT	Strength, weaknesses, opportunities, threats
TALENs	Transcription activator-like effector nucleases
TCA	Tricarboxylic acid
T-DNA	Transfer DNA
THF	Tetrahydrofuran
TILLING	Targeting Induced Local Lesions in Genomes
TSO	Tobacco seed oil
UA	United Airlines
USP	Ultrasonic pre-treatment
VBS	Vector Backbone Sequences
VR	Virus resistance
WEP	Wet oxidation pre-treatment
WWTP	Waste water treatment plant
XTH	Xyloglucan transglycosylase/hydrolase
XyG	Xyloglucans
ZFNs	Zinc finger nucleases

About the Editors

Palmiro Poltronieri is researcher at the Agri-Food Department of the National Research Council, Italy. He holds a PhD in Molecular and Cellular Biology and Pathology, obtained from Verona University in 1995. In 1996–1997, he was Japanese Society for Promotion of Science postdoctoral fellow at University of Tsukuba, Japan. Since 1999, as a researcher at the Institute of Sciences of Food Productions, he studied plant protease inhibitors and their applications, and plant signaling in response to abiotic stresses in legumes. He is cofounder of Biotecgen Company, involved in several European projects, and has supported as supervisor the biotech company, Bioesplora. He is associate editor to BMC Research Notes and editor-in-chief of the journal Challenges, MDPI, Basel, Switzerland.

Dr. Oscar Fernando D'Urso obtained a BSc in Biological Science in 2005 from the University of Salento and a PhD in Physiology in 2010. For 6 years, he has been the head of the laboratory at Biotecgen in Lecce, Italy. He founded a small company, Bioesplora, San Michele Salentino, Italy, in 2011, working in the field of novel diagnostic kits. Bioesplora counts on two divisions: biotechnology and engineering. Bioesplora's core business relies on molecular biology services, and in the development and commercialization of microelectronics and electronics devices. Bioesplora has a strong expertise in the development of microarray and real-time-based protocol in clinical and environmental fields.

About the Contributors

Cristóbal Noé Aguilar González is Dean of the School of Chemistry at Universidad Autónoma de Coahuila, México. He is Chemist (1992) by the Universidad Autónoma de Coahuila (Mexico); his MSc Program in Food Science and Biotechnology (1995) was held in the Universidad Autónoma de Chihuahua, México. His PhD program in Fermentation Biotechnology (2000) was developed in the Universidad Autónoma Metropolitana, Mexico. He was working on a postdoc in Biotechnology and Molecular Microbiology (2001) in the IRD, Marseille, France. Dr. Aguilar has published more than 130 papers published in indexed journals, more than 35 articles in Mexican journals; 16 book chapters, 8 Mexican books, 4 editions of international books, 34 proceedings, and more than 250 contributions in scientific meetings. Dr. Aguilar is a member, Level III of SNI (Mexican National System of Researchers) from 1998, receiving several prizes and awards, among which the most important are: National Prize of Research 2010 of the Mexican Academy of Sciences, the prize "Carlos Casas Campillo" (2008) of the Mexican Society of Biotechnology and Bioengineering, National Agro-Bio Prize (2005), and the Mexican Prize in Food Science and Technology. Consejo Nacional de Ciencia y Tecnologia (CONACYT)-Coca Cola México, 2003. Prof. Aguilar has developed more than 21 research projects, including 4 international exchange projects Evaluation-orientation de la COopération Scientifique Nord (ECOS-Nord), alpha network European Project; Marie Curie European project for mobility of researchers. He has been advisor of 13 PhD theses, 19 MSc theses, and 40 BSc theses. He is now a regular member of the Mexican Academy of Science (2014).

Lilia Arely Prado Barragán is full-time Professor at the Universidad Autónoma Metropolitana-Iztapalapa (UAM-I). She graduated in Food Engineering at the Universidad Autónoma Metropolitana–I (UAM-I) (Mexico) and performed technical studies on flavor release, Master's degree and PhD in Biochemistry and Food Science at the University of Nottingham (England), performed a research program (2003) at the University of Victoria (British Columbia, Victoria Canada) and in the Biotechnology and Molecular Microbiology Laboratory in the Institute for the Research and Development, IRD, Marseille, France (2004). She published more than 45 papers in indexed journals and 8 book chapters, and edited 2 national books, 15 proceedings, and more than 90 contributions in scientific meetings; she is co-author of 3 patents on the production of biomolecules by solid state fermentation processes.

She is an expert in fermentation processes for the valorization of bioactive molecules from agroindustrial by-products. She has experience in isolation of native strains with biotechnological potential and has developed solid state fermentation processes for the production, isolation, and characterization of biomolecules with enzymatic, antioxidant, antimicrobial, and hypoglycemic activities. Dr. Prado has been awarded the "Sergio Sanchez-Esquivel" prize by the Mexican Society of Biotechnology and Bioengineering (2009), has been recognized for the desirable profile for full-time teachers by the Ministry of Education Programa Mejoramiento del Profesorado (PROMEP), and belongs to the National System of Researchers, Level II (SNI). She is member of the Commission on Graduate Biotechnology, a member of the Mexican Society of Biotechnology and Bioengineering; she belongs to the Research Group of Bioprocess, Fermentations and Natural Products; and she is the Coordinator of the Editorial Committee of Biological Science Faculty (UAM-I) for the development of textbooks and teaching materials.

José Juan Buenostro Figueroa is a food engineer. He obtained Master's and doctoral (2008) degrees in Food Science and Technology by the Universidad Autónoma de Coahuila and accomplished postdoctoral research at the Biotechnology Department of the Universidad Autonoma Metropolitana, Iztapalapa. He is coauthor of 2 bioprocess patents, 16 indexed papers, and 28 contributions in scientific events, a member of Mexican Society of Biotechnology and Bioengineering, and belongs to the Research Group of Bioprocess, Fermentations and Natural Products.

Ana Carolas is lab leader and fermentation scientist at Biotrend. She holds an MSc in Biological Engineering (University of Lisbon, Portugal) and has performed research at NIZO Food Research BV, Ede, The Netherlands. Ana oversees all fermentation trials at Biotrend and has an impressive experience of having personally run hundreds of fermentation trials in controlled bioreactors with diversified biological systems, including yeasts, bacteria, and filamentous fungi.

Steven De Meester holds a Master's in Bioscience Engineering with a specialization in Environmental Technology at Ghent University. After his Master's, he worked with a PhD fellowship at Ghent University on the use and development of the Life Cycle Analysis (LCA) methodology in the biorefinery sector, and has now a PhD degree. He is author or coauthor of several scientific articles on environmental sustainability assessment. Within Organic Waste Systems (OWS), he worked as Sustainability Assessment Services project manager and is currently working at Environmental Organic Chemistry and Technology (EnVOC) group at Ghent University, Belgium.

Thomas Dietrich is an engineer in biotechnology, and project manager and researcher at Tecnalia Research & Innovation (Spain) in the area of bioprocesses and food ingredients. He participated in several European-funded research projects and coordinated the biomass valorization projects HELICAS and TRANSBIO. After receiving his engineering degree from Technical University of Berlin (Germany) in 2000, he started his career as project manager in ttz Bremerhaven (Germany). In November 2004, he took over the responsibility for the bioprocess engineering group at the food department of ttz. In the end of 2011, he moved to the applied research center LEIA CDT (Spain), mainly working in projects related to food ingredients and preservation, as well as utilization of microalgae for different approaches. Additionally, he coordinated the European project AIRTV, aiming at environmental technology verification of air emission abatement technologies, and was involved in EU projects: FERMATEC, continuous bioreactor for bioethanol; CRUSTAMEL, preventing browning (melanosis) of shrimps; FISHNOSE, electronic nose to detect quality of smoked fish, SO2SAY, prevention of polyphenol oxidase-induced browning in fruits and vegetables; and BreadGuard. Since 2011, he is part of Tecnalia, working in biomass valorization as well as new sensor systems for food applications.

Pedro J. Echeverría is an agronomist engineer; his degree was obtained at U.P.N.A University in Navarra, Spain. He is Occupational Safety and Health Senior Technician, and Master's in Environmental Management Engineering. He has more than 15 years of experience in the agro-alimentary sector, basically devoted to the technical, environmental, and the R&D areas. Between 1999 and 2007, he worked in different projects related to enterprise management, always within the scope of CONSEBRO (enterprise association of agro-alimentary industries that has a unit of services of enterprise management), and since 2008, he is Technical Director of the company TRASA, Tratamiento de Subproductos Agroalimentarios, San Adriàn, Navarra, Spain. The aim of the company is the treatment and valorization of subproducts from the agro-food sector. During these last years, he participated in several R&D projects in collaboration with other companies, technological centers, and universities related to the remainders/agro-alimentary by-products and their possibilities of valorization (BISOSTAD, CARBIO, VALBIO, VALSOST^{+}).

Bruno Sommer Ferreira is co-founder and CEO of the BIOTREND company, located in Cantanhede, Portugal, specializing in bioprocess development, optimization, and scale-up, and he is co-founder of Silicolife, a leading computational biology company. He was visiting scientist at the Institute of Biomaterials and Biomedical Engineering, University of Toronto, Canada, and researcher at the Center for Biological and Chemical Engineering, Technical University of Lisbon, Portugal. Bruno holds a PhD in Biotechnology, University of Lisbon, and an MSc in Chemical Engineering, Instituto Superior Técnico, Portugal. Bruno has been Vice-President of the Portuguese Bioindustries Association. He is a member of the expert panel of ERA-Net Industrial Biotechnology and has contributed to advisory groups for the Portuguese Government, the European Commission, and the OECD on the topic of the bioeconomy. He advises public and private initiatives aiming to bridge the gap between biotechnology research and commercial applications.

Diogo Figueira is downstream processing scientist at Biotrend, specializing in the development of process operations for the recovery of biomolecules from fermentation products, both at laboratory and at pilot scale. He holds an MSc in Biotechnology (University of Aveiro, Portugal) and has past research experience in the brewery industry.

Ricardo Franco-Duarte is a researcher from Center of Molecular and Environmental Biology, in University of Minho, Portugal. He holds a Master's in Molecular Genetics (2008), and a PhD in Biology (2014). His research interests are focused mainly in yeast diversity and genomics, molecular biotechnology, computational science, bioinformatics, metabolomics, and phenomics. He participated in several national and international projects, such as INNOYEAST, WineMetrics, PhenoMet, and Transbio, to name a few.

Fedra Francocci holds a PhD and MD in Industrial and Environmental Biotechnologies, obtained at Sapienza University of Rome in 2006 and in 2010. She performed a PhD in Botanical Sciences and Plant Biology, a collaborative thesis under "La Sapienza" University in Rome and the Universidad Autónoma de Madrid, in which she studied the role of the cell wall in plant development and defense against microbial pathogens. Her research interests are pectin and cell wall recalcitrance to enzymatic hydrolysis during saccharification, and processes to facilitate higher fermentative biomasses for biofuel production. She collaborated with the Royal Holloway, University of London, to improve sugar release from plant biomass by pretreating the plant material with a white rot fungus. She has several high impact publications and is inventor the international patent EP2204451A1, "Use of plants with reduced levels of de-esterified homogalacturonan in the cell wall or portions thereof for improving the saccharification of plant biomasses."

Caroline Hennigs, Engineer in Biotechnology, has been working for Naturstoff-Technik GmbH in Osterholz-Scharmbeck, Germany, since 2015, where she is responsible for the business development. After receiving her engineering degree from Hochschule für Angewandte Wissenschaften Hamburg (Germany) in 1998, she started her career in the "cosmed" department of Beiersdorf AG in Hamburg, Germany. In 2001, she became project manager in ttz Bremerhaven (Germany). In 2005, she took over the responsibility for the natural resources group at the food department of ttz. She has more 15 years experience in applied research in bioprocesses and food ingredients using renewable resources. She has elaborated, coordinated, and participated in more than 15 European and national research projects, as well as direct R&D contracts within the European food industry.

Izaskun Marañon is researcher at Foundation TECNALIA Research & Innovation, Spain. She has 15 years of experience in developing R&D projects in macromolecular chemistry, working with synthetic and natural polymers. Her last 12 years of experience related to food ingredients development and micro- and nanoencapsulation technologies for different applications: food ingredients, nutraceutics, enzymes, yeasts, and bacteria. Since 2002, she is active as member of the Bioencapsulation Research Group. She has five international contributions, and she is an inventor in two international patents.

Nike Mortier holds a Master's of Science in Bioscience Engineering with specialization in Environmental Technology at Ghent University. She is project manager for compostability and ecotoxicity testing and has been working at OWS for 7 years.

Srinivasan Narasimhan, recipient of the Tamil Nadu Scientist award, obtained his PhD in Chemistry from the University of Madras. Subsequently, he became associated with Prof. Herbert C. Brown, Nobel Laureate in Chemistry (1979) in Purdue University, Lafayette, Indiana, USA. He returned to India and served as Senior Scientist in IDL-Nitro Nobel Basic Research Institute, Bangalore, and as Associate Director, Spice Science Foundation. He is currently the Founder/Director of Asthagiri Herbal Research Foundation, an institution that achieved accreditation from Department of Scientific and Industrial Research (DSIR), Ministry of Science and Technology; Ayurveda, Yoga, Unani, Siddha and Homoeopathy (AYUSH), Ministry of Health and Family Welfare; and University of Madras. He works in interdisciplinary areas involving natural products, biotechnology, entomology, microbiology, and analytical chemistry. His research has resulted in the development of several crop protection formulations, health care products, and novel lead molecules. His innovative technology on the application of pheromone technology to control sugarcane borers earned him the Tamil Nadu Scientist Award and citation given by the Tamil Nadu State Council for Science and Technology. He has supervised researchers, leading to PhD degrees from the University of Madras, Sri Ramachandra Medical College and University, Bharathiar University, and Bharath University. He has so far guided 36 PhDs, and 14 more are currently pursuing research in chemistry and biotechnology. He has published more than 125 papers in national and international journals, applied for 15 patents, and conducted 3 international conferences. He has also established collaborations with Indian and foreign institutions for conducting research. A few of them are: Vel's University (Department of Biotechnology); Greenwich University, UK; Nag Research Labs, Fremont, USA. He has generated a number of projects funded by Department of Science and Technology (DST), Department of Biotechnology (BDT), AYUSH, Indian Institute of Chemical Technology (CSIR), and a few private sponsors. He has been serving as a member in the advisory committees of several institutions. He also served as the task force member of the Department of Biotechnology, Ministry of Science and Technology, India.

Célia Pais holds a PhD in Biology and a postgraduation in Plant Pathology by Imperial College, London. She is an Associate Professor at the University of Minho, Braga, Portugal, where she has been Vice-President of the School of Sciences, Head of the Biology Department, and Coordinator of the doctoral program in Molecular and Environmental Biology. She supervised numerous PhD and masters theses, and she has been involved in over 20 research projects and networks, European and national, as principal investigator or participant. Presently, she develops her research at the CBMA, University of Minho, focusing mainly on yeast biotechnology, molecular biology, and taxonomy.

Carlota Peral is a researcher at TECNALIA Research & Innovation (Spain) in the area of bioprocesses. She has 20 years of experience in developing R&D projects in the chemistry industry, the last 5 years of experience being related with food ingredients development, biomolecules extraction from agricultural by-products, and encapsulation technologies for different applications. She has participated in several European-funded research projects. Since 2011, she is part of Tecnalia, working in biomass valorization, mainly in the development of pretreatment and downstreaming operations for the recovery of biomolecules from fermentation products.

Ana Mercedes Pérez Carvajal holds a PhD in Food Science from the University of Montpellier, France, obtained in 1992. She is researcher at the National Center of Food Science and Technology (CITA) of the University of Costa Rica (UCR) since 2001, and she has the coordination of the research area. Also, she is currently the Director of the PhD Program in Sciences from the UCR. Her research interests are mainly the development of functional foods, antioxidant activity assessment, analysis of bioactive compounds, and membrane technologies applications in foods. She has published as coauthor more than 30 papers in high-impact factor journals, and she is co-inventor in three patents.

Palmiro Poltronieri is researcher at the Agri-Food Department of the National Research Council, Italy. He holds a PhD in Molecular and Cellular Biology and Pathology, obtained from Verona University in 1995. In 1996–1997, he was Japanese Society for Promotion of Science postdoctoral fellow at Tsukuba University, Japan. Since 1999, as a researcher at the Institute of Sciences of Food Productions, he studied plant protease inhibitors and their applications, and the plant signaling in response to abiotic stresses in legumes. He is cofounder of Biotecgen Company, is involved in several European projects, and has supported as supervisor the biotech company, Bioesplora. He is associate editor to BMC Research Notes, and editor-in-chief of the journal Challenges, MDPI, Basel, Switzerland.

Bianca Pop, environmental science engineer, is Senior Project Manager at Tritecc, Romania. She is founder of the company, which, apart from being a member in various project consortia since FP7, is also supporting other Romanian Small Medium Enterprises (SMEs) to have access to funding for their innovative ideas. Ms. Pop is a former scholarship holder of the German Environmental Foundation (DBU) and now a member of the scholarship awarding committee. During the 1-year scholarship, she specialized in Germany through various research projects.

Ida Barbara Reca is postdoctoral fellow at CNR-ISPA Institute in Lecce, Agri-Food Department of the National Research Council, Italy. She graduated in Biology and biotechnology from "La Sapienza" University in Rome in 2004. She holds a PhD in Plant biotechnology obtained from "La Tuscia" University, Viterbo, and a PhD in Nutrition, molecular and cellular aspects, from the "Paul Cezanne" University, Aix-Marseille. She worked from 2009 to 2013 at the Great Lake Bioenergy Research Center, Plant Research Laboratory, Michigan State University, on plan cell wall and hemicellulose biosynthesis.

Presently, she studies pathogen recognition and immunity conferred Leucine Rich Repeat (LRR) proteins in plants.

Luis Víctor Rodríguez Durán is a biochemical engineer by the Universidad Autónoma de Tamaulipas (Mexico) and holds a Master's degree in Food Science and Technology by the Universidad Autónoma de Coahuila (Mexico). He is author or coauthor of 12 scientific papers in international peer-reviewed journals, 4 books chapters, and 17 contributions to congresses, co-author of a Mexican patent on the extraction of biomolecules by solid state fermentation, and has participated in several projects funded by the National Council of Science and Technology (Mexico). His research interests include the production, purification, characterization, and application of enzymes involved on the phenolic compounds release. His research is focused on the application of solid state fermentation for the production of fungal enzymes such as tannases, feruloyl esterases, and chlorogenate hydrolases.

Alexandru Rusu (PhD) is the Leader of Strategic Research Department in Biozoon GmbH, Germany. He has extensive background in different areas of biotechnology and food research. From the combined experience in academia and industry, he also gained a strong professional know-how in managing national and international collaborative Research and Technical Development (RTD) projects. Thus, he possesses a successful track record, both as a participant and as coordinator, in managing numerous European and national RTD projects.

Paula Sampaio is a biochemist and holds a PhD in Molecular Microbiology by the University of Minho, Braga, Portugal. She is an assistant professor at the same university and a researcher at the Center of Molecular and Environmental Biology (CBMA), University of Minho. During the last decade, she has focused her research on yeast molecular biology and taxonomy, as well as on the yeast–host interactions. She supervised numerous PhD and masters theses, and has been involved in several European and national research projects and networks.

Lasse Six holds a Master's of Science in Bioscience Engineering with a specialization in Environmental Technology at Ghent University. He has been working as a SACS Sustainability Assessment Service & Compensations project manager at OWS for 2 years.

Bikshandarkoil Sivaraj Srikanth is a Management Consultant/Facilitator with 17 years experience in market and business research, strategy planning and execution (including merger and acquisitions/joint ventures), feasibility studies, operational audits, performance improvement, market development across consumer packaged goods (CPG), retail, agro-services, industrials (automotive and light engineering goods), infrastructure, and consumer services. He had the opportunity to work with a few ministries of the Government of India, state government departments like tourism, agriculture, and other government/quasi-government bodies like Confederation of Indian Industries (CII), Federation of India Chambers of commerce (FICCI), National Disaster Management Authority (NDMA), National Institute for Smart Government (NISG), etc., in numerous capacities involving policy papers, capacity building, and communication programs. His association with the educational ecosystem has been through management/business magazines publications (prepared articles/case studies), British Council/English programs in India, preparing and running small programs in marketing management, modern retailing, and strategic management with some notable B-schools in India. In the agriculture domain, he had the opportunity to work with teams in some of the leading agriculture firms, working on strategy and execution programs for "horticulture, organic seeds, fruits and vegetables, farm implements, and inputs, etc.," over the years. This involved interface with R&D, strategy, and operations teams.

Carmela Velázquez is a full professor at UCR and currently the General Director of CITA of this university since 2005. She holds a Master's Degree in Food Science from Iowa State University, United States, from 1998 and an Engineering Degree in Food Technology from the UCR. Her main research has been focused in the use of agro-industrial wastes as biomass for industrial fermentations in lines like lactic acid, ethanol, and recently, poly-2-hydroxybutyrate (PHB). She has worked with the scaling-up process from lab to pilot plant facilities with a special interest in the full use of the biomass for different industrial applications. She also has developed a large amount of new products with the local industry through extension programs developed by CITA and UCR. Se has also specialized in different models of cooperation between academia and industry, and has started a new spin-off between CITA and the Chamber of Food Industry in Costa Rica to develop a platform of cooperation for the production of value added products with scientific background. She has participated in several academic activities and represented academia in several industrial and government organizations.

Filip Velghe is a project engineer at OWS and is responsible for anaerobic digestion lab tests and follow-up of full-scale AD plants, and has 9 years experience with anaerobic digestion. He is author or coauthor of several scientific articles on anaerobic digestion. Before OWS, he worked for 2 years as coordinator and technological advisor of Biogas-E, a nonprofit organization promoting the use of anaerobic digestion in agriculture. He is Master of Science in Bioscience Engineering with specialization in Environmental Technology at Ghent University and Master of Science in Physical Land Resources with specialization in Soil Science.

Steven Verstichel holds a Master's in Bioscience Engineering with a specialization in Environmental Technology at Ghent University. He has been working for 15 years at OWS and is the head of the Biodegradability, Compostability and Ecotoxicity (BCE) department. He is the author or coauthor of several scientific articles on biodegradation and compostability. Within the company, he is responsible for the organization of European projects.

María del Carmen Villaran Velasco holds a PhD in Chemistry from the University of Basque Country (Spain). She is the Director of Bioprocesses Area in Fundacion TECNALIA Research & Innovation, a Technological Center, Miñano, Alava, in the North of Spain. She has more than 25 years' expertise in developing and management of R&D projects at national and international levels, related to the development of drying and dehydration processes for the food industry, and the development of processes for food ingredients development and stabilization. Since 2002, she is a member of the Bioencapsulation Research Group. She is inventor in two international patents. She has collaborated in more than 55 projects subsidized by the public organism at the national and international level, and more than 50 contracts with private companies. She has more than 30 contributions to congresses and 31 papers in national and international journals. She has tutored 3 PhD fellows.

Jessica Wildner graduated in Industrial Biology at the University of Applied Sciences in Bremen (Germany). She had the opportunity to work for different research organizations in Europe in the fields of food science and microbiology. Since about the past 10 years, she has performed applied research and project management at ttz Bremerhaven (Germany), an independent provider of market-driven research and development services in the areas of food, environment, and health. Her main research interests include food microbiology and biotechnology, valorization of by-products, integrated fermentation, and biorefinery concepts.

Introduction

The bioeconomy and sustainable growth are top topics, both in Europe and around the world. There is a need to exploit all the added value in the present productions and to unravel new markets and technologies for the future. An important topic in plant biotechnology showing good prospective and economic potential is the exploitation of plants with modified cell walls for improved processes and industries to be applied also to biofuel production.

This book presents an evaluation on plant species better exploitable for a particular transformation. Crops are already covering large parts of cultivable soils, so it is not conceivable to try to extend the cultures beyond the limit of available soils. A further increase in productivity is also difficult to obtain. Agriculture for food products covers the largest part of crop productions, while feeds represent the second largest market. A smaller part is devoted to production of maize for biofuel. Life cycle assessment (LCA) and CO_2 footprint assessment evaluate the production processes as sustainable or not. Transport of small distances (within 100 km), and through shipment transfer, are the most economic means to deliver the waste and by-products. Nowadays, the food industry residues are recycled by composting and by biomass incorporation in the soil to sustain farmyard manure and fertilizers in order to increase the organic matter in the soil. Composting is presently the most used method for waste treatment, the largest applicative transformation in the management of differentiated wastes. However, there is a need to digest enzymatically complex fiber-rich polysaccharides to obtain simple sugars suitable for biofermentations. Further advances in technology and plants design are supporting the exploitation and valorization of vegetable and fruit by-products through fermentation (feed-batch liquid fermentation, solid state fermentation) in biobased biochemicals and industrial enzymes production. The need to substitute the decreased reserves of petroleum with chemicals from fermentative processes and the request for an increase in crop yields pose new challenges to scientists in Europe and in the world. New production systems on an industrial scale exploit tobacco plants (seeds) for biofuel, while the residual proteins are processed as feeds for livestock. A chapter is dedicated to the biotechnology attempts to improve the efficiency of second-generation bioethanol production from lignocellulosic plant material. Fours chapters describe the activities done within the framework of the TransBio project (www.transbio.eu).

Chapter 1 describes how a suitable biomass feedstock for conversion to energy or refining to biobased products is the first step of a biobased supply chain. It is a prerequisite before establishing any conversion facility to review the four Ws in the overall approach: **W**hat type of biomass, in **W**hat quantities it is available, at **W**hat time of the year, and **W**hat properties it has. Proper logistic solutions must be implemented to establish economic feasible and environmental sustainable value-added chains using crop and plant biomass as feedstock. Afterward, valuable compounds can be extracted and/or biomass can be used as feedstock for fermentation processes.

In Chapter 2, the fruit production and processing in Central America and Mexico are discussed, considering the volume of production, the competitiveness that these crops have in the region and world, and some sociocultural differences between the countries that help to explain the different aspects of the agroindustry. Central American countries are world leaders in agricultural production in tropical fruits like banana and pineapple, and Mexico dedicates importance to agriculture also and is a world leader in products like avocado, coffee, and citrus fruits. The composition of these major tropical fruits are analyzed and compared, considering not only nutrients, but also biological activities related to major molecules of special interest.

Chapter 3 discusses food processing industries and awareness that waste and by-products offer new market opportunities. Research addresses the issue of mango wastes and by-products in India, as one of the most growing Brazil, Russia, India and China (BRIC) countries, and in Europe by taking a comprehensive approach at looking for sustainable processes to turn food processing by-products and nonfood plant fibers into valuable industrial products (fabrics and textiles, vegetal leather, elements for the automotive and building industry, cases for electronics products or household appliances, composites, bioresins for plastic materials, and lightweight structural composites based on biobased polymers, reinforced with cellulose pulp fibers). There are many additional applications for plant wastes, including the use of wood fibers to produce lower carbon, energy efficient construction materials with antiseismic properties, or the use of biobased and biodegradable agricultural mulch films for improving agricultural productivity. The bioeconomy must take into account the balance of industrial use of biomass resources for multiple uses including food, feed, chemicals, fibers, and fuels. To achieve this, the concept of smart and sustainable use of biomass should be applied. In this chapter, several types of plant by-products are discussed, and case studies of companies exploiting new production lines for textiles, lightweighted insulating components, and biobased biochemicals are presented. A Strengths, Weaknesses, Opportunities, Threats (SWOT) analysis of the most significant products is presented.

In Chapter 4, organic waste is discussed, along with the treatment options developed and installed to deal with this waste. This chapter focuses on two commonly used technologies: composting and anaerobic digestion. Landfilling and incineration are not discussed in this chapter, as they are less sustainable than composting and anaerobic digestion, and are therefore being faded out in waste management solutions.

In Chapter 5, agricultural wastes are discussed as the most abundant and renewable sources in the world. These wastes, commonly called lignocellulosic biomass, are composed mainly of cellulose, hemicellulose, and lignin. Both the cellulose and hemicellulose fractions are polymers of sugars and thereby a potential source for fermentation processes. Lignin can be used for many purposes, including production of chemicals, heat, and energy. Lignocellulosic biomass is recalcitrant to hydrolysis with cellulolytic enzymes. Therefore, a pretreatment step is necessary for the production of fermentable sugars. Various pretreatments methods, such as physical, chemical, physicochemical, biological, and a combination of them are available. Selection of the best pretreatment depends on such factors as foreseen utilization of the main biomass components, efficiency in separation, environmental concerns, compatibility with subsequent fermentation, energy requirements, and costs that must be taken into account for commercial implementation.

In Chapter 6, tobacco is described as a model plant used to study the effect of genetic interventions on biomass production, lignification, and plant growth. In addition, tobacco has been studied for its ability to produce seed oil in established and new varieties devoid of nicotine. One of these health-friendly varieties, Solaris, has been patented for the high amount of seeds and seed oil of good quality for transformation into biofuels. Feedstock biomasses and plant species are briefly introduced. The bioeconomics of biofuels, biokerosene, syngas, and different types of biofuels for aircrafts are presented. The chapter gives an overview on the economic measures and debate on biofuels, and second-generation alternative fuels in Europe and around the world. Sustainability issues in this field are presented. Finally, Solaris pilot plants in South Africa are discussed from a sustainable economy point of view. Chapter 7 summarizes the main aspects of the fermentative methods and its application to the biomolecules' production. Traditionally, most enzyme production was implemented by submerged fermentation (SmF); however, in the from 1990 up to 2015, an increasing trend toward the use of the solid state fermentation (SSF) started to figure in the enzymes and other microbial metabolites production. Title comparisons between SSF and SmF processes must be done under the same fermentation and analytical parameters, and care should be taken in comparing the same enzymatic units. There exist advantages and disadvantages in both processes; however, the best process will be dictated by means of availability in infrastructure, substrates, and strain, and also to be considered is the downstream process. Then, it shows how it is not a matter of a "better that" statement; it is rather what can be achieved with the available chain of recourses.

In Chapter 8, plant cell walls are described as the most abundant renewable source of energy of the earth. For thousands of years, their use mainly concerned animal feed and heat power, as well as large employ in the paper industry. This abundant resource is today considered a promising source of energy that can be converted into biofuels for transport and products for electricity and heat production, as well as biogas. Pathways in the biosynthesis of fibers, sugars, and metabolites are presented with the focus on the life cycle of plant species.

Cellular structures and organization of cell walls are analyzed to show polysaccharides favoring fermentative processes. Cellulose, the main component of the plant cell wall, is embedded in a matrix of other polysaccharides, and together they form the skeleton of the plant tissue. This structure is naturally recalcitrant to the deconstruction carried out by microbes and enzymes; therefore, breakthrough technologies are needed to overcome barriers to developing cost-effective processes for the conversion of biomass into fuels and chemicals. For all these reasons, the interest to deeply understand cell wall structure and composition of plants is becoming even more essential. The progress made in understanding plant cell wall biosynthesis and remodeling and the biotechnological approaches to reducing biomass recalcitrance are here described.

In Chapter 9, the biotechnological production of biobased dicarboxylic acids is introduced as a hot topic in industrial biotechnology, with many investments involved in the development, piloting, and validation at demonstration scale of diverse processes using renewable raw materials. This chapter will review the main markets and applications of some of the most important dicarboxylic acids, and will briefly present their current chemical and biotechnological production processes. Then, the chapter will mainly focus on the particular case of succinic acid, and the microbial platforms that have been proposed will be reviewed, with emphasis on yeast strains. Next, the basic requirements for setting up and scaling the bioprocess and the required purification strategy to obtain an economically feasible process yielding a product meeting the required specifications will be presented. Then, the specific challenges of using very low-cost raw materials, such as agroindustrial residues, will be highlighted.

Chapter 10 describes the problem of industrial exploitation of fermentation products, introducing the markets for microbial enzymes. Enzyme production and stabilization methods, with a focus on the techniques used, such as liposome encapsulation, are presented. Industrial applications are described, such as proteases and lipases for laundry and dish powder applications, and use of immobilized enzymes in food industry, hygiene products, cosmetics, and medical and pharmaceutical applications.

Chapter 11 discusses the development and selection of sustainable waste management solutions, as global waste production rates are expected to continue rising. The chapter focuses on environmental sustainability considerations for organic waste management, comparing the principles of the waste hierarchy to results obtained in comparative case studies following the LCA method, and reviewing the competitiveness of existing organic waste valorization systems. It can be concluded that the waste hierarchy is not sufficient for decision-making in the field of organic waste valorization, and that LCA studies are needed in order to avoid misconceptions. Furthermore, the referenced studies show that certain bio-waste-based products are capable of competing with conventional processes, but these results cannot be generalized and should be evaluated on a case-by-case basis, as seemingly small differences can drastically change environmental impacts.

Chapter 12 presents an overview on transgenesis, cisgenesis, and alternative and innovative methods of genetic modification to obtain novel plant products, as well as nongenetically modified methods based on mutagenesis. Genetically modified organism studies and genetic improvements using genome editing are discussed, as well as novel techniques to achieve new varieties and organisms for novel foods and feed applications. Finally, the various assessment regulations around the world are discussed, as well as the need for harmonization among different states. Novel plant products, in many cases, could go through a simplified authorization process due to the assimilation to traditional breeding and mutagenesis techniques.

Acknowledgment

The research carried on in the project "TransBio: Bio-Transformation of by-products from fruits and vegetables processing industry into valuable Bioproducts" has received funding from the European Union's Seventh Framework Program (FP7/2007–2013) under grant agreement n° 289603.

CHAPTER 1

Crop and Plant Biomass as Valuable Material for BBB. Alternatives for Valorization of Green Wastes

T. Dietrich, M.C. Villaran Velasco
Tecnalia, Miñano (Alava), Spain

P.J. Echeverría
Tratamiento Subproductos Agroalimentarios S.L., San Adrián (Navarra), Spain

B. Pop
TRITECC SRL, Cluj Napoca, Romania

A. Rusu
Biozoon GmbH, Bremerhaven, Germany

Abbreviations

DM Dry matter
EU27 European Union (27 member States) Belgium (BE), Denmark (DK), France (FR), Germany (DE), Greece (EL), Ireland (IE), Italy (IT), Luxembourg (LU), Netherlands (NL), Portugal (PT), Spain (ES), United Kingdom (UK), Austria (AT), Finland (FI), Sweden (SE), Cyprus (CY), Czech Republic (CZ), Estonia (EE), Hungary (HU), Latvia (LV), Lithuania (LT), Malta (MT), Poland (PL), Slovakia (SK), Slovenia (SI), Bulgaria (BG) and Romania (RO)
EU15 European Union (15 member states) Belgium (BE), Denmark (DK), France (FR), Germany (DE), Greece (EL), Ireland (IE), Italy (IT), Luxembourg (LU), Netherlands (NL), Portugal (PT), Spain (ES), United Kingdom (UK), Austria (AT), Finland (FI) and Sweden (SE)
FVW Fruit and vegetable waste
SFE Supercritical fluid extraction

1.1 Introduction

Until about 15 years ago, bulk agricultural raw materials such as wheat, rice, and corn have been continuously low in price because of increasing agricultural yields. This tendency has drastically changed with competition between biomass for food versus biomaterials, chemicals, or biofuels becoming a societal issue (Vandamme, 2009). Despite increasing world production and yields, world prices are rising. Fig. 1.1 shows the world production increase of wheat, coarse grain, and rice versus price changes.

Biotransformation of Agricultural Waste and By-Products. http://dx.doi.org/10.1016/B978-0-12-803622-8.00001-X

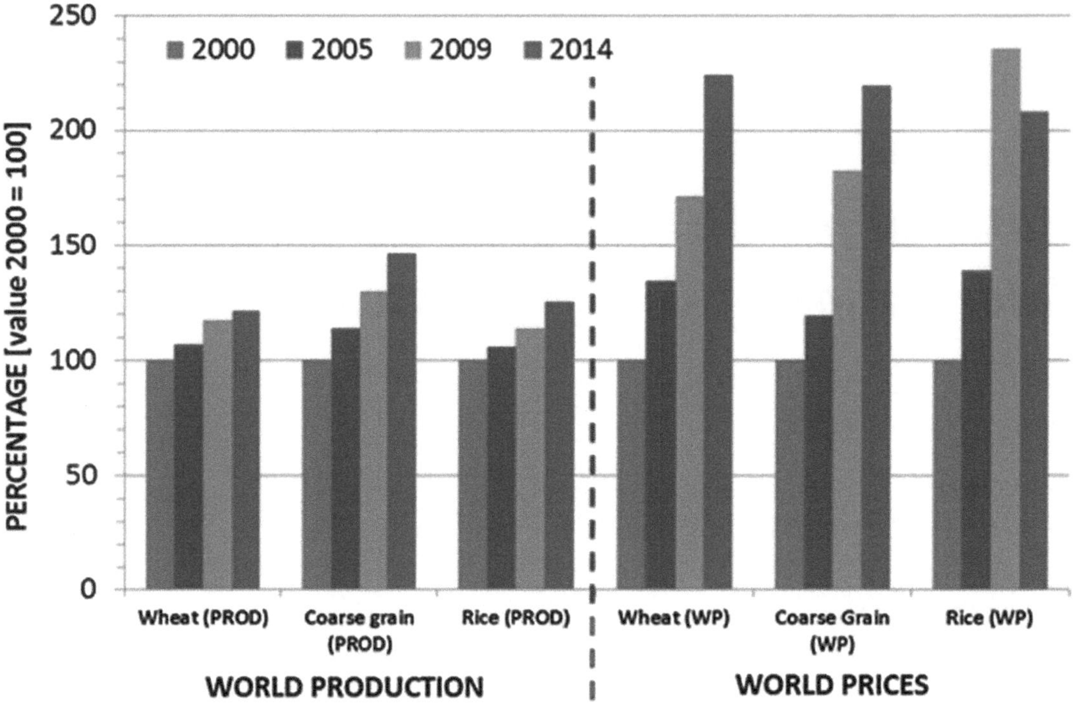

Figure 1.1
World production and world price indices (value 2000 = 100) (Database FAO/OECD Agricultural Outlook 2014–2023).

Currently, around 12% of the global land surface is used for crop cultivation (Foley et al., 2005; Ramankutty et al., 2008). In order to stay in the planetary boundaries, Rockström et al. (2009) propose that no more than 15% of the global ice-free land surface should be converted to cropland. Especially in EU27, more than 40% of available lands were utilized as agricultural area (Eurostat, 2015). Therefore, further increase of agricultural area is not an option, as transformation of forests, grasslands, wetlands, and other vegetation types to agricultural land is the driving force behind the serious reductions in biodiversity and impacts in water flows as well as biogeochemical cycling of carbon, nitrogen, phosphorous, and other important elements. Nitrogen and phosphorus are both essential elements for plant growth, so fertilizer production and application is the main concern (Rockström et al., 2009). Besides others, anthropogenic perturbation levels of land-system change and biochemical flows exceed already proposed planetary boundaries (Steffen et al., 2015). Therefore, new alternatives must be identified to ensure sufficient food supply as well as sustainable production of biomaterials, biochemicals, or biofuels. New approaches must consider the use of all available biomass resources.

1.2 Plant Biomass as Feedstock for Industrial Applications: General Considerations

Significant amounts of agricultural residues and by-products are generated from agricultural crop production and food processing such as straw, corncobs, corn stover, or cutoffs from fruit and vegetable processing (Figs. 1.2–1.4).

These residues and by-products are rich in lignocellulosic materials or other value-added compounds, which can be valorized via physical, chemical, and biotechnological treatments. Most of the crop residues are available in the countries with a large agricultural sector and high agricultural production as well as food processing industry. Scarlet et al. (2010) calculated that the total amount of crop residues produced in EU27 is estimated at 258 million dry

Figure 1.2
Asparagus waste.

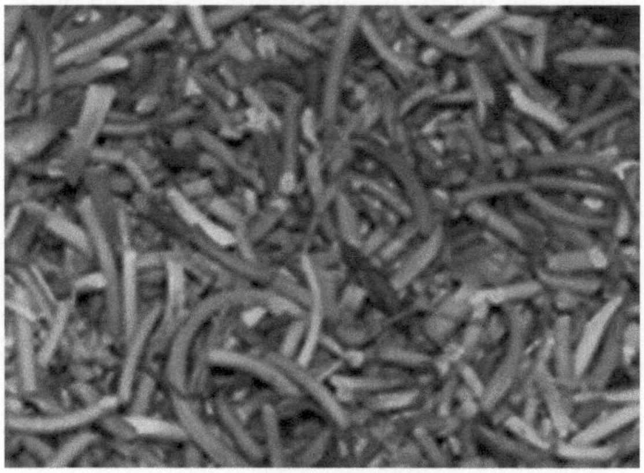

Figure 1.3
Garlic residues.

Figure 1.4
Green bean by-products.

ton per year (sunflower, wheat, rye, rapeseed, rice, maize, oats, barley). Of these residues, approximately 109 million dry ton per year could be collected and used as feedstock for other applications, if environmental and harvesting constraints are taken into account. Furthermore, fruit and vegetable waste should be considered as valuable feedstock, as FVW are produced in large quantities and are available throughout the world. According to FAO (2011), around 45 percent of fruit and vegetable biomass is wasted during agriculture, postharvest, processing, distribution, and consumption. The organized sector of fruit and vegetable processing, packaging, distribution, and consumption in India, the Philippines, China, and the United States of America generate for example a total of approximately 55 million tons of FVW. A large proportion of these wastes are dumped in landfills or rivers, causing environmental hazards (FAO, 2013). Table 1.1 is showing total vegetable waste amount in EU28 and key producing countries for agricultural production and manufacture of food products.

Nevertheless, general statistics are not sufficient in analyzing the potential of crop and plant biomass resources as feedstock for valuable applications, as the use requires accurate data on their availability by crop and biomass type, but also on their local and annual variability. The

Table 1.1: Vegetable Waste Amount in 2012 (Eurostat, 2015).

	Vegetable Waste from Agriculture, Forestry, and Fishing (t/year)	Vegetable Waste from Manufacture of Food Products, Beverages & Tobacco Products (t/year)
EU28	6,600,000	15,670,000
Italy	49,950	2,903,042
Poland	388,635	1,430,485
France	2693	1,241,609
Spain	653,604	723,994
Germany	77,765	565,365
Romania	3,192,623	346,825

estimation of the by- and co-products available as feedstock for valorization purposes provides also information on the best locations for a biorefinery plant location as well as plant size.

1.2.1 Temporal Availability and Compositions of Biomass

The selection of crop and plant residues as feedstock for valorization strategies is primarily concerned with the challenge of sustainable and sufficient year-round supplies (Parajuli et al., 2015). Therefore, a specific region must be assessed carefully. During the European funded research project TRANSBIO (grant agreement n° 289603); the Ebro Valley with fruit vegetable transforming industry was selected as a test case. A first analysis showed that the region provides around 80,000–120,000 tons of different types of by-products per year. Nevertheless, this amount was reduced to around 58,000 tons, considering only the most promising subproducts (Fig. 1.5).

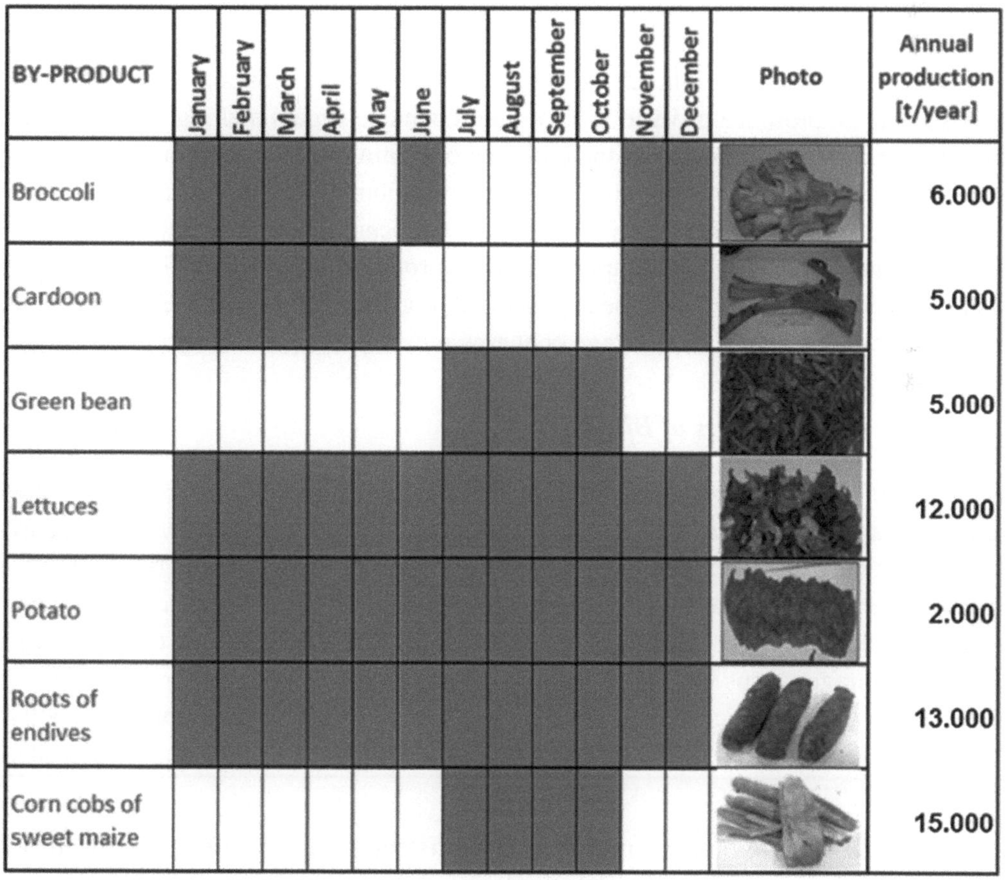

Figure 1.5
By-product calendar of subproducts available in the Ebro valley (Spain) (Personal communication with TRASA S.L.).

Furthermore, chemical compositions of different types of by-products must be considered since these compositions are the basic elements that undergo transformation or extraction processes to deliver valuable products. A key barrier is the lack of knowledge on how to deal with natural differences in plant components and characteristics from one plant to the next within the same species for feedstock considerations. Agricultural residues such as straw, corncobs, corn stover, etc., constitute an important category of potential biorefinery feedstock that is not in confrontation with food availability. These lignocellulosic materials have three basic constituents: cellulose, hemicellulose, and lignin. A typical composition of agricultural residues is cellulose, 40–50%; hemicellulose, 25–30%; and lignin, 15–20% (Ghatak, 2011); however, the components of fruit and vegetable by-products are more diverse. Potato by-product from peeling industry consist of up to 50% starch per DM, and also corncob residues from sweet corn processing have a significant starch content of up to 29% DM. In root residues of endives after hydroponic cultivation, the polymeric inulin, stored in the roots, is depolymerized into mono- and oligosaccharides. Therefore, up to 12% DM of fructose can be found in the by-products very useful for industrial fermentation processes. The cellulose content varies between 10% and 33%; hemicellulose, 5% and 33%; and lignin, 3% and 20%. One important issue represents the physical barrier that lignocelluloses possess such as complex structures, presence of various hexose and pentose sugars making fermentation process more complicated, as well as the presence of lignin or other compounds inhibiting the fermenting organisms (Galbe et al., 2007). These limitations are further related with higher energy demand, having thus the potential to increase processing costs (Parajuli et al., 2015) and must be considered by selecting appropriate by-products as feedstock for value chains. Once suitable biomass sources have been identified, logistic requirements as well as quality control, preprocessing needs, and storage concerns must be thought through.

1.2.2 Logistic Requirements of Biomass

Beside year-round availability and appropriate compositions, several other aspects related to environmental sustainability and economic feasibility should be considered. Scattered availability and bulkiness can be impediment for the collection of the waste and by-product streams. Another critical issue is the prevention of spoilage of the by-products before being processed. Residual moisture of by-products determines further processing (e.g. drying) and suitability for storage versus biological stability. Many types of waste material already contain large numbers of microbes and/or will be altered quickly through microbial activity (Russ and Schnappinger, 2007). Therefore, a properly developed storage and pretreatment plan is required.

Collection and transport of biomass resources from distant places can affect significantly the economic viability of green biomass. Delivered cost of the agricultural residues would include harvesting cost and logistic costs. Especially, the logistic costs account for as much as 90% of total costs, while harvesting costs sum up only to 10%. Therefore logistic costs play a major role in the overall profitability of a biorefinery (Ghatak, 2011). Technical literature provides a range of figures for the delivery costs of agricultural residues as biorefinery feedstocks (Stephen et al., 2010; Thorsell et al., 2004; Nilsson, 1999). The European project AWARENET calculated transport costs for waste and by-products of 7–20 € per ton (Fig. 1.6).

During the European project TRANSBIO (Biotransformation of by-products from fruit and vegetable processing industry into valuable bioproducts (Grant agreement no: 289603)), a detailed analysis regarding logistic aspects for the Ebro Valley was performed. The average distance to the collection and pretreatment facility is around 35 km, and the transport is performed by trucks with a loading of 12 tons, resulting in transport and provision costs of around 8,75 €/ton by-products. These data are based on calculation of the company TRASA, Tratamiento Subproductos Agroalimentarios S.L., and are fully in line with calculations of the EU Project AWARENET. Nevertheless, in order to establish appropriate logistic solutions, the role of different actors in the supply chain must be clarified. Although some biomass conversion facilities will have the capability to handle all aspects of biomass logistics from the farm or processing facility to the valorization and transformation facility, most biomass projects will not have the in-house expertise on all aspects. Therefore, the identification of support operations needed and who should complete those activities is a key requisite. This will have also significant impact in the business model of the final biomass transformation facility. Additionally, some biomass sources, such as fruit and vegetable by-products, must be pretreated and stabilized in order to avoid degradation. After these challenges are solved, plant biomass is available for industrial applications.

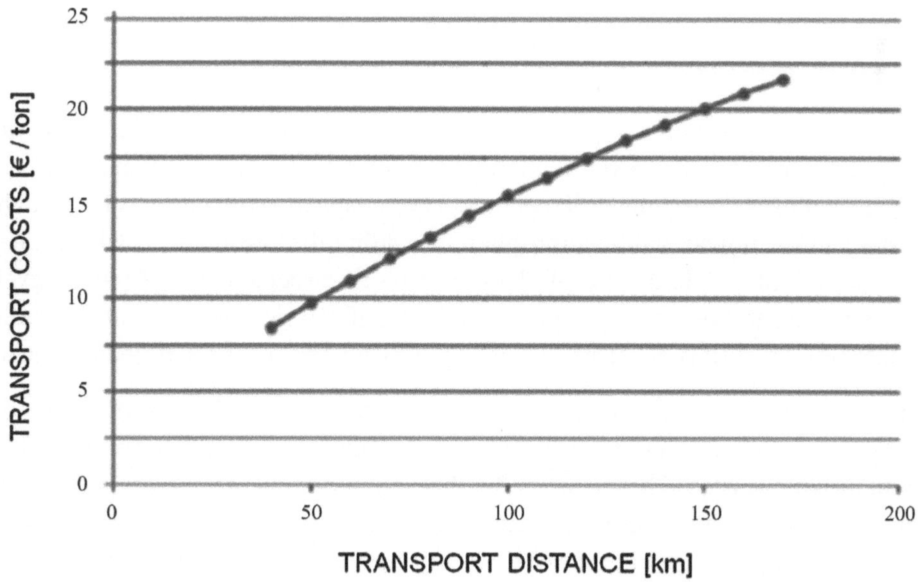

Figure 1.6
Transport costs of waste and by-products versus transport distance (AWARENET).

1.3 Plant Biomass: A Sustainable Feedstock for Industrial Applications

Biomass has been defined to be "any material, excluding fossil fuel, which was a living organism that can be used as a fuel either directly or after a conversion process" (ASTM, 1995). There is increasing concern about the impact that chemicals, fuels, and energy obtained from fossil-based feedstocks (i.e. natural gas, coal, and oil) are having on the earth's environment, particularly in view of the increasing demand from rapidly expanding industrial nations such as India and China. There are several reasons to look for alternative raw materials: reserves are finite and nonrenewable, the environmental risks related with their transport and use, especially when they are combusted to energy production, and the increasing volatility in the price of crude oil alternative, renewable feedstocks from which key chemicals, fuels, and energy sources can be isolated are required in the longer term, in order to maintain both current demands and facilitate future worldwide industrial expansion.

One renewable source of all of these materials is plant material (biomass), and there is increasing research activity surrounding the use of plant-based feedstocks in industrial processing as a source of these key commodities (NNFCC, 2007).

A complete biorefinery can be established from plant biomass, and several value products can be obtained from it (Charlton et al., 2009): extracts with food and pharmaceutical applications, platform chemicals such as organic acids from fermentation (Willke and Vorlop, 2004), biofuels (Huber et al., 2006), energy sources, and fiber residues that have applications in biocomposite materials (Fowler et al., 2006).

1.3.1 Current Applications

For years, the two most common applications of plant biomass have been the production of biomaterials and bioenergy. However, the development of oil from fossil fuel replaced many of these traditional uses during the 20th century. The prospect of oil supplies being exhausted and concern about the impact on the atmosphere of adding the carbon in fossil fuels have resulted in renewed interest in the use of plant biomass as direct sources of bioenergy, biomaterials, biochemicals, and other biocompounds (Henry, 2010) in order to create a biorefinery based on renewable materials as alternative to the conventional refinery based on petrol.

Among other resources of renewable materials, the wastes generated in agrofood sector can be a good resource of plant biomass. It is an economic important sector in Europe with high generation of organic wastes. These wastes, by their composition, have high potentiality to be used for bioactive compounds recovery or as substrate to obtain, chemicals, polymers, additives, etc., through biological processes. According to the European project AWARENET, only in fruit and vegetable processing industries in EU15, the production of subproducts achieve 192 million tons.

As a consequence of their biological instability, their potentiality pathogen nature, their high water content, their rapid autoxidation, and their high enzymatic activity, the use and disposal of these wastes is very difficult. They have commonly been used as direct feed for animals. However, their use as feed should be avoided since it may cause environmental and food safety risks.

There are other developments to use this biomass as compost. However, this application has low value for companies.

More interesting is the use of this biomass as an energy resource and as raw material to obtain bioactive compounds. Although the energy production is interesting, this application does not consider other potential applications of this biomass in the areas of food, health, or chemistry.

These last valorization routes are more innovative alternatives with high potential and interest considering the composition of this biomass. This is a less studied alternative, which requires more technical effort to its implantation.

The European Directive 1999/31/CE demands to the member states to reduce by 65% the biodegradable wastes in dumping sites before 2016 (in relation with levels of 1995). Therefore, also keeping in mind environmental reasons, it is necessary to exploit the valorization of plant biomass.

1.3.2 Resource for Value-Added Compounds

There are several studies in literature that deal with the biorefinery concept. Due to the well-known problem of land competition between the use of crops for food or feedstock applications, there are a growing number of studies related to the use of subproducts as a resource for energy and raw materials production. Both agriculture and food processing sectors are considered as relevant sectors for biorefineries (Ghatak, 2011; Kammand Kamm, 2004), especially due to the potential use of food residues, e.g. Mahro and Timm, 2007.

Besides the current uses of these residues, materials produced from the food industry have a high potential as resources of compounds of high value for the food, pharmacy, and chemical industry. Food by-products can be an important and alternative raw material to obtain molecules as antioxidants, carotenoids, sterols, etc., which can be used, depending on their activity, as ingredients in conventional food products or as functional ingredients for nutraceuticals and functional foods production (Herrero et al., 2006). For example, phenols and carotenoids from fruit by-products could be applied as natural food or beverage preservatives since they extend the shelf-life of the product (Oreopoulou and Tzia, 2007). Pectin could be utilized as a gelling agent in confectionary or as fat replacement in meat products (Galanakis et al., 2010b). Water insoluble fibers (i.e. hemicelluloses) are able to improve intestinal regulation (Rodríguez et al., 2006) and thereby are destined to supplement food products or ready meals.

Food industries produce large amount of vegetable and fruit waste with high potential for obtaining value-added compounds (Mirabella et al., 2014). Food wastes are composed of complex ingredients, which have been discharged from the original material. These wastes can be generated during the agricultural production or during postharvest handling and storage. The scientists have focused their research in looking for the functional compounds present in by-products generated in agricultural and food processing activities. These resources are abundant, but they can have some complications related to their seasonality and the requirements of stabilization to avoid losses of desired compounds or microbiological deterioration. However, an adequate application of stabilization technologies as drying and the implementation of a viable strategy of valorization can solve these challenges.

Among other plant biomass, cereals, roots and tubers, and fruits and vegetables are the most promising resources for added-value production (Galanakis, 2012). Among crops, wheat and rice are the most important, depending the geographical area. Grinding of wheat and dehulling of rice generates by-products like bran or straw, which are rich in high nutritional proteins, dietary fibers, and particularly glucuronoarabinoxylans (Hollmann and Lindhauer, 2005). Oat mill waste has been studied as resource for β-glucan extraction with advanced gelling properties (Patsioura et al., 2011).

Among roots and tubers, potato is the largest crop. The wastes of potatoes processing are peels and cull potatoes. Potato peels have been investigated for the extraction of phenols (Oreopoulou and Tzia, 2007). In the case of oil crops, sunflower and soybean seeds are used for the extraction of phytosterols (Copeland and Belcher, 2001). Other important oil crop is the olive. Olive mill pomace has been valorized to obtain bioactive polyphenols and pectin (Galanakis, 2010a; Obied et al., 2005).

However, fruit and vegetables are the most studied biomass for obtaining added value, especially antioxidants and dietary fibers. Pomace and its extracts, the most important wastes of the apple manufacturing industry, have a great potential as a resource of valuable compounds. Apple pomace is constituted of simple sugars (glucose, fructose, and sucrose) and is a rich source of carbohydrate, pectin, crude fiber, proteins, vitamins, and minerals, and as such, is a good source of nutrients, worth being recovered. It can be a good resource for pectin and fiber. Other important wastes from apple juice production are the peels. Several studies have been carried out for their valorization due their high content of phenolic compounds and antioxidants with high possibilities for their use in the food and nutraceutical industry (Huber and Rupasinghe, 2009; Henríquez et al., 2010). Ribes, raspberries, and blackberries have a high content of antioxidants, phenolic acids, flavonoids, polyphenols, and fibers. They are often used in the food industry for juice, jams, etc., and the wastes produced have been widely studied for recuperation of high value compounds (Bakowska-Barczak et al., 2009). Citrus fruits are perhaps the most studied fruits as resources of added value molecules. In contrast with other fruits, they have a small edible portion. In consequence, during their processing, high amounts of biomass is produced, as peels and seeds, etc., with high possibilities for their valorization. Residues of citrus juice production are a source of several compounds, principally water, soluble sugars, fiber, organic acids, amino acids and proteins, minerals, oils, and lipids, and they also contain flavonoids and vitamins (Fernández-López et al., 2004). Some studies have been carried out to obtain fiber, polyphenols, antioxidants, oils, etc., from these subproducts for their use as food ingredients (Fernández-López et al., 2004; Crizel de Moraes et al., 2013), compounds for resins, or soil-amendment products, among other applications (Kato-noguchi and Tanaka, 2006).

The industry of exotic fruits processing also produces high amount of wastes such as peels, seeds, and flesh. Exotic fruits are rich in bioactive compounds, phenols, carotenoids, vitamins, and fibers (Mirabella et al., 2014). Several studies have been carried out for recovering these kinds of compounds and their application as antioxidants, antimicrobials, flavoring, colorants, and texturizer additives (Ayala-Zavala et al., 2010). Among other studies, Canteri et al. (2010) studied the extraction of pectin from passion fruit industry, and Upadhyay et al. (2012) studied the extraction of bromelain from pineapple. Ajila et al. (2010) found that mango peels are a good source of bioactive compounds such as polyphenols, carotenoids, vitamins, enzymes, and dietary fibers; its extracts have also good antioxidant properties.

Morais Ribeiro da Silva et al. (2014) quantified the levels of resveratrol, coumarin, and other bioactives in pulps and by-products of 12 tropical fruits from Brazil (pineapple, acerola, mombin, cashew apple, guava, soursop, papaya, mango, passion fruit, Surinam cherry, sapodilla, and tamarind pulps, as well as their by-products: peel, pulp's leftovers, and seed) for future application in the food industry.

Not only is the fruit biomass a good resource to added value compounds production, some vegetables have molecules with high functional activity. However, with the exception of some plants such as tomatoes, the rest of the vegetables are been less studied as a raw material to obtain compounds of high value.

Tomatoes are rich in bioactive and valuable compounds such as carotenoids, mainly lycopene (80–90%) and β-carotene, in addition to vitamin C, vitamin E, and various phenolic compounds (Dumas et al., 2003). The majority of the studies published in literature focus on characterization of carotenoids, especially for lycopene and β-carotene (Riggi and Avola, 2008). However, peels can be a resource of fiber and macronutrients (proteins, ash, total available carbohydrates, and soluble sugars) (García Herrera et al., 2010). Concerning other vegetables, it is interesting to consider the studies carried out with plants of the Brassicaceae family as cauliflower and broccoli. Cauliflower is a vegetable rich in bioactive compounds and nutrients, especially fiber and protein. Broccoli is a good resource of bioactive compounds such as glucosinolates, phenolic acids, and flavonoids and nutrients (vitamin C, minerals, and trace elements) (Domínguez-Perles et al., 2010).

Lante et al. (2011) evaluated red chicory leaf residue as a natural substitute for synthetic antioxidants for the food and feed industry. Horse chestnut extracts are used in pharmacy and cosmetic applications by their content in oleane-type saponins, and their seeds have an important content in flavonoids (Kapusta et al., 2007).

The by-products of asparagus are also rich in many of the phytochemicals, such as phenols (flavonoids and hydroxycinnamic acids) and saponins, located in the edible part of the spears (Fuentes-Alventosa et al., 2013).

Peas and other subproducts from legumes production are being studied for protein and fiber production for food use and for developing of new packaging systems (Leguval project, 2014). Table 1.2 summarizes valuable compounds contained in different biomass sources.

Economically viable strategies for the revalorization of this plant biomass requires the definition of processes to obtain its complete valorization, considering the separation and obtaining of molecules of interest by different extraction technologies and to complete the total use of biomass from high to low added-value products with different applications such as pharmacy, food, chemicals or energy production, in a similar strategy than the chain planned in the TRANSBIO project (Transbio, 2012).

To recover the above-mentioned compounds from plant biomass, it is necessary to define efficient technologically and economically viable extraction technologies. The selection of the most appropriate extraction technology will be done in function of raw material, compound to be recovered, and the final application of the recovered compound. Actually, there are several technologies for added-value compounds extraction from plant biomass: solid–liquid extraction, Soxhlet extraction, ultrasound-assisted extraction (UE), microwave-assisted extraction (MAE), supercritical fluid extraction (SFE), pressurized solvent extraction (PSE), pulsed electric field extraction, and enzyme-assisted extraction (Baiano, 2014).

Table 1.2: Valuable Compounds From Fruit and Vegetable Biomass.

Plant Biomass Resource	Valuable Compounds
Apple	Antioxidants, fiber, flavonoids, polyphenols, phytochemicals
Berry	Fiber, polyphenols
Grapes	Fiber, polyphenols
Citrus fruits	Cattle feed, substrate, sugar, cellulose, flavonoids, vitamins, fiber, polyphenols, antioxidant, antimicrobials
Exotic fruits	Additive, antioxidant, antimicrobials, polyphenols, bioactive compounds (coumarin, resveratrol, bromelain, etc), pectin, substrate
Potatoes	Antioxidant, substrate
Oil crops (sunflower, soybean)	Phytosterols, albumin, glucuronoarabinoxylans
Olive	Herbicide, pesticide, various sugars, carbon sequestration, lipids
Tomatoes	Antioxidant, carotene, lycopene, carotenoids
Other vegetables	Substrates, flavonoids, bioactive compounds (glucosinolates, phenolic acids, etc), pectin, vitamins antioxidant

Solid–liquid extraction allows soluble components to be removed from solids using aqueous organic solvents, alone or mixed together. The selection of solvent will be carried out to obtain high yield and to avoid interferences with the desired compound. The disadvantage of this technology is the use of solvents, which in numerous cases are toxic and flammable, and the long times required. Solid–liquid extraction has a great number of industrial applications. For example, in herbal and other food preparation industries, it is currently applied when a vegetable matrix requests extraction for further processing. The actual and future research direction is represented by the use of cheap and nontoxic solvents (i.e. water or ethanol) to be combined with other mild extraction techniques.

The advantages of Soxhlet extraction compared with conventional solid–liquid extraction is the possibility of application for a high number of food matrixes and the possibility to carry out repeated washing of the matrix with fresh solvent, the higher possible solubilization of the analyte due to the use of hot solvent. The disadvantages of this technology are the same as mentioned for solid–liquid extraction, plus the possible degradation of temperature sensitive compounds. So, it is not a good option to recover thermosensitive compounds. The conventional Soxhlet extraction has been combined with other technologies such as microwave, ultrasound, etc., in order to reduce the operation times.

Pressurized solvent extraction is similar to Soxhlet extraction using solvent near to their supercritical region. In these high temperature conditions, the solute solubility and diffusion in solvent is higher, reducing solvent amounts and the process time.

In supercritical fluid extraction (SFE), temperature and pressure are higher than the supercritical values of solvent. Carbon dioxide is the most used supercritical fluid, sometimes modified by co-solvents such as ethanol that change its polarity. The choice of supercritical CO_2 is due to its moderate critical conditions (31.1°C and 73.8 MPa), absence of toxicity, and chemical stability. Some of the advantages of SFE are (Baiano, 2014): solvating powers similar to liquid organic solvents, high solute diffusivities, lower viscosity, lower surface tension, and the possibility of adjusting the solvating power by changing pressure or temperature. Some disadvantages are the high capital investment required and the high operational costs. Therefore, this technology is only applied to obtain products with high added value.

Ultrasound-assisted extraction has been used to extract bioactive compounds such as antioxidants and tocols, essential oils, steroids, and lipids from plants. Ultrasounds allow a greater penetration of solvent in the material, improving mass transfer and consequently, the extraction yield and kinetic. Ultrasound does not suppose a reduction of temperature and pressure in relationship with the conventional extraction process. A disadvantage of this method is that it must be developed and scaled up for each application. However, the application of ultrasounds improves energy efficiency of the equipment; it has ease of installation and/or of retrofitting and has competitive energy costs and low maintenance costs (Patist and Bates, 2008).

Microwave-assisted extraction supposes the application of electromagnetic waves during the extraction process. Microwave-assisted extraction depends on the dielectric susceptibility of both solvents and matrices, and is particularly promising for the extraction of compounds of medium to high polarity from a solid matrix (Eskilsson and Björklund, 2000). This is a good alternative to extract bioactive compounds from plants due to some reasons: reduction of both extraction time and solvent usage and improved extraction yields. It is comparable to SFE but at lower cost. However, its cost is higher than ultrasound-assisted extraction. The microwave-assisted extraction is an adequate technology at laboratory scale, but its use at industrial level is limited.

Pulsed electric field extraction allows the extraction of intracellular compounds. The application of a pulsed electric field produces the pores' formation in the cellular membrane. Generation of high voltage pulses having sufficient peak power is the limitation in processing large quantities of fluid economically.

Enzyme-assisted extraction allows the cell disruption and extraction of active compounds from plant cells. It is a promising alternative to conventional solvent-based extraction methods. It is considered an environmentally friendly extraction method (Gardossi and Halling, 2009). The use of enzymes as pectinases, cellulases, hemicellulases, etc. hydrolyze cell wall components, increasing the extraction yield of targeted compounds such as polysaccharides, oils, natural pigments, flavors, antioxidants, and medicinal compounds (Moura et al., 2008; Puri et al., 2012). Enzymes can be also used in a raw material pretreatment step in order to reduce extraction times, minimize use of solvents, and provide increased yield and quality. Some limitations of this technique are related to the enzyme costs, the shortage of enzymes to achieve the complete hydrolysis of plant cell walls, and the difficulty to scale up enzyme-assisted extraction to industrial scale.

1.3.3 Feedstock for the Fermentation Industry

The EU fermentation industry relies on agricultural carbohydrate feedstocks. Depending on the crop, sugars can originate directly from the crop itself or from starch or cellulosic material. Starch and cellulosic material require hydrolysis and enzymes to break the material down into sugars before the fermentation step. Currently, fermentation is based on processing C6 sugars ($C_6H_{12}O_6$), while the commercial processing of C5 sugars is still in early stages of development. The costs of carbohydrates represents up to 50–70 percent of the net sales value of the fermentation industry's products. Therefore, flexibility of raw material choice and access at world market prices are the key factors for the global competitiveness of the fermentation industry (CSG, 2008; Deloitte, 2014). Different types of carbohydrates are currently used in industrial fermentation processes. Table 1.3 shows some industrial examples.

Nevertheless, given the limited availability of arable and fertile land and the use of sugars for fuel, bio-based chemicals and materials using first generation fermentation technology raises

Table 1.3: Currently Used Carbohydrate Carbon Sources in Industrial Fermentation.

Carbohydrate	Product Example	Company Example
Starch	Succinic acid	Reverdia, www.reverdia.com
Glucose	Succinic acid	Myriant, www.myriant.com
	Citric acid	Jungbunzlauer www.jungbunzlauer.com
Dextrose	Succinic acid	Bioamber, www.bio-amber.com
	PHBV	TianAn biologic materials, www.tianan-enmat.com
Lactose	Lactic acid	Cellulac, http://cellulac.co.uk/en/
Sugar cane	PHB	PHB Industrial S.A., www.biocycle.com.br
Sucrose	Bioethanol	Nordzucker AG, www.nordzucker.de
Wood hydrolysate	Bioethanol	Borregaard, www.borregaard.com

concern. Taking into account that already 12 percent of global land surface is used for crop production (Foley et al., 2005; Ramankutty et al., 2008), but no more than 15 percent should be converted to cropland in order to not undermine human well-being and long-term sustainability (Rockström et al., 2009), new carbon-containing feedstocks for fermentation must be identified.

Significant research is performed, for example, agro–industrial or even agro–waste streams, as feedstock for fermentation. Cesário et al. (2014) realized a high-productivity fermentation of P(3HB), and Patel et al. (2006) produced bioethanol using wheat straw as cheap and readily available agricultural residue. Hass et al. (2015) implemented successfully a P(3HB) fermentation process using chicory roots (*Cichorium intybus*) after hydroponic cultivation as a carbon source. Pandey et al. (2001) showed that for the production of citric acid, subproducts such as pineapple waste, carrot processing waste, okara (soy residue), corncobs, or kiwi fruit peel could be used. During the European project TRANSBIO several sub-products from transforming industry (e.g. sweet corn cobs, potato peelings, banana pulp) could be successfully used to produce succinic acid, P(3HB), as well as enzymes (lipase, protease). An additional huge range of literature is available showing the potential of this kind of feedstock for the fermentative production of food ingredients, biomaterials, and platform chemicals, and to replace first-generation carbohydrate sources, ensuring sustainable growth of the fermentation industry.

1.4 Conclusion

Growing demand on biofuels and biomaterials makes it inevitable to establish proper sustainability criteria beyond environmental performance. To ensure feeding of the population as well as providing feedstocks for industries, new approaches are needed. Existing raw materials must be analyzed and considered in a sustainable manner. Nevertheless, the switch to agro–industrial by-products or waste streams requires solving the bottleneck of releasing fermentable sugars left in the lignocellulosic matrix (Zhang, 2008; Sarath et al., 2008; Vandamme, 2009) in an economic feasible way, as well as establishing appropriate logistic solutions for sustainable biomass supply chains. Merely stating that there is low-cost biomass available for conversion to a carbon source for fermentation, ingredients, or feedstock for energy production will not result in a well-developed biomass supply chain. It is a prerequisite that comprehensive planning for the biomass supply chain starts before or at the same time that the extraction or conversion technology is discussed. The large number of biomass feedstocks, possible industrial use, and conversion technologies makes almost every biomass valorization project somewhat unique, and a unique set of skills are needed for a specific biomass supply and valorization chain (Tallaksen, 2011).

Acknowledgment

The research leading to these results has received funding from the European Union's Seventh Framework Program (FP7/2007–2013) under grant agreement n° 289603.

References

Ajila, C.M., Aalami, M., Leelavathi, K., Rao, U.J.S.P., 2010. Mango peel powder: a potential source of antioxidant and dietary fiber in macaroni preparations. Innovative Food Science and Emerging Technologies 11, 219–224.

ASTM, 1995. Standard E1705, 1995. Standard Technology Relating to Biotechnology ASTM International. Westconshohocken, PA. www.astm.org.

AWARENET, 2004. Handbook for the Prevention and Minimization of Waste and Valorization of By-products in European Agro-food Industries. published by European Network - Agro-food wastes minimisation and reduction network. (Contract No. G 1 RT-CT-2000-05008; Project No. GRD 1-T2000-28033). Legal Deposit: BI 223–2004 Oficina Depósito Legal Vizcaya.

Ayala-Zavala, J.F., Rosas-Domínguez, C., Vega-Vega, V., González-Aguilar, G.A., 2010. Antioxidant enrichment and antimicrobial protection of fresh-cut fruits using their own byproducts: looking for integral exploitation. Journal of Food Science 75, R175–R181.

Baiano, A., 2014. Recovery of biomolecules from food wastes—a review. Molecules 19, 14821–14842.

Bakowska-Barczak, A.M., Schieber, A., Kolodziejczyk, P., 2009. Characterization of Canadian black currant (*Ribes nigrum* L.) seed oils and residues. Journal of Agricultural and Food Chemistry 57, 11528–11536.

Canteri, M.H.G., Scheer, A.P., Wosiacki, G., Ginies, C., Reich, M., Renard, C.M.C.G., 2010. A comparative study of pectin extracted from passion fruit rind flours. Journal of Polymers and the Environment 18, 593–599.

Cesário, M.T., Raposo, R.S., de Almeida, M.C.M.D., van Keulen, F., Ferreira, B.S., da Fonseca, M.M.R., 2014. Enhanced bioproduction of poly-3-hydroxybutyrate from wheat straw lignocellulosic hydrolysates. New Biotechnology 31 (1), 104–113.

Charlton, A., Elias, R., Fish, S., Fowler, P., Gallagher, J., 2009. The biorefining opportunities in wales: understanding the scope for building a sustainable, biorenewable economy using plant biomass. Chemical Engineering Research and Design 87, 1147–1161.

Copeland, D., Belcher, M.W., 2001. Methods for Refining Vegetable Oils and Byproducts Thereof. US Patent, US 6172247.

Crizel de Moraes, T., Jablonski, Rios de Oliveira, A., Rech, R., Flôres Hickmann, S., 2013. Dietary fiber from orange byproducts as a potential fat replacer. LWT-Food Science and Technology 53 (1), 9–14.

CSG—Carbohydrates sector Group, 2008. The European fermentation industry—a case study. In: Presentation Within the Framework of the HLG on Chemicals 25th June 2008.

Deloitte, 2014. Opportunities for the Fermentation-based Chemical Industry—An Analysis of the Market Potential and Competitiveness of North-West Europe.

Domínguez-Perles, R., Martínez-Ballesta, M.C., Carvajal, M., García-Viguera, C., Moreno, D.A., 2010. Broccoli-derived by-products a promising source of bioactive ingredients. Journal of Food Science 75, C38–C392.

Dumas, Y., Dadomo, M., Di Lucca, G., Grolier, P., 2003. Effects of environmental factors and agricultural techniques on antioxidant content of tomatoes. Journal of the Science of Food and Agriculture 83, 369–382.

Eskilsson, S.C., Björklund, E., 2000. Review: analytical-scale microwave-assisted extraction. Journal of Chromatography A 902, 227–250.

Eurostat, 2015. Land Use by NUTS 2 Regions. http://ec.europa.eu/eurostat/data/database.

FAO, 2011. Global Food Losses and Food Waste—Extent, Causes and Prevention. Rome.

FAO, 2013/04. Utilization of Fruit and Vegetable Wastes as Livestock Feed and as Substrates for Generation of Other Value-added Products. RAP Publication.

Fernández-López, J., Fernández-Ginés, J.M., Aleson-Carbonell, L., Sendra, E., Sayas-Barberá, E., Pérez-Alvarez, J.A., 2004. Application of functional citrus byproducts to meat products. Trends in Food Science and Technology 15, 176–185.

Foley, J.A., DeFries, R., Asner, G.P., Barford, C., Bonan, G., Carpenter, S.R., Chapin III, F.S., Coe, M.T., Daily, G.C., Gibbs, H.K., Helkowski, J.H., Holloway, T., Howard, E.A., Kucharik, C.J., Monfreda, C., Patz, J.A., Prentice, I.C., Ramankutty, N., Snyder, P.K., 2005. Global consequences of land use. Science 309, 570–574.

Fowler, P.A., Hughes, J.M., Elias, R.M., 2006. Biocomposites: technology, environmental credentials and market forces. Journal of the Science of Food and Agriculture 86, 1781–1789.

Fuentes-Alventosa, J.M., Jaramillo-Carmona, S., Rodríguez-Gutiérrez, G., Guillén- Bejarano, R., Jiménez-Araujo, A., Fernández-Bolaños, J., Rodríguez-Arcos, R., 2013. Preparation of bioactive extracts from asparagus by-product. Food and Bioproducts Processing 91 (2), 74–82.

Galanakis, C.M., Tornberg, E., Gekas, V., 2010a. A study of the recovery of the dietary fibres from olive mill wastewater and the gelling ability of the soluble fibre fraction. LWT-Food Science and Technology 43, 1009–1017.

Galanakis, C.M., Tornberg, E., Gekas, V., 2010b. Dietary fiber suspensions from olive mill wastewater as potential fat replacements in meatballs. LWT-Food Science and Technology 43, 1018–1025.

Galanakis, C.M., 2012. Conventional, emerging technologies and commercialized applications. Trends in Food Science and Technology 26, 68–87.

Galbe, M., Sassner, P., Wingren, A., Zacchi, G., 2007. Process Engineering Economics of Bioethanol Production. In: Olsson, L. (Ed.), Biofuels. Springer, Berlin Heidelberg, pp. 303–327.

García Herrera, P., Sánchez-Mata, M.C., Cámara, M., 2010. Nutritional characterization of tomato fiber as a useful ingredient for food industry. Innovative Food Science and Emerging Technologies 11, 707–711.

Gardossi, L., Halling, P.J., 2009. Guidelines for reporting of biocatalytic reactions. Trends in Biotechnology 28, 171–180.

Ghatak, H.R., 2011. Biorefineries from the perspective of sustainability: feedstocks, products, and processes. Renewable and Sustainable Energy Reviews 15, 4042–4052.

Haas, C., Steinwandter, V., Diaz De Apodaca, E., Maestro Madurga, B., Smerilli, M., Dietrich, T., Neureiter, M., 2015. Production of PHB from chicory roots—comparison of three *Cupriavidus necator* strains. Chemical and Biochemical Engineering Quarterly 29 (2), 99–112.

Henríquez, C., Speisky, H., Chiffelle, I., Valenzuela, T., Araya, M., Simpson, R., Almonacid, S., 2010. Development of an ingredient containing apple peel, as a source of polyphenols and dietary fiber. Journal of Food Science 75, H172–H181.

Henry, R.J., 2010. Evaluation of plant biomass resources available for replacement of fossil oil. Plant Biotechnology Journal 8, 288–293.

Herrero, M., Cifuentes, A., Ibañez, E., 2006. Sub- and supercritical fluid extraction of functional ingredients from different natural sources: plants, food-by-products, algae and microalgae a review. Food Chemistry 98, 136–148.

Hollmann, J., Lindhauer, M.G., 2005. Pilot-scale isolation of glucuronoarabinoxylans from wheat bran. Carbohydrate Polymers 59, 225–230.

Huber, G.W., Iborra, S., Corma, A., 2006. Synthesis of transportation fuels from biomass: chemistry, catalysis and engineering. Chemical Reviews 106, 4044–4098.

Huber, G.M., Rupasinghe, H.P.V., 2009. Phenolic profiles and antioxidant properties of apple skin extracts. Journal of Food Science 74, C693–C700.

Kamm, B., Kamm, M., 2004. Principles of biorefineries. Applied Microbiology and Biotechnology 64, 137–145.

Kapusta, I., Janda, B., Szajwaj, B., Stochmal, A., Piacente, S., Pizza, C., Franceschi, F., Franz, C., Oleszek, W., 2007. Flavonoids in horse chestnut (Aesculus hippocastanum) seeds and powdered wastewater byproducts. Journal of Agricultural and Food Chemistry 55, 8485–8490.

Kato-noguchi, H., Tanaka, Y., 2006. Potential of citrus junos fruit waste from the food processing industry for weed management. Horticultural Science 41 (6), 1516–1517.

Lante, A., Nardi, T., Zocca, F., Giacomini, A., Corich, V., 2011. Evaluation of red chicory extract as a natural antioxidant by pure lipid oxidation and yeast oxidative stress response as model systems. Journal of Agricultural and Food Chemistry 59, 5318–5324.

Leguval Project, 2014. Legume Subproducts for Bio-packaging Applications. http://leguval.eu.

Mahro, B., Timm, M., 2007. Potential of biowaste from the food industry as a biomass resource. Engineering in Life Sciences 7, 457–468.

Mirabella, N., Castellani, V., Sala, S., 2014. Current options for the valorization of food manufacturing waste: a review. Journal of Cleaner Production 65, 28–41.

Morais Ribeiro da Silva, L., Teixeira de Figueiredo, E.A., Pontes Silva Ricardo, N.M., Gusmao Pinto Vieira, I., Wilane de Figueiredo, R., Montenegro Brasil, I., Gomes, C.L., 2014. Quantification of bioactive compounds in pulps and byproducts of tropical fruits from Brazil. Food Chemistry 143, 389–404.

Moura, J.M.L.N., Campbell, K., Mahfuz, A., Jung, S., Glatz, C.E., Johnson, L., 2008. Enzyme-assisted aqueous extraction of oil and protein from soybeans and cream de-emulsification. Journal of the American Oil Chemists' Society 85, 985–995.

Nilsson, D., 1999. SHAM—a simulation model for designing straw fuel delivery systems. Part 2: model applications. Biomass Bioenergy 16, 39–50.

NNFCC- National Non Food Crops Centre, Smith, W., 2007. Literature Review-State of the Art Biorefinery Development (NFC 07/008).

Obied, H.K., Allen, M.S., Bedgood, D.R., Prenzler, P.D., Robards, K., Stockmann, R., 2005. Bioactivity and analysis of biophenols recovered from olive mill waste. Journal of Agricultural and Food Chemistry 53, 823–837.

Oreopoulou, V., Tzia, C., 2007. Utilization of plant by-products for the recovery of proteins, dietary fibers, antioxidants, and colorants. In: Oreopoulou, V., Russ, W. (Eds.), Utilization of By-products and Treatment of Waste in the Food Industry. Springer Science Business Media, New York, pp. 209–232.

Pandey, A., Soccol, C.R., Rodriguez-Leon, J.A., Nigam, P., 2001. Production of organic acids by solid state fermentation. In: Solid State Fermentation in Biotechnology-fundamentals and Applications. Asiatech Publishers, New Delhi, pp. 113–126.

Parajuli, R., Dalgaard, T., Jørgensen, U., Adamsen, A.P.S., Trydeman Knudsen, M., Birkved, M., Gylling, M., Kofod Schjørring, J., 2015. Biorefining in the prevailing energy and materials crisis: a review of sustainable pathways for biorefinery value chains and sustainability assessment methodologies. Renewable and Sustainable Energy Reviews 43, 244–263.

Patel, J.S., Onkarappa, R., Shobha, K.S., 2006. Study of ethanol production from fungal pretreated wheat and rice straw. The Internet Journal of Microbiology 4, 1.

Patist, A., Bates, D., 2008. Ultrasonic innovations in the food industry: from the laboratory to commercial production. Innovative Food Science and Emerging Technologies 9, 147–154.

Patsioura, A., Galanakis, C.M., Gekas, V., 2011. Ultrafiltration optimization for the recovery of β-glucan from oat mill waste. Journal of Membrane Science 373, 53–63.

Puri, M., Sharma, D., Barrow, C.J., 2012. Enzyme-assisted extraction of bioactives from plants. Trends in Biotechnology 30, 37–44.

Ramankutty, N., Evan, A.T., Monfreda, C., Foley, J.A., 2008. Farming the planet: 1. Geographic distribution of global agricultural lands in the year 2000. Global Biogeochemical Cycles 22, 1.

Rockström, J., Steffen, W., Noone, K., Persson, Å, Chapin III, F.S., Lambin, E., Lenton, T.M., Scheffer, M., Folke, C., Schellnhuber, H., Nykvist, B., De Wit, C.A., Hughes, T., van der Leeuw, S., Rodhe, H., Sörlin, S., Snyder, P.K., Costanza, R., Svedin, U., Falkenmark, M., Karlberg, L., Corell, R.W., Fabry, V.J., Hansen, J., Walker, B., Liverman, D., Richardson, K., Crutzen, P., Foley, J., 2009. Planetary boundaries: exploring the safe operating space for humanity. Ecology and Society. 14 (2), 32. http://www.ecologyandsociety.org/vol14/iss2/art32/.

Russ, W., Schnappinger, M., 2007. Waste related to the food industry: a challenge in material loops. In: Oreopoulou, V., Russ, W. (Eds.), Utilization of By-products and Treatment of Waste in the Food Industry. Springer Science Business Media, New York, pp. 1–13.

Riggi, E., Avola, G., 2008. Fresh tomato packinghouses waste as high added-value biosource. Resources, Conservation and Recycling 53, 96–106.

Rodríguez, R., Jiménez, A., Fernández-Bolaños, J., Guillén, R., Heredia, A., 2006. Dietary fibre from vegetable products as source of functional ingredients. Trends in Food Science and Technology 17, 3–15.

Sarath, G., Mitchel, R.B., Satler, S.E., Funnel, D., Pedersen, J.F., Graybosch, R.A., Vogel, K.P., 2008. Opportunities and roadblocks in utilizing forages and small grains for liquid fuels. Journal of Industrial Microbiology and Biotechnology 35 (5), 343–354.

Scarlat, N., Martinov, M., Dallemand, J.F., 2010. Assessment of the availability of agricultural crop residues in the European Union: Potential and limitations for bioenergy use. Waste Management 30, 1889–1897.

Steffen, W., Richardson, K., Rockström, J., Cornell, S.E., Fetzer, I., Bennett, E.M., Biggs, R., Carpenter, S.R., de Vries, W., de Wit, C.A., Folke, C., Gerten, D., Heinke, J., Mace, G.M., Persson, L.M., Ramanathan, V., Reyers, B., Sörlin, S., 2015. Planetary boundaries: guiding human development on a changing planet. Science 347 (6223). http://dx.doi.org/10.1126/science.1259855.

Stephen, J.D., Sokhansanj, S., Bi, X., Sowlati, T., Kloeck, T., Townley-Smith, L., et al., 2010. The impact of agricultural residue yield range on the delivered cost to a biorefinery in the Peace River region of Alberta, Canada. Biosystems Engineering 105, 298–305.

Tallaksen, J., 2011. Chapter 6: Guidelines for developing a sustainable biomass supply chain. In: Final Report to the USDA Rural Development Grant 68–3A75-5-232.

Thorsell, S., Epplin, F.M., Huhnke, R.L., Taliaferro, C.M., 2004. Economics of a coordinated biorefinery feedstock harvest system: lignocellulosic biomass harvest cost. Biomass Bioenergy 27, 327–337.

TRANSBIO project, 2012. Biotransformation of by-products from fruit and vegetable processing industry into valuable bioproducts. http://www.transbio.eu.

Upadhyay, A., Chompoo, J., Araki, N., Tawata, S., 2012. Antioxidant, antimicrobial, 15–LOX, and AGEs inhibitions by pineapple stem waste. Journal of Food Science 77, H9–H15.

Vandamme, E.J., 2009. Agro-industrial residue utilization for industrial biotechnology products. In: Singh nee'Nigam, P., Pandey, A. (Eds.), Biotechnology for Agro-Industrial Residues Utilization. Springer Science+Business Media B.V, pp. 3–11.

Willke, T., Vorlop, K.-D., 2004. Industrial bioconversion of renewable resources as an alternative to conventional chemistry. Applied Microbiology and Biotechnology 66, 131–142.

Zhang, Y.H.P., 2008. Reviving the carbohydrate economy via multi-product lignocellulose biorefineries. Journal of Industrial Microbiology and Biotechnology 35 (5), 367–375.

CHAPTER 2

Fruit Processing in Central America and Mexico

C. Velázquez, A.M. Pérez Carvajal
CITA-University of Costa Rica, San José, Costa Rica

L.A. Prado Barragán
Metropolitan Autonomous University-Iztapalapa, México City, México

2.1 Introduction

Central America is a very diverse region with an intense agriculture activity and an area with a persistent inequity that tends to concentrate its production and income opportunities in a few segments of the population. Greater integration into the world economy enabled the region to diversify its production base, but also made it more sensitive to adverse international market cycles. Public policies have shown little possibilities of action to mitigate external shocks and limited opportunities to generate endogenous capacities based on productivity and innovation (PEN, 2014).

The Human Development Index (HDI) is used to quantify and compare the level of welfare of a country (United Nations, 2014). In Central America, it has historically been below the average of the other Latin American countries and of the world. All the countries in the region have increased this index in a continuous way over the past 5 years. In the region, there are two blocks: Panama, Costa Rica, and México with a level of high human development index (HDI), and the rest of the countries with a medium HDI, showing the main gaps in education and income levels (PEN, 2014; GPG, 2015).

The total area for cultivation in the region varies between countries. Costa Rica is the one that has a lower available arable land per person, 0.05 Ha/person, while other countries fluctuate between 0.10 (Guatemala) and 0.32 (Nicaragua) Ha/person. Honduras and Panamá are the countries with the lowest percentage of territory occupied by agricultural land, about 23%. Costa Rica, Guatemala, and Nicaragua use about 40%, and El Salvador, the smallest country in Central America, uses 74%.

Central America has experienced a slowdown in foreign trade over the past two years. After the international financial crisis, the region lived two years of commercial dynamism. In 2010

Biotransformation of Agricultural Waste and By-Products. http://dx.doi.org/10.1016/B978-0-12-803622-8.00002-1

and 2011, exports grew (13% and 23%, respectively), and imports grew as well (19% and 24%, respectively). However, in 2012 and 2013, the growth of business activity fell to 4.5% and −1.4% in exports. This is the first negative change in exports since the global financial crisis of 2009. According to the Competitiveness Index (WEF, 2015), the region has two of the three most competitive countries in Latin America, Costa Rica and Panamá, only less than Chile, which occupies the first position (PEN, 2014), while Mexico, when considered in the Central America block, it is cited in the fifth position.

Roache (2007) indicates that the cycle of business in Central America is very dependent on the United States; 40% of the exports of the region go to this market. El Salvador and Costa Rica are the two most sensitive to this economic cycle. The increment in productivity through new technologies and the introduction of international standards requested by different markets are the major drivers to have a more competitive agricultural activity and some leeway to mitigate the impact of the contraction in the US economy of the previous years. The decision by the United States and European Union to negotiate free trade agreements with Central American countries as a whole helped create and strengthen many of the conglomerates that exist today in the Isthmus and revitalize government bodies responsible for foreign trade. Exports to the region went to countries which have signed Free Trade Agreements (FTAs) since the early 1990s; Mexico, Dominican Republic and Chile where the Central American countries showed a better performance with their total external sales.

The region shows important agricultural activity, especially in the production of tropical fruits for the export market (PEN, 2009). The production of fresh pineapples, bananas, melons, mangos, and watermelons have been a major and differentiating activity that develops an integrated chain of economical and social development for the region. This activity generates an opportunity for the valorization of the second-quality fruit through agro–industry and the whole use of its potential with well-funded bioactive molecules that are driven by the international markets that look for foods as a major factor in wellness and health of the consumers.

2.2 Fruit Production in the Region

World production of tropical fruits, excluding bananas, is currently about 56 million tons, most of which are absorbed domestically in the producing countries, either in fresh or processed forms. Global demand for fresh pineapples has increased significantly due to the combination of increased distribution networks through supermarkets and increased consumer awareness. The rise in population, improved distribution networks, and higher disposable incomes have caused the explosion of import demand in many countries in recent years, particularly for fresh pineapples.

2.2.1 Costa Rica

Costa Rica is a small country of about 50,000 km^2, which dedicates about 40% of its territory to agriculture. This productive activity needs to compete with the protected areas dedicated to national parks and rainforest reserves, which occupy almost a quarter of the total extension of the country and compete also with the increase in urban areas. The agro-climatic conditions of Costa Rica are optimal for the production of a variety of tropical and exotic fruits. It is one of the major exporters in the world of pineapple, banana, and melon.

The major agricultural product of the country in the period of 2009–2013 was sugar cane with 21,540,000 tons, followed by pineapple with 12,029,410 tons, and banana with 10,251,450 tons. In the fruit area, the exports of banana and pineapple are the major products over a diverse production of other fruits, just as indicated in Table 2.1 (Procomer, 2015).

Pineapple

Pineapple (*Ananas comosus*) is a nonclimacteric fruit native to Brazil and Paraguay, which belongs to the Bromeliaceae family (Dull, 1971). This crop has spread through the tropical regions, and it is available all year round. Its texture is juicy and fibrous, and the taste is sweet and sour (Dembitsky et al., 2011).

In Costa Rica, pineapple has been cultivated for many decades. This activity began around the 1950s in Buenos Aires and Puntarenas, and the predominant varieties were initially of the Cayenne group, such as Smooth Cayenne and Champaka. Later, the varieties Sugarloaf and Monte Lirio were grown at a smaller scale and for national consumption. Currently, the MD-2 hybrid that was patented by Fresh Del Monte and is known commercially as Golden, Yellow, Gold, or Golden, is grown with great commercial success because of its sweet taste and bright golden color, constituting one of the main exports in Costa Rica (Brenes Gamboa, 2005). Costa Rica has become the leading exporter country of fresh pineapple in the world (in metric tons), followed by the Philippines and Ecuador (Ti, 2000).

Costa Rica is the largest exporter of fresh pineapple to the US and the European markets, and production area has tripled since 2002 with associated economic benefits and environmental concerns. The pineapple, based on income, is the second agriculture export product in Costa

Table 2.1: Main Seven Exported Fruits Form Costa Rica for the Year 2014, Expressed in Thousands of US Dollars and in Tons.

Product	Value (In Thousands of US, $)	Weight (Tons)
Banana	905,066.38	2,171,384.33
Pineapple	865,087.28	2,070,540.11
Melon	70,884.22	131,850.61
Watermelons	16,316.21	36,043.43
Mangos	4,844.74	6,322.21
Papaya	3,948.03	4,781.10
Plantain	1,811.92	3,208.99
Total	**1,938,399.71**	**4,481,120.86**

Adapted from Procomer, 2015. Estadísticas del sector exportador.http://www.procomer.com/contenido/sector-alimentario.html.

Rica, only behind banana. In 2014, it generated US $865 million, just behind bananas with US $905 million. The pineapple export activity is developed by 170 exporters and 72 packaging houses, distributed in three major areas of the country: the North with 47%, the Atlantic with 31%, and the Pacific with 22%. It generates about 26,600 direct employments. The main markets for export are Europe with 52% and the United States with 48%, according to reports of 2014. In Costa Rica, the production area has increased in the past 5 years, generating from a total export in 2011 of about 2.07 million tons to 2.20 million tons in 2014. The area dedicated to pineapple production in 2014 was reported at a peak of 45,000 Ha.

This tendency of growth is showing a decrease, and for the first semester of year 2015, there is an estimate of at least a 10% reduction due to climatic problems associated with an excess of rain in the North and the Caribbean that started in July 2014 and of the high production costs associated with currency exchange rate and labor. The area is estimated to be reduced to 40,000 Ha in 2015 (CANAPEP, 2015).

The leadership of Costa Rica in the world pineapple production has been accompanied by massive organizational changes: the pineapple sector is increasingly driven by large downstream actors who derive their power from the ability to impose their definition of quality to the whole chain and from economies of scale and scope. Even if the innovation presented in varieties throughout the last 10 years, commercial and logistical innovations are very important to define this success. The major concern on this system is the risk associated with small producers to be left behind or marginalized.

World production of tropical fruits, excluding bananas, is currently about 56 million tons; most of this production is consumed in the local country markets of the producing countries. The major crops are mangoes and mangosteen (43.3 million tons) followed by pineapples. The world pineapple production is about 24.7 million tons, of which about 30% is used for processing and the rest for fresh consumption. The major producer in the world is Costa Rica (2.7 million tons), which accounted for almost 11% of world production in 2013, followed closely by Brazil, Philippines, and Thailand, which accounted for between 9% and 10% each (Vagneron et al., 2009).

Banana

Costa Rica's agricultural sector has been and is largely influenced and dependent on banana cultivation. Banana was the first agricultural export product in Costa Rica in 2014. According to CORBANA (2015), for the year 2014, Costa Rica exported 110 million boxes of bananas, 18.14 kilos each, around 2.17 million tons; this constituted 46.5% of the total fresh produce exports of the country for that year. In the world, 8 out of 10 bananas are exported, and Latin America is the largest export area. The main producers in Latin America are Ecuador, Costa Rica, Colombia, and Honduras. Costa Rica ranks as the fourth largest exporter in the world.

The introduction of bananas to Central American Region was a result of early Spanish exploration. In Costa Rica, it was initiated in 1870 by Minor Keith, an American entrepreneur. Keith was commissioned by the Costa Rican government to construct a transnational railroad and began planting alongside the railroad. Keith's company merged with Boston Fruit Company in the 1890s and became the United Fruit Company, which held a monopoly over the banana industry until the 1950s. The production evolved, and now about 52% of the production is in hands of big companies: Chiquita Brands (formerly United Fruit), Dole, and Del Monte (Sauma, 2014).

Bananas produced in Costa Rica come from two wild species: *Musa acuminata* (AA) and *Musa balbisiana* (AB). The breeding between them has generated a large amount of cultivars that belong to the Musaceae family. In Costa Rica, the largest plantations are located in the Caribbean side of the country, where bananas are produced for export and are mainly of the variety *Musa cavendish* (AAA). The dessert banana or table bananas (AAA) are derived from the pure acuminata group. These are sweeter and less starchy than the AAB (*Musa sapientum,* or true banana; *Musa paradisiaca*, or plantains), ABB bananas (cooking banana), or pure BBB ("Saba," or cooking banana). The variety that is cultivated in Central America for export activity is the triploid AAA of *M. acuminata*, referred to as Cavendish group bananas.

The good soil conditions and the humid and hot climate make the Costa Rican and Honduran areas the best in all of the Central American region for the production of bananas. The specific and optimal conditions required to have a successful activity require a flat terrain and deep (no more than 1.20 m), well-structured, and well-drained soils with a high balance of nutrients, especially potassium, and pH between 6 and 7.5.

The main markets for Costa Rica's bananas are North America and the European Union; these two capture over 60% of world imports, while Japan and the USSR import about 12%. In global commerce, by far the most important cultivars belong to the Cavendish group.

Costa Rica devotes around 44,000 Ha to the cultivation of bananas. The international production has expanded considerably in recent years, and this is leading to a more saturated market. Countries like Ecuador, Philippines, Guatemala, and Cameroon are offering low prices and impacting the world market. Costa Rica has higher laborer wages than countries in Africa and Latin America; this makes it very difficult to compete with price. Big efforts have been organized by Costa Rica to let the consumers know about the value added by ethical social and environmental policies. The initiatives with the protection of the environment are being coordinated with the Environmental Banana Commission (CAB); most of the producers in the country are certified by Global Gap, ISO 14001, and Rainforest Alliance. The country is working with the EU for the designation of a protected geographical indication to Costa Rica's bananas (Sauma, 2014).

2.2.2 Other Central American Countries

The main Central American agriculture product is sugar cane, followed by palm oil and fruits like pineapple and banana. The regional production in agricultural products is shown in Table 2.1.

As indicated by FAO in 2013, Honduras is a major producer of palm oil; Guatemala, El Salvador and Nicaragua of sugar cane; and Costa Rica of palm oil and pineapple (FAO, 2014).

The production in the major crops of the region, after the world crisis of 2009, started to show an increase in the period of 2009–2013, as reported in Table 2.2. This situation was especially important with the production of pineapple due to Costa Rica's increase in production areas. The only crop that showed a very small change in the last years was the banana production and this was attributed to climate factors and special commercial situations with European markets. The trends in the production in this period are shown in Fig. 2.1.

In the world, the American region in this period of 5 years produced an average of 36.3% of the pineapple produced in the world, following Asia, which produced 46.1%. In the Central American region, a total of 4,010,739 tons were produced in 2013 (FAO, 2014). Also in this period from 2009 to 2013, a growing trend is shown with an average growth rate of 5.92% in the region. In regards to banana, the Latin American region is the second largest producer in

Table 2.2: Agricultural Production in Tons for Central American Countries in the Year 2013.

Country	Pineapple	Banana	Sugar cane	Palm oil
Costa Rica	2,685,131	2,174,986	4,850,000	5,348,750
Honduras	139,000	4,164,711	6,082,000	9,206,803
Guatemala	243,570	3,188,050	26,334,667	6,000,000
El Salvador	1,708	0	7,162,995	0
Nicaragua	60,091	37,227	7,041,332	15,000

FAO Stat Database In: http://faostat3.fao.org/browse.

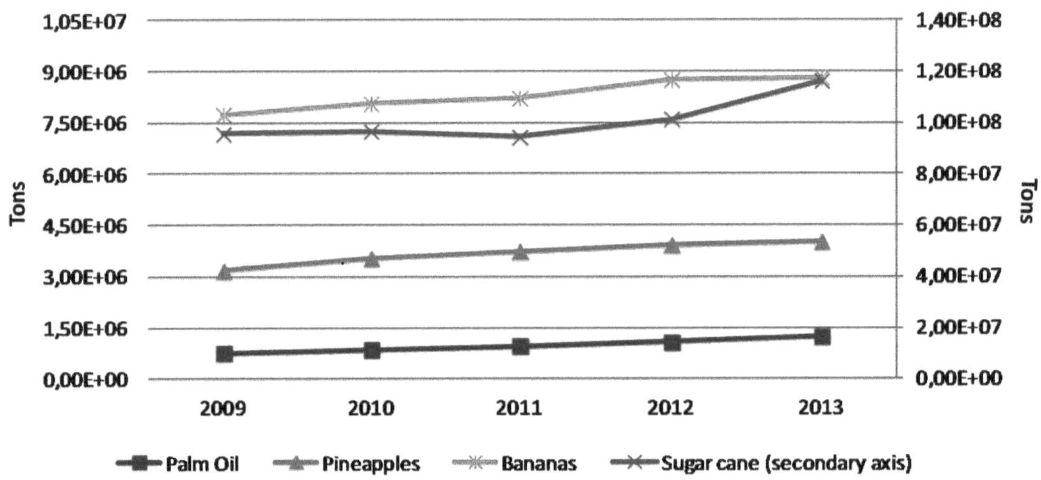

Figure 2.1
Central American major crops produced in the period 2009–2013. Sugar cane data reported against secondary axis. *FAO Stat* http://faostat3.fao.org.

the world with the leadership of Ecuador and Colombia, but these followed the production of Costa Rica and Honduras, who are major actors of these crops in the Central American region.

2.2.3 Mexico

Mexico, a 1,972,550 km^2-sized country, dedicates 55% of the land to agriculture. This productive activity and the forest-protected areas are in a hard fight to urban areas' growth. In Mexico, the cultivation of fruit and vegetables accounts for about 45% of the agricultural land. The climate variation from tropical to template encompasses a vast array of geographical soil types, climates, and microclimates, oscillating from arid pasturelands in the north, humid farmlands in the midsection of the country, and hot, humid, tropical zones in the south. Data from Mexico's Ministry of Agriculture (the *Secretaría de Agricultura, Ganadería, Desarrollo Rural, Pesca y Alimentación*, or SAGARPA) shows that the country produces more than 350 distinct varieties of farm products each year. A wide spectrum of fruit varieties and almost any vegetable can be grown; however, around 20% of fruit and vegetable production is exported. Mexico's vegetable production is concentrated in Sinaloa, Zacatecas, Guanajuato and Chihuahua. Among the major products are avocado, beans, corn, coffee beans, citrus, grass, and sorghum. However, the major exported products are avocados, citrus (lime, limes, lemons, and grapefruit), bell peppers, coffee, mango, papaya, pineapple onion, and tomatoes.

According to the Ministry of Agriculture in 2012, Mexico was the largest exporter of avocados worldwide, with about 52.8% of the global total, and it also accounted for 23.6% of global sales of tomatoes. Mexico leads in global onion exports with 14.7% of the global market, while its papayas accounted for 31.8%. Mango and guavas each have about 20% of the global market share, and Brussels sprouts, 26.6%. Pineapple output was 15.8 million tons, nearly 26% of world tropical fruit production. Mexico is also successful in the citrus category, as the country is the fourth largest exporter, especially of limes, although it also exported oranges and grapefruit, totaling 470,000 tons (FAO, 2012).

2.3 Fruit Industrialization in the Region

2.3.1 Costa Rica

The agro–industry in Costa Rica has more than 100 years of experience in the development of value-added products from its important produce. In the 1970s, the local industry made an important development in the implementation of international certifications, such as ISO, Kosher, USDA Organic, and Fair Trade, among others that guarantee the commitment with quality, health, and environment.

The major fruits produced in Costa Rica are processed as juices, purees, pastes, and marmalades, frozen for the development of other products and dehydrated through innovative technologies. One important activity is the preparation of extracts from byproducts of the banana and orange juice industry. Pineapple, banana, and mango are the main fruits used in the country for the production of juice concentrates for export activities (Procomer, 2015).

Regarding the industrialization of pineapple, although only one-third of its output is utilized for processing, pineapple products account for more than two-thirds of the trade in pineapples, by value. If about 90% of the produced fruit is exported as a fresh product, this leaves a large amount of second-quality product that is used as a raw material for industrialization, about 0.24 million tons for 2014 production reported previously. The major products generated from this fruit are juice and frozen chunks.

The use of banana byproducts rejection and commercial exploitation of fresh fruit is a national concern in these countries. In Costa Rica for example, considering that the average rejection of fresh fruit in packaging facilities runs between 10 and 15% and that the total production of fruits in 2014 represented about 2.17 million tons, the amount of fruit dedicated to processing is about 0.48 million tons (CORBANA, 2015; Murillo, 2013).

The possibilities of products produced from bananas are very large. In Costa Rica, there are big agro–industrial facilities for the production of frozen purees and aseptic pulps, and they absorb about 75% of the fruit available for processing. For the rest of the fruit, the companies compete to get this fruit and use it for the production of frozen pieces using IQF facilities, smaller facilities that produce dehydrated pieces from the ripe fruit, and fried snacks from the plantains and green bananas. A large amount of this fruit is used also for the production of flours from the green banana because of its high content of resistant starch. The processing of bananas undergoes a large variety of possibilities, and in the country, the use of the fruit is intense. Almost all of this fruit is used for agro–industry, but if some is left behind, it will be used in animal feeding for pork production, mainly.

2.3.2 Central America

As for the production of canned pineapple, Central America stands as the second region in the world with 12.3% of world production, following Asia, which produced 76.9%, highlighting the contribution of Thailand and Indonesia. Central American production accounted in 2013 for 64,856 tons, with Costa Rica being the largest producer in the region with 49,200 tons. The industrialization of canned pineapple chunks for pulp, juice, or frozen product is often associated with producing countries because the export of fresh fruit produces between 5% and 8% of a second-quality fruit that does not meet the strict international standard, so it is used with no problem for industrialization. With transport being the biggest economic problem for industrialization, the processing facilities are often located nearby (within 100 km) the production areas. That is why countries like Thailand, Brazil, and Costa Rica have an important industrialization activity around pineapple products. Costa Rica is the number one world producer of pineapple juice (°Brix <20) (Procomer, 2015).

Central America is a large producer of pineapple juice because of Costa Rica's production. When considering the imports of this juice to the European Union, Costa Rica (42%) and Thailand (36%) account for more than three-quarters of a reported 232,000 tons of pineapple juice send to the EU for the year 2011 (AIJN, 2012). For the year 2013, a total of 360,000 tons were sent to the EU, showing an important 50% increase in this flavor. Pineapple is in the top five fruit juice flavors consumed in the EU, and this is primarily due to the traditional propensity of this flavor on the Spanish market (AIJN, 2014). This market trend is used by the Central American region, which is increasing considerably the amount of pineapple production.

Associated with the industrialization of pineapples, a major byproduct is generated from the agro–industrial activity, composed of the heart, the crown, and peels of the fruit, which could add up to 65% of the fruit weight (Quesada et al., 2005). These byproducts could be divided in two groups: solids, like the ones mentioned before, and liquids.

The liquids come from the extraction of juices of the peels and residues of the pulp and contain mainly sugars like sucrose, glucose, and fructose and in a minor amount, dextrins, raffinose, and galactose. If these are not used, they can become a major contaminant, not only because of the high amount generated, but also because of their high organic content. These do have a special importance as biomass for fermentation activities. Much research has been developed to use this biomass considering the high fermentable sugar content and the stringent environmental regulations regarding waste disposal. The literature show interesting applications for the production of lactic acid (Idris and Suzana, 2006; Araya-Cloutier et al., 2010, 2012) and biofuels (Choonut et al., 2014; Rani and Nand, 2004), among others.

The solid byproducts produced at the field include the stubble ("rastrojo") are mainly composed of the crown and leaves of the plant. It has been determined that per hectare of pineapple cultivated, about 300 tons of stubble are generated (Quesada et al., 2005). In countries like Costa Rica where there are about 45,000 Ha, these will produce about 13.5 thousand tons of this material. This material can be a big problem in the fields because of its large volume and slow degradation. If it is left in the field, it generates the proliferation of the *Stomoxys calcitrans* fly, which causes disease in cattle and also generates human health risks. Others who treat this product use herbicides to dry it up and then burn it, causing severe impact to the environment. The producers need to generate friendly protocols to handle this byproduct or develop technological applications for its valorization, such as dehydration for year-round storage and to use it a biomass for fuel or fermentation activities due to its high fibers and complex sugar content; some experiences of this application have been developed for Costa Rica and Thailand.

Currently, due to an oversupply of fruit in the international market driven by Ecuador (one of the largest producers of bananas), the drop in international prices, and the implications of free trade agreements, Costa Rica and Honduras have had a significant volume of surplus and export rejections.

Each hectare of banana crop generates nearly 220 tons of plant residual waste that consists mainly of lignocellulose material. Although these materials can be converted to biogas or compost, transportation makes this practice uneconomical. Most of the residual waste produced is discarded by farmers, and in some cases, could be an environmental concern. In recent years, there has been significant interest in efficient use of agro–industrial residues. Several processes have been developed based on these materials as substrates in the bioprocess for production of single cell proteins, organic acids, ethanol, mushrooms, enzymes, and biologically important secondary metabolites. The main residual wastes of the banana crop are leaves and pseudostems, both containing high levels of lignocelluloses. Major analysis of the potential of banana fibers has been developed due to the better tensile strength and its use in engineering materials. The quality and quantity of biological macromolecules, especially the cellulose, hemicellulose, lignin, wax, engineering, and mechanical properties of banana biofibre have been discussed and analyzed in the world (Pappu et al., 2015). The use of these macromolecules as biomass for fermentations (Reddy et al., 2003) and applications in Costa Rica for the production of paper, biomass as fermentation substrate, and others related to the integral use of all these materials, which are produced either in the field or as byproducts from the agro–industry.

2.4 Physicochemical Composition and Biological Activity of Fruits of the Region

2.4.1 Physicochemical Composition of Fruits of Costa Rica

Table 2.3 shows the physicochemical composition of the main fruits grown in Costa Rica. Gold pineapple cultivar from Hawaii was found to have significantly higher ($p < 0.0001$) pH, °Brix/acid ratio, yellow color, total carotenoid content, and vitamin C than Hawaiian Smooth Cayenne (Ramsaroop and Saulo, 2007). For Gold pineapple fruits grown in Costa Rica, Brix value of about 13 was obtained, indicating high natural sugar content. Furthermore, acidity found for this variety is lower than data reported in the literature (Dull, 1971; Flath, 1980) for other varieties of pineapple (0.5–1.62%), which promotes its acceptance and consumption. Costa Rican Gold pineapple have a °Brix/acid ratio slightly higher than Gold fruits cultivated in Hawaii but lower than those of Smooth Cayenne from the same origin.

Banana has the highest protein content among the Costa Rican fruits studied, as well as the highest value of °Brix and total sugars. USDA (2015a) reports a similar value for protein content (1.09 g/100 g) without specifying the variety of bananas. However, Forster et al. (2002) determined a protein content of 1.61% for the Grand Naine variety grown in Tenerife, Spain. These authors also report dietary fiber content within the same ranges, as seen in Table 2.3.

In general, a great variability is found in data reported in the literature regarding the physicochemical composition and nutritional value of fruits evaluated, with respect to the data reported for Costa Rica crops (Table 2.3). These results may be due to different factors such as variety, crop growing conditions, maturity, and postharvest handling.

Table 2.3: Physicochemical Composition of Cultivated and Exported Fruits From Costa Rica and Central America.

Component	Banana[a,b] *Musa* AAA, Cavendish Variety, cv. Gran Naine	Pineapple[a,b,c] *Ananas comosus*, MD-2 Hybrid or *Gold* Cultivar	Mango[b] *Mangifera indica*, cv. Tommy Atkins	Cantaloupe[a,b,d] *Cucumis melo* Cantaloupe Variety, cv. Veracruz	Watermelon[e] *Citrus lanatus* cv. Micky Lee
Moisture (g/100 g FW)	75.6 ± 1.2[f]	87 ± 1[l]	84 ± 1[l]	93 ± 1[i]	92.4 ± 1.3[l]
Acidity (%)	0.38 ± 0.00[k,m]	0.56 ± 0.04[l,m]	0.11 ± 0.05[k,n]	0.12 ± 0.02[k,n]	0.073 ± 0.005[l,m]
Protein (g/100 g FW)	1.0 ± 0.1[g]	0.4 ± 0.2[l]	0.45 ± 0.05[k]	0.5 ± 0.0[k]	0.6 ± 0.0[l]
Fat (g/100 g FW)	0.12 ± 0.00[g]	0 ± 0.0[l]	0.05 ± 0.05[k]	0 ± 0.0[k]	<0.1[l]
Dietetic fiber (g/100 g FW)	1.8 ± 0.4[g]	1.8 ± 0.1[l]	N.D.	N.D.	0.53 ± 0.05[l]
Ash (g/100 g FW)	0.73 ± 0.06[g]	0.22 ± 0.02[k]	0.31 ± 0.01[k]	0.6 ± 0.1[k]	0.24 ± 0.01[l]
Total carbohydrates (g/100 g FW)	20 ± 1[l]	10.9–13.3[j]	12.8 ± 0.5[k]	7.3 ± 0.3[k]	7.7 ± 0.2[l]
Sucrose (g/100 g FW)	6 ± 2[f]	6 ± 1[l]	8.1 ± 0.4[k]	3.05 ± 0.05[k]	2.07 ± 0.09[l]
Glucose (g/100 g FW)	9 ± 1[i]	2.5 ± 0.5[l]	0.55 ± 0.05[k]	2.0 ± 0.2[k]	1.39 ± 0.04[l]
Fructose (g/100 g FW)	8 ± 2[h]	2.4 ± 0.3[l]	4.4 ± 0.3[k]	2.3 ± 0.3[k]	2.97 ± 0.04[l]
°Brix	22.4 ± 0.1[g]	13 ± 1[k]	14.8 ± 0.3[k]	8.8 ± 1[h]	9.2 ± 0.2[l]
pH	4.93 ± 0.03[g]	3.7 ± 0.1[k]	4.41 ± 0.08[k]	6.6 ± 0.3[f]	5.37 ± 0.05[l]
°Brix/acid ratio	58.9	23.2	134.5	73.3	126.0

[a]Badilla, M., 2005. Caracterización de compuestos volátiles en jugos pulposos y clarificados de frutas de exportación. Project Graduation. Food Engineering. Universidad de Costa Rica.
[b]Pérez, A.M., 2005. Caracterización físico-química de frutas de exportación: piña, mango, melón y banano. Estudio sobre su contenido en pigmentos y compuestos volátiles. Informe final de proyecto. Centro Nacional de Ciencia y Tecnología de Alimentos, Universidad de Costa Rica. Proyecto MICITT-CONICIT 147-01-FI, 26 pp.
[c]Quesada, C., 2013. Determinación de las condiciones óptimas del tratamiento enzimático acoplado a un proceso de microfiltración tangencial para la obtención de jugo clarificado de piña en una empresa procesadora de frutas y vegetales. Project Graduation. Food Engineering. Universidad de Costa Rica.
[d]Chaverri, M., 2003. Elaboración de un jugo de melón clarificado por medio de microfiltración tangencial. Project Graduation. Food Engineering. Universidad de Costa Rica.
[e]CITA, 2015. Reporte de Análisis Químico. Laboratorio de Química. 4 pp.
[f]Mean of 6 samples.
[g]Mean of 5 samples.
[h]Mean of 8 samples.
[i]Mean of 7 samples.
[j]Range of 4 samples.
[k]Mean of 2 samples.
[l]Mean of 3 samples.
[m]expressed as malic acid.
[n]expressed as citric acid, FW, Fresh weight; N.D., not determined.

Vitamin C

Pineapple is a fruit that is rich in bioactive compounds, such as vitamin C and bromelain (digestive enzyme of protease type), with beneficial health effects. Pineapple had the highest vitamin C content among the analyzed fruits from Costa Rica (Table 2.4). Ascorbic acid is the main biologically active form of vitamin C, and as shown in Table 2.4, the values determined for the pineapple of the Golden variety grown in Costa Rica are similar to those of citrus such as orange (containing ascorbic acid 46 mg/100 g) (Proteggente et al., 2002; Hernández et al., 2006), although higher than that reported for other pineapple varieties as Smooth Cayenne. For this variety, Dull (1971) found a vitamin C content of 10–25 mg/100 g FW. Hassimotto et al. (2005) reported the following value for ascorbic acid in pineapple: 22.4 ± 0.9 mg/100 g. Meanwhile, Almeida et al. (2011) assessed a lower value in pineapple for ascorbic acid content (13.0 ± 0.9 mg/100 g) for fruits from Brazil, and Ramsaroop and Saulo (2007), a content of 26 mg/100 g for Hawaiian Smooth Cayenne pineapple. These authors reported, for the Gold variety cultivated in Hawaii, a value of vitamin C of 91 mg/100 g, otherwise Valente et al. (2011) indicated a content of 61 mg/100 g for the Golden fruits from Costa Rica (Table 2.4) and 11 mg/100 g for Smooth Cayenne, which originated in Azores Islands.

Table 2.4 shows the vitamin C content in bananas from Costa Rica was higher (14 mg/100 g) that those reported in literature (Leong and Shui, 2002; Hernández et al., 2006; Valente et al., 2011; Almeida et al., 2011). Also, Kanazawa and Sakakibara (2000) and Forster et al. (2002) determined vitamin C values for bananas of 5.4–10 mg/100 g and 9.56 mg/100 g, respectively. These results are similar for Costa Rican mango and melon fruits (Table 2.4), considering that

Table 2.4: Vitamin C Content (mg/100 g) of the Main Cultivated Fruits in Costa Rica and Central America.

References	Fruit Content (mg/100 g) Fresh Weight, Origin, and Variety				
	Banana	Pineapple	Mango	Melon	Watermelon
Pérez (2005)[a]	14 ± 1 Costa Rica cv. Grain naine	50 ± 4 Costa Rica cv. Gold	29 ± 1 Costa Rica cv. Tommy Atkins	26 ± 2 Costa Rica cv. Cantaloupe	7 ± 0.2 Costa Rica cv. Micky Lee
Leong and Shui (2002)[a]	2.1 ± 0.8 Philippines	54.0 ± 7.9 Malaysia	19.7 ± 9.1 Philippines	3.7 ± 0.2 Malaysia	3.7 ± 0.2 Malaysia
Hernández et al. (2006)[a]	2.15 ± 1.2 Canary Islands cv. Grain naine	26.9 ± 2.7 Canary Islands cv. roja española	55 ± 8 Canary Islands cv. Keitt	N.R.	N. R.
Valente et al. (2011)[a]	1.42 ± 0.08 Portugal cv. Williams	61.0 ± 0.1 Costa Rica cv. Sweet Gold	40.9 ± 0.6 Brazil cv. Palmer	N.R.	N. R.
Mélo et al. (2006)[b]	9.8 ± 0.7 Brazil cv. Comprimida	25.0 ± 0.7	83.9 ± 0.3 Brazil cv. espada	2.3 ± 0.2 Brazil cv. espanhol	N. R.

[a]HPLC method.

[b]AOAC method with dichloroindophenol; N. R, Not reported.

Shofian et al. (2011) reported a lower ascorbic acid content in mango of 8.36 ± 2.33 mg/100 g and melon of 2.24 ± 0.35 mg/100 g. However, Table 2.4 shows that vitamin C content of mango in other studies was significantly higher than our data (Pérez, 2005).

The vitamin C content of watermelon cultivated in Costa Rica was analyzed by CITA of University of Costa Rica, and this value of 7.0 ± 0.2 mg/100 g was higher than the one found by Leong and Shui (2002) and by Shofian et al. (2011) in fruits from Malaysia (3.7 ± 0.2 and 1.75 ± 0.37 mg/100 g, respectively), but lower than those reported by Tlili et al. (2011) in five cultivars from Tunisia (11.3–20.6 mg/100 g). On the other hand, values ranging from 3.82 to 6.98 mg/100 g FW were assessed by Leskovar et al. (2004) for different watermelon cultivars grown under 100 percent evapotranspiration irrigation. Higher values attaining 57.62 mg/100 g FW were reported by Mélo et al. (2006) available on the market in Recife, Brazil. According to Leskovar et al. (2004), the variations are probably due to differences in genotypic and cultural practices.

Phenolic compounds content (TPC)

Bananas had the highest total polyphenol content and melon the lowest among the analyzed fruits of Costa Rica (Table 2.5). TPC values reported by other researchers for selected tropical fruits varied from the values obtained in our study. A higher total phenolic content was reported by Kumar Reddy et al. (2010) in mango and watermelon, but similar in pineapple and lower in banana (Table 2.5). These authors do not indicate the origin or variety of fruit, and moreover do not eliminate the interference analysis with the Folin–Ciocalteu reagent, as reported by González et al. (2012) in the case of the presence of vitamin C and reducing sugars. On the other hand, Patthamakanokporn et al. (2008) found a higher value in mango and banana (Table 2.5). Meanwhile, a higher value was assessed by Shofian et al. (2011) in mango (99.69 mg GAE/100 g FW) and watermelon (29.3 mg GAE/100 g FW), but lower in melon (16.71 ± 1.40 mg GAE/100 g FW).

Hassimotto et al. (2005) reported the phenolics in pineapple as a content of 67.2 ± 0.6 mg of GAE/100 g FW (mg Gallic Acid Equivalents (GAE)/100g Fresh Weight (FW)), Almeida et al. (2011) determined a value of 38.1 ± 0.7 mg of GAE/100 g, and that of honey pineapple was 34.7–54.7 GAE/100 g, according to Alothman et al. (2009).

According to Yi et al. (2006), who were cited by Dembitsky et al. (2011), the obtained results of the contents of phenolic compounds in peel, core, and pulp of pineapple by HPLC (high-performance liquid chromatography) showed that there were four phenolic compounds in the peel of pineapple: gallic acid, catechin, epicatechin, and ferulic acid. While in pulp and core of pineapple, only low concentrations of epicatechin and ferulic acid were detected.

Tlili et al. (2011) reported total phenolics in five cultivars of watermelon from Tunisia. At the red-ripe stage, Giza ranked first for total phenol content (26.02 mg GAE/100 g FW), followed by Dumara (24.64 mg GAE/100 g FW), P403 (20.08 mg GAE/100 g FW), P503 (18.34 mg GAE/100 g FW), and Crimson Sweet (13.72 mg GAE/100 g FW) cultivars. A much higher

Table 2.5: Phenolic Compounds Content (mg GAE/100 g) of the Main Cultivated Fruits.

	Fruit Content (mg GAE/100 g) Fresh Fruit Weight, Origin, and Variety				
References	**Banana**	**Pineapple**	**Mango**	**Melon**	**Watermelon**
Pérez (2005)	66 ± 2 Costa Rica cv. Grain naine	159 ± 35 Costa Rica cv. Gold	45 ± 2 Costa Rica cv. Tommy Atkins	48.6 ± 0.7 Costa Rica cv. Cantaloupe	15.1 ± 0.2 Costa Rica cv. Micky Lee
Alothman et al. (2009)	72.2 ± 2.0 Malaysia cv. Lady finger	54.1 ± 0.8 Malaysia	N.R.	N.R.	N.R.
Fu et al. (2011)	57.1 ± 3.6 China	94.0 ± 1.5 China	37.0 ± 2.1 China	31.5 ± 1.3 China Cantaloupe	24.7 ± 1.07
Brat et al. (2006)	51.5	32.7–61.6	68.1	7.8	11.6
Lako et al. (2007)	11 Fiji	15 Fiji	48 Fiji cv. Yalewa	N.R.	14 Fiji
Kumar Reddy et al. (2010)	43 ± 2.6 India cv. Lady finger	48 ± 7 India	307 ± 17 India	N.R.	26 ± 2.5
Patthamakanokpom et al. (2008)	78.3 ± 0.7 Thailand cv. Nam-wa	N.R.	113.0 ± 5.5 Thailand cv. Nam-dok-mai	N.R.	N.R.
González (2011)	53.0 ± 2.2 Costa Rica cv. Grain naine	52.5 ± 0.7 Costa Rica cv. Gold	40.9 ± 1.0 Costa Rica cv. criollo	5.4 ± 0.2 Costa Rica cv. Cantaloupe	21.3 ± 0.3 Costa Rica
Wolfe et al. (2008)	54.8 ± 1.3 United States	78.1 ± 0.8 United States	62.6 ± 4.2 United States	16.0 ± 0.4 United States	14.1 ± 0.3 United States

GAE, gallic acid equivalents; *N.R.*, not reported.

content of phenols, ranging between 87 and 91 mg GAE/100 g FW was reported by Perkins-Veazie et al. (2012) in red-fleshed watermelon cultivars grown in Oklahoma. Although, according to Tlili et al. (2011), the phenol contents in watermelons are lower than in other fruits such as cranberry and pomegranate, the large consumption of this fruit in the Mediterranean and American diet may considerably contribute to the daily intake of such compounds.

Such differences in the results of TPC compared with other researchers may be related to different varieties of fruits and the varying antioxidant extraction methods used (Shofian et al., 2011). Moreover, factors such as fruit maturity, agroclimate, and postharvest storage conditions are known to affect the content of polyphenols in fruits.

Antioxidant activity

Due to the potential of antioxidants to decrease the risk of different chronic diseases, it is important to be able to measure antioxidant activity using biologically relevant assays (Wolfe et al., 2008). Because they are rapid and cost-effective, chemical assays provide much information about the antioxidant potential of food (Gonzalez et al., 2012). The H-ORAC (hydrophilic oxygen radical absorbance capacity) assay is one of the most popular methods that have been used (Wu et al., 2004), and many databases have been created for food products (Gonzalez

Table 2.6: Antioxidant Capacity Assessed by ORAC Method (TE Micromol/g) of the Traditional Tropical Fruits.

References	Fruit Content (TE Micromol/g) Fresh Weight, Origin, and Variety				
	Banana	Pineapple	Mango	Melon	Watermelon
Pérez (2005)	10.7 ± 1.0 Costa Rica cv. Grain naine	13.3 ± 0.5 Costa Rica cv. Gold	6.96 ± 0.03 Costa Rica cv. Tommy Atkins	2.03 ± 0.05 Costa Rica cv. Cantaloupe	2.9 ± 0.4 Costa Rica cv. Micky Lee
Wu et al. (2004)	8.1 ± 2.0 Markets USA	7.6 ± 2.1 Markets USA	9.88 Markets USA	3.0 ± 0.6 Markets USA	1.2 ± 0.2 Markets USA
González (2011)	5.61 ± 0.25 Costa Rica cv. Grain naine	7.28 ± 0.23 Costa Rica cv. Gold	2.55 ± 0.08 Costa Rica cv. criollo	3.09 ± 0.13 Costa Rica cv. Cantaloupe	2.2 ± 0.1 Costa Rica
Wolfe et al. (2008)	5.65 ± 0.18 United States cv. Williams	10.6 ± 0.8 United States cv. Sweet Gold	11.6 ± 1.6 United States cv. Palmer	2.4 ± 0.2 United States	13.9 ± 0.1 United States
Kevers et al. (2007)	7.8 ± 2.6 Markets Belgium	N.R.	N.R.	3.8 ± 0.4 Markets Belgium	N. R.

TE, trolox equivalents; *N.R.*, not reported.

et al., 2012). The ORAC method was applied in order to determine the antioxidant activity in main cultivated fruits in Costa Rica and Central America (Table 2.6). In general, pineapple presented the highest ORAC value and melon the lowest one among the analyzed fruits.

2.4.2 Physicochemical Composition of Fruits Produced in Mexico

Coffee

Coffee is one of the most consumed beverages around the world and is the second largest traded commodity after petroleum (Mussatto et al., 2011a). In 2014, the production of coffee in Mexico was 1,166,025.82 tons (SIAP, 2014). The coffee cherries are composed by two coffee beans, silver skin, pulp, and husk. However, the chemical composition of coffee cherries show more than 1000 chemical compounds (Clarke, 2012).

The composition of green coffee depends of variety, origin, processing, climate, and also the genetic aspects. A good revision related to the components of coffee was published by Belitz et al. (2009). Caffeine is the most recognized component of coffee beans, and its content changes between varieties. In Arabica coffee, the values of caffeine can be found between 0.8% and 1.4% (w/w), while for Robusta coffee, the values vary between 1.7% and 4.0% (w/w). Nevertheless, the coffee beans are constituted by several compounds such as cellulose, minerals, sugars, lipids, tannins, and polyphenols (Table 2.7).

Although caffeine is the most known compound of coffee, there are many other bioactive compounds that can be obtained from byproducts of coffee industry. Coffee silver skin (CS) and the spent coffee grounds (SCG) are the main residues obtained from coffee industry.

Table 2.7: Composition of Green Coffee of Varieties Arabica and Robusta, Produced in Mexico.

Component	Arabica[a]		Robusta[a]	Constituents
Soluble carbohydrates	9–12.5		6–11.5	
Monosaccharides		0.2–0.5		Fructose, glucose, galactose, arabinose (traces)
Oligosaccharides	6–9		3–7	Sucrose (>90%), raffinose (0–0.9%), stachyose (0–0.13%)
Polysaccharides		3–4		Polymers of galactose (55–65%), mannose (10–20%), arabinose (20–35%), glucose (0–2%)
Insoluble polysaccharides	46–53		34–44	
Hemicelluloses	5–10		3–4	Polymers of galactose (65–75%), arabinose (25–30%), mannose (0–10%)
Cellulose, $\beta(1-4)$ mannan	41–43		32–40	
Acids and phenols				
Volatile acids		0.1		
Nonvolatile aliphatic acids	2–2.9		1.3–2.2	Citric acid, malic acid, quinic acid
Chlorogenic acid	6.7–9.2		7.1–12.1	Mono-, dicaffeoyl-, and feruloylquinic acid
Lignin		1–3		
Lipids	15–18		8–12	
Wax		0.2–0.3		
Oil		7.7–17.7		Main fatty acids: 16:0 and 18:2 (9,12)
N compounds		11–15		
Free amino acids		0.2–0.8		Main amino acids: Glu, Asp, Asp-NH_2
Proteins		8.5–12		
Caffeine	0.8–1.4		1.7–4.0	Traces of theobromine and theophylline
Trigonelline	0.6–1.2		0.3–0.9	
Minerals		3–5.4		

[a]Percentage in dry-weight basis.

Adapted from Belitz, H.-D., Grosch, W., Schieberle, P., 2009. Food chemistry. In: Coffee, Tea, Cocoa, Springer-Verlag Berlin Heidelberg, pp. 938–970.

The CS is a byproduct obtained through the roasting process and possesses a high concentration of dietary fiber (86%) and high antioxidant capacity due to the presence of phenolic compounds in coffee beans (Borrelli et al., 2004). The fibrous composition of CS corresponds to the presence of cellulose and hemicellulose. Glucose, xylose, galactose, mannose, and arabinose are the monosaccharides present in CS. Also, proteins are present in significant amounts (Mussatto et al., 2011b).

The SCG is a residue with high moisture (80–85%) and sugars (up to 73.4%) content. For example, it contains hemicellulose (36.7%), arabinose (1.7%), galactose (13.8%), and mannose (21.2%) (Mussatto et al., 2011b). Also, it possesses high antioxidant activity by the presence of tannins such as catechin, gallocatechin, chebulic acid, and gallocatechin-3-O-gallate (Low et al., 2015).

CS and SCG are considered as highly pollutant residues due to the presence of organic materials. The content of caffeine, tannins, and polyphenols present in the byproducts materials confer a toxic nature to them (Mussatto et al., 2011b); however, the added value biomolecules that may be recovered from such materials widely open the field of opportunities to reevaluate the byproducts from the coffee industry, together with the reduction of the impact of pollution to the environment.

Pomegranate

The pomegranate (*Punica granatum*) is a shrub or small tree that grows to heights between 5 and 8 m and bears highly colored fruit with many juicy seeds inside. The edible portion of pomegranate fruit, called an aril, is comprised of hundreds of seed coats. Seeds are sometimes soft or hard, depending on the cultivar. The juice obtained can be light pink to dark red and in some varieties can appear yellow or clear (MacLean et al., 2014).

In Mexico, the pomegranate is consumed and used as an ornamental tree. Despite its good adaptation, high productivity, and good quality, the fruits are used sporadically and are seasonal. Usually, the pomegranate is consumed as a fresh fruit or as garnish in national dishes. In 2014, the production of pomegranate in Mexico was of 4362.82 tons, obtained from the cultivation of 716.02 Ha (SIAP, 2014). In the last 15 years the production has increased due to the introduction of new products, for example: juice, tea, wines, liquors, and food supplements (Mercado et al., 2011). Also, the number of publications reporting bioactive molecules from pomegranate has increased the interest in the consumption of pomegranate (Aguilar et al., 2008; Aguilera-Carbo et al., 2008; Madrigal-Carballo et al., 2009).

The pomegranate composition is affected by factors such as genetic heterogeneity, varietal factors, climate conditions, soil structure, agricultural practices, water–heat stress, harvesting time, and storage conditions (Karaaslan et al., 2014). The physicochemical composition of seeds and husks was reported by Robledo et al. (2008). High values of total and reducing sugars were found in seeds and husks. Total protein and crude fat were higher in pomegranate seeds. Polyphenols were found in pomegranate husk with high content of hydrolyzable polyphenols (Table 2.8).

Pomegranate is rich in ellagitannins, such as punicalagin, and its monomers, punicalagin, gallagic acid, ellagic acid; and glycosides such as anthocyanins, gallotannins, hydroxybenzoic acids, and hydroxycinnamic acids (Fischer et al., 2011; Mena et al., 2012; Seeram et al., 2005). The ellagitannins are highly appreciated because of their biological properties; it is well-known that ellagitannins are promoters of good health, due to their antioxidant, antimutagenic, anticarcinogenic, and antiviral properties (Ascacio-Valdés et al., 2011).

The byproducts generated from the processing of pomegranate fruits represent an alternative and low-cost material to obtaining commercial products, including new nutritional products with potential health benefits. Particularly the pomegranate husk has been employed for the

Table 2.8: Physicochemical Characterization of Pomegranate Seeds and Husk in g/100 g of Fruit Produced in Mexico.

Dry Base Analysis	Seed (g)	Husk (g)
Dry matter	74.33	94.45
Moisture	25.66	5.5
Ash	3.62	3.59
Crude fat	10.33	3.57
Total protein	10.42	1.26
Crude fiber	12.12	17.75
Total sugars	10.13	16.08
Reducing sugars	4.67	4.34
Total hydrolysable phenols	0.002	6.11
Ellagic acid	0.00048	0.26
WAI (g gel/g dry sample)	1.62	4.84
CHP (%)	64.55	22.08
pH	5.37	5.70
Aw	0.156	0.189

WAI, Water absorption index; *CHP*, Critical humidity point; *Aw*, Water activity.
Adapted from Robledo, A., Aguilera-Carbo, A., Rodriguez, R., Martinez, J.L., Garza, Y., Aguilar, C.N., 2008. Ellagic acid production by Aspergillus niger in solid state fermentation of pomegranate residues. Journal of Industrial Microbiology and Biotechnology 35 (6), 507–513.

purification and releasing of ellagitannins and ellagic acid and for the production of enzymes with potential application in the food industry (Ascacio-Valdés et al., 2014; Buenrostro-Figueroa et al., 2014; De la Cruz et al., 2014; Huang et al., 2007; Sepúlveda et al., 2014; Wang et al., 2013).

Grapefruit

The production of citrus is one of the most important crops in the world. Orange, lemons, mandarins, and grapefruit account approximately 98% of the entire industrialized crops (Wang et al., 2015). The grapefruit (*Citrus paradisi*), a subtropical citrus tree originated in Barbados by the accidental cross between sweet orange (*Citrus sinensis*) and pomelo (*Citrus grandis*). The grapefruit is also named the forbidden fruit or pomelo and is known for its sour to semisweet flavor (Carrington and Fraser, 2003).

In Mexico, the production of grapefruit was 424,678.08 tons, according to data reported by SIAP (2014). The grapefruit in Mexico is destined for the consumption as fresh fruit, juices, and for confectionary. For industry, 20 percent of production is destined for the elaboration of juices, nonalcoholic beverages, and jams. Essential oil is extracted from the husk, which is used in the elaboration of fragrances (Pérez et al., 2010).

Grapefruit is a rich source of nutrients such as vitamin C and pectin fiber, and the pink and red hues contain the beneficial antioxidant lycopene (Fellers et al., 1990). The grapefruit helps to lower cholesterol and glycemic index, helps to metabolize of fat, and the seeds have antioxidant properties (Murphy et al., 2014). According to USDA (2015b) the grapefruit contains (per 100 g): 90.48 g of water, 0.69 g of protein, 0.10 g of total lipid, 0.33 g of ash, 8.41 g of carbohydrate, and 1.1 g of dietary fiber. It also contains high amounts of potassium (148 mg), vitamin C (33.3 mg of ascorbic acid), and flavonoids such as naringenin (21.3 mg).

The volatile fractions that represent the flavor of grapefruit are still inexplicit. Studies on the volatile compounds of grapefruit indicated that it possesses a unique aromatic profile (Cheong et al., 2011a,b). The identified compounds were 100 volatiles in grapefruit peel and 50 aroma active compounds (mainly unsaturated aliphatic aldehydes, terpene aldehydes, esters, terpene alcohols, and nootkatone), identified as the compounds that provide the odor and flavor to grapefruit (Cheong et al., 2011a,b).

2.5 Biological Activities of Fruits From Costa Rica and the Central America Region

2.5.1 Banana

Therapeutic and medicinal effects of different parts of banana were described by Dixit et al. (2014). Antiulcerogenic activity and hypocholesterolemic effect in rats was reported by these authors in the unripe, green Musa fruits, and the cholesterol-lowering effect is probably related to soluble and insoluble fibers. On the other hand, Chen et al. (2011) found a low glycemic index for banana fruits (35.1 ± 5.6), perhaps because of their resistant starch content (4 g/100 g), which was determined by Murphy et al. (2014) in ripe banana.

In a clinical study, Emery et al. (1997) concluded that banana flakes reduced the severity of diarrhea in critically ill tube-fed patients, and this product can be used as a safe, cost-effective treatment.

Banana contained dopamine at high levels in both the peel and pulp (Kanazawa and Sakakibara, 2000). According to these authors, dopamine levels ranged from 80 to 560 mg/100 g in peel and 2.5–10 mg/100 g in pulp, even in ripened bananas ready to eat. Dopamine is a strong, water-soluble antioxidant that has a greater antioxidant activity than glutathione or food additives such as butylated hydroxyanisole and hydroxytoluene.

A high concentration of serotonin 5-hydroxytryptamine (5-HT), which is a monoamine neurotransmitter, is found in fruits such as pineapple, 17.0 μg/g, and banana, 15.0 μg/g, (Dembitsky et al., 2011).

2.5.2 Mango

Parts of the mango, such as stem bark, leaves, and pulp are known for various biomedical applications, including antioxidative and free radical scavenging (Ajila et al., 2007; Rocha Ribeiro et al., 2007), antiinflammatory (Hernandez et al., 2007), and anticancer (Percival et al., 2006) activities.

The antioxidant and antiproliferative properties of flesh and peel of mango (*Mangifera indica* L.) were investigated by Kim et al. (2010). In this work, the cytoprotective effects of mango flesh and peel extracts on oxidative damage induced by H_2O_2 in a human hepatoma cell line, HepG2, were determined. According to these authors, treatment of HepG2 cell with mango peel extract prior to oxidative stress was found to inhibit DNA damage. The results also showed that the antiproliferative activity of mango flesh and peel extracts correlated with their phenolic contents. Thus, mango peel, a major byproduct obtained during the processing of mango products, exhibited good antioxidant activity, and it is a potential source of phenolics with antitumor activity.

The major polyphenols in mango in terms of antioxidative capacity and/or quantity were: mangiferin, catechins, quercetin, kaempferol, rhamnetin, anthocyanins, and gallic and ellagic acids, among other phenolic compounds (Dembitsky et al., 2011).

Mangiferin is a natural plant polyphenol with various pharmacological activities, and the mango tree (*M. indica*) is one of the primary sources (Matkowski et al., 2013). Mangiferin has strong antiradical and antioxidant properties; it is also an efficient iron chelator, therefore preventing the generation of hydroxyl radicals. In vitro and in vivo pharmacological studies demonstrated other activities of mangiferin, such as analgesic; antidiabetic; antisclerotic; antimicrobial; and antiviral; cardio-, hepato-, and neuroprotective; antiinflammatory; antiallergic; as well as radioprotective against X-ray, gamma, and UV radiation. Several studies indicated also its ability to inhibit cancerogenesis and cancer cells growth by apoptosis induction in vitro and in vivo.

2.5.3 Pineapple

Pineapple has been used as a medicinal plant in several native cultures, and these therapeutic qualities of pineapple are attributed to bromelain, which is a pharmacologically active compound, belonging to a group of protein-digesting enzymes obtained commercially from the fruit or stem of pineapple (Pavan et al., 2012; Dembitsky et al., 2011). In vitro and in vivo studies demonstrate that bromelain exhibits various fibrinolytic, antiedematous, antithrombotic, and antiinflammatory activities. The antitumor-initiating effects of bromelain in the mouse skin tumorigenesis model were reported (Bhui et al., 2009). Bromelain accounts for many therapeutic benefits such as the treatment of angina pectoris, bronchitis, sinusitis, surgical trauma, and thrombophlebitis, debridement of wounds, and enhanced absorption of drugs, particularly antibiotics. It also relieves osteoarthritis, diarrhea, and various cardiovascular disorders. Bromelain also possesses some anticancerous activities and promotes apoptotic cell death. Most of the in vitro and in vivo studies on anticancer activity of bromelain are concentrated on mouse and human cells, both cancerous and normal, treated with bromelain preparations (Bhui et al., 2009).

Several sterols were detected in the juices of pineapple; ergostanol and stigmastanol were found to be the sterol markers for pineapple juice (Ng and Hup, 1998).

2.5.4 Watermelon

Watermelon juice is naturally rich in L-citrulline, a nonprotein amino acid, and the benefits derived from its intake include improving athletic performance. This compound accelerates lactic acid removal, and according to Tarazona-Díaz et al. (2013), supplying 500 mL of natural watermelon juice significantly reduced muscular soreness 24 h after exercise. Also, citrulline malate was shown to increase levels of arginine and ornithine (which are important for muscle growth) and influence levels of growth hormone (Sureda et al., 2010).

A beneficial effect of watermelon pomace juice as a functional food was found by Wu et al. (2007). These authors reported increasing arginine availability, reducing serum concentrations of cardiovascular risk factors, improving glycemic control, and ameliorating vascular dysfunction in obese animals with type-II diabetes. Most of the L-citrulline released into blood by the liver is taken up by the kidney for the biosynthesis of arginine. It is also an essential amino acid for young mammals and adults with trauma, burn injury, massive small bowel resection, and renal failure (Jayaprakasha et al., 2011).

Citrulline could play a pivotal role in maintaining protein homeostasis and is a promising pharmaconutrient in nutritional support strategies for malnourished patients, especially in aging and sarcopenia (Bahri et al., 2013).

2.6 Agro–Industrial Fruit Wastes and Its Potential as Biomass

The fruit that is not exported exhibits a high potential to be used as feedstock for fermentation for its high content of fermentable sugars. The ripe banana has a total carbohydrate content of 21.8%, composed almost entirely of sucrose, glucose, and fructose (Chacon et al., 1987), and is a good source of vitamins C and Group B, along with various minerals (potassium, magnesium, phosphorus, manganese). Given these potential fermentable sugars and nutrients, these fruits are converted into a rich substrate with a high potential for fermentation processes.

The green bananas are subjected to a maturation process in order to bring them to a maturity grade 7 (yellow with initial stage black spots), wherein the conversion of starch to glucose and fructose generates about 180 g/L fermentable sugars. During banana ripening starch hydrolysis occurs, as well as accumulation of sugars such as glucose and fructose; these simple sugars increase due to strong enzymatic activity that presents this fruit, specifically invertase, which converts sucrose into a molecule of glucose and one of fructose. Overall, bananas are consumed in Costa Rica when they reach a level of maturity of 7, have a pH between 4.8 and 5.2 and Brix are between 22.5 and 23.5. After ripening, bananas receive a blanching step, pulping, and 50% dilution in order to prepare banana pulp for the enzymation process with pectinases.

For the preparation of a proper biomass suitable for fermentation, the bananas receive mechanical and enzymatic treatments before the fermentation process in order to generate monosaccharides (glucose and fructose) and disaccharides (sucrose) that can be used by the bacterium that

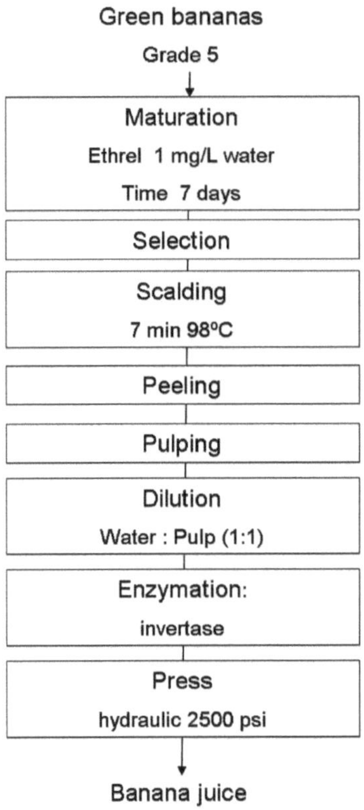

Figure 2.2
Flow diagram of the preparation of bananas to be used as carbon source in lactic acid fermentation process.

will undergo the fermentation process. In the case of the production of lactic acid, for example, the bacteria responsible for the transformation process will be *Lactobacillus casei* subsp. *rhamnosus*. The physical and enzymatic treatment will involve a pectinase to help the juice extraction, mainly. The pulp is then filtered or pressed in order to produce the juice that will be the carbon source in the medium for the liquid fermentation. This treatment is necessary to eliminate any excess solids that will more difficult later on in the purification and recovery during the downtream processing. The preparation of bananas to be used as carbon source in lactic acid fermentation is shown in Fig. 2.2 (CITA-CENIBIOT-CORBANA, 2012).

2.7 Conclusions

Central America and Mexico are regions with a high production of banana, pineapples, citrus fruits, and avocadoes, especially because of a very intense production from Costa Rica and Mexico. The major tropical fruits that are exported as high-quality fresh fruits generate an important amount (10–15%) of second-quality fruits that stay in the countries and are used by the agro–industry for processing and value added products. Even this industrialization produces a large amount of byproducts that are an excellent source of biomass for fermentation and biotechnological applications. The nutritional composition and the presence of bioactive molecules give these fruits a high-quality profile that gives them an advantage over other foods. These molecules are related to functional properties that need to be addressed to stimulate welfare of the population and that the region needs to incorporate through innovative processing technologies and science that supports all the potential functional claims of commercial value. The opportunities need to be exploited in order to generate positive economical and social impacts.

References

Aguilar, C.N., Aguilera-Carbo, A., Robledo, A., Ventura, J., Belmares, R., Martinez, D., Contreras, J., 2008. Production of antioxidant nutraceuticals by solid-state cultures of pomegranate (*Punica granatum*) peel and creosote bush (*Larrea tridentata*) leaves. Food Technology and Biotechnology 46 (2), 218–222.

Aguilera-Carbo, A., Augur, C., Prado-Barragan, L.A., Favela-Torres, E., Aguilar, C.N., 2008. Microbial production of ellagic acid and biodegradation of ellagitannins. Applied Microbiology and Biotechnology 78 (2), 189–199.

AIJN, European Fruit Juice Association, 2012. Liquid Fruit, Market Report. https://revipackonline.files.wordpress.com/2013/01/aijn-2012-liquid-fruit-market-report.pdf.

AIJN, European Fruit Juice Association, 2014. Liquid Fruit, Market Report. In: http://aijn.org/files/default/aijn2014-full.pdf.

Ajila, C.M., Bhat, S.G., Prasada Rao, U.J.S., 2007. Valuable components of raw and ripe peels from two Indian mango varieties. Food Chemistry 102, 1006–1011.

Almeida, M.M.B., de Sousa, P.H.M., Arriaga, A.M.C., do Prado, G.M., Magalhães, C.E.C., Maia, G.A., de Lemos, T.L.G., 2011. Bioactive compounds and antioxidant activity of fresh exotic fruits from northeastern Brazil. Food Research International 44, 2155–2159.

Alothman, M., Bhat, R., Karim, A.A., 2009. Antioxidant capacity and phenolic content of selected tropical fruits from Malaysia, extracted with different solvents. Food Chemistry 115, 785–788.

Araya-Cloutier, C., Rojas-Garbanzo, C., Velázquez-Carrillo, C., 2010. Síntesis de ácido poliláctico a través de la hidrólisis enzimática simultánea a la fermentación de un medio a base de un desecho de piña (*Ananas comosus*), para uso como materia prima en la elaboración de ácido poliláctico. Revista Iberoamericana de Polímeros 11 (7), 407–416.

Araya-Cloutier, C., Velázquez-Carriilo, C., Rojas-Garbanzo, C., 2012. Effect of initial sugar concentration on the production of L (+) lactic acid by simultaneous enzymatic hydrolysis and fermentation of an agro-industrial waste product of pineapple (*Ananas comosus*) using *Lactobacillus casei* subspecies *rhamnosus*. International Journal of Biotechnology for Wellness Industries 1 (1), 91–100.

Ascacio-Valdés, J.A., Buenrostro, J.J., De la Cruz, R., Sepúlveda, L., Aguilera, A.F., Prado, A., Aguilar, C.N., 2014. Fungal biodegradation of pomegranate ellagitannins. Journal of Basic Microbiology 54 (1), 28–34.

Ascacio-Valdés, J.A., Buenrostro-Figueroa, J.J., Aguilera-Carbó, A., Prado-Barragán, A., Rodriguez-Herrera, R., Aguilar, C.N., 2011. Ellagitannins: biosynthesis, biodegradation and biological properties. Journal of Medicinal Plants Research 5 (19), 4696–4703.

Badilla, M., 2005. Caracterización de compuestos volátiles en jugos pulposos y clarificados de frutas de exportación. Project Graduation. Food Engineering. Universidad de Costa Rica.

Bahri, S., Zerrouk, N., Aussel, C., Moinard, C., Crenn, P., Curis, E., Chaumeil, J.C., Cynober, L., Sfar, S., 2013. Citrulline: from metabolism to therapeutic use. Nutrition 29, 479–484.

Belitz, H.-D., Grosch, W., Schieberle, P., 2009. Food chemistry. In: Coffee, Tea, Cocoa. Springer-Verlag Berlin Heidelberg, pp. 938–970.

Bhui, K., Prasad, S., George, J., Shukla, Y., 2009. Bromelain inhibits COX-2 expression by blocking the activation of MAPK regulated NF-kappa B against skin tumor initiation triggering mitochondrial death pathway. Cancer Letters 282, 167–176.

Borrelli, R.C., Esposito, F., Napolitano, A., Ritieni, A., Fogliano, V., 2004. Characterization of a new potential functional ingredient: coffee silverskin. Journal of Agricultural and Food Chemistry 52 (5), 1338–1343.

Brat, P., George, S., Bellamy, A., Du Chaffaut, L.D., Scalbert, A., Mennen, L., Arnault, N., Amiot, M.J., 2006. Daily polyphenol intake in France from fruit and vegetables. Journal of Nutrition 136, 2368–2373.

Brenes Gamboa, S., 2005. Caracterización vegetativa y productiva del cultivar MD-2 de piña (*Ananas comosus*) bajo las condiciones climáticas de Turrialba. Inter Sedes 6 (11), 27–34.

Buenrostro-Figueroa, J., Huerta-Ochoa, S., Prado-Barragán, A., Ascacio-Valdés, J., Sepúlveda, L., Rodríguez, R., Aguilar, C.N., 2014. Continuous production of ellagic acid in a packed-bed reactor. Process Biochemistry 49 (10), 1595–1600.

Cámara Nacional de Productores y Exportadores de Piña—CANAPEP, 2015. Statistics. http://canapep.com/canapep/estadisticas/.

Carrington, S., Fraser, H.C., 2003. Grapefruit. A~Z of Barbados Heritage. Macmillan Caribbean, pp. 90–91.

Chacón, S.I., Viquez, F., Chacón, G., 1987. Escala físico-química de maduración de banano. Physico-chemical banana ripening scale. Fruits 42, 95–102.

Chaverri, M., 2003. Elaboración de un jugo de melón clarificado por medio de microfiltración tangencial. Project Graduation. Food Engineering. Universidad de Costa Rica.

Chen, Y.-Y., Wu, P.Ch., Weng, S.F., Liu, J.F., 2011. Glycemia and peak incremental indices of six popular fruits in Taiwan: healthy and type 2 diabetes subjects compared. Journal of Clinical Biochemistry Nutrition 49 (3), 195–199.

Cheong, M., Liu, S., Yeo, J., Chionh, H., Pramudya, K., Curran, P., Yu, B., 2011a. Identification of aroma-active compounds in Malaysian pomelo (*Citrus grandis* (L.) Osbeck) peel by gas chromatography-olfactometry. Journal of Essential Oil Research 23 (6), 34–42.

Cheong, M.-W., Loke, X.-Q., Liu, S.-Q., Pramudya, K., Curran, P., Yu, B., 2011b. Characterization of volatile compounds and aroma profiles of Malaysian pomelo (*Citrus grandis* (L.) Osbeck) blossom and peel. Journal of Essential Oil Research 23 (2), 34–44.

Choonut, A., Saejong, M., Sangkharak, K., 2014. The production of ethanol and hydrogen from pineapple peel by *Saccharomyces cerevisiae* and *Enterobacter aerogenes*. Energy Procedia 52, 242–249.

CITA, 2015. Reporte de Análisis Químico. Laboratorio de Química. 4 p.

CITA-CENIBIOT-CORBANA, 2012. Development of an Scaling up Process for the Fermentation of L(+) Lactic Acid Using Banana Fruit (*Musa Cavendish*) as Substrate. Final Report of Project. .

Clarke, R., 2012. Coffee, vol. 1. Chemistry: Springer Science & Business Media.

CORBANA, 2015. Banana Export Data. In: https://www.corbana.co.cr.

De la Cruz, R., Ascacio, J.A., Buenrostro, J.J., Sepúlveda, L., Rodríguez, R., Prado-Barragán, A., Aguilar, C.N., 2014. Optimization of ellagitannase production by *Aspergillus niger* GH1 by solid state fermentation Preparative Biochemistry and Biotechnology 45, 617–631.

Dembitsky, V., Poovarodom, S., Leontowicz, M., Vearasilp, S., Trakhtenberg, S., Gorinstein, S., 2011. The multiple nutrition properties of some exotic fruits: biological activity and active metabolites. Food Research International 44, 1671–1701.

Dixit, P., Kumar, V., Shukla, R., 2014. Therapeutic and medicinal effects of different parts of *Musa sapientum*. Vivechan International Journal of Research 5 (1), 62–68.

Dull, G.G., 1971. The Pineapple: General. In: Hulme, A.C. 1971. The Biochemistry of Fruits and Their Products. Academic Press, London. 620 p.

Emery, E.A., Ahmad, S., Koeth, J.D., Skipper, A., Perlmutter, S., Paskin, D.L., 1997. Banana flakes control diarrhea in enterally fed patients. Nutrition in Clinical Practice 12, 72–75.

FAO, 2012. The State of Food and Agriculture. www.fao.org/docrep/017/i3028s/i3028s.pdf.

FAO, 2014. Fao Stats Database. In: http://faostat3.fao.org/browse.

Fellers, P., Nikdel, S., Lee, H., 1990. Nutrient content and nutrition labeling of several processed Florida citrus juice products. Journal of the American Dietetic Association 90 (8), 1079–1084.

Fischer, U.A., Carle, R., Kammerer, D.R., 2011. Identification and quantification of phenolic compounds from pomegranate (*Punica granatum* L.) peel, mesocarp, aril and differently produced juices by HPLC-DAD-ESI/MS(n). Food Chemistry 127 (2), 807–821.

Flath, R.A., 1980. Tropical and Subtropical Fruits. Avi Publishing, Westport.
Forster, M.P., Rodríguez Rodríguez, E., Díaz Romero, C., 2002. Differential characteristics in the chemical composition of bananas from Tenerife (Canary Islands) and Ecuador. Journal of Agricultural and Food Chemistry 50, 7586–7592.
Fu, Li, Xu, B.-T., Xu, X.-R., Gan, R.-Y., Zhang, Y., Xia, E.-Q., Li, H.-B., 2011. Antioxidant capacities and total phenolic contents of 62 fruits. Food Chemistry 129, 345–350.
González, E., 2011. Efecto del tratamiento térmico y la digestión in vitro sobre la capacidad antioxidante de un jugo de fruta a nivel químico y celular. Master Thesis in Food Science, Universidad de Costa Rica.
González, E., Vaillant, F., Pérez, A., Rojas, G., 2012. In vitro cell-mediated antioxidant protection of human erythrocytes by some common tropical fruits. Journal of Nutrition and Food Sciences 2 (3), 139.
GPG, 2015. Economic Competitiveness Rating—Mexico Compared to Continent. In: http://www.globalpropertyguide.com/Latin-America/Mexico/competitiveness.
Hassimotto, N.M.A., Genovese, M.I., Lajolo, F.M., 2005. Antioxidant activity of dietary fruits, vegetables, and commercial frozen fruit pulps. Journal of Agricultural and Food Chemistry 53 (8), 2928–2935.
Hernández, Y., Lobo, G.M., González, M., 2007. Determination of vitamin C in tropical fruits: a comparative evaluation of methods. Food Chemistry 96, 654–664.
Hernandez, P., Rodriguez, P.C., Delgado, R., Walczak, H., 2007. Protective effect of *Mangifera indica* L. polyphenols on human T lymphocytes against activation-induced cell death. Pharmacology Research 55, 167–173.
Huang, W., Niu, H., Li, Z., Lin, W., Gong, G., Wang, W., 2007. Effect of ellagitannin acyl hydrolase, xylanase and cellulase on ellagic acid production from cups extract of valonia acorns. Process Biochemistry 42 (9), 1291–1295.
Idris, A., Suzana, W., 2006. Effect of sodium alginate concentration, bead diameter, initial pH and temperature on lactic acid production from pineapple waste using immobilized *Lactobacillus delbrueckii*. Process Biochemistry 41 (5), 1117–1123.
Jayaprakasha, G.K., Chidambara Murthy, K.N., Patil, B.S., 2011. Rapid HPLC-UV method for quantification of L-citrulline in watermelon and its potential role on smooth muscle relaxation markers. Food Chemistry 127, 240–248.
Kanazawa, K., Sakakibara, H., 2000. High content of dopamine, a strong antioxidant, in Cavendish banana. Journal of Agricultural and Food Chemistry 48, 844–848.
Karaaslan, M., Vardin, H., Varlıklıöz, S., Yılmaz, F.M., 2014. Antiproliferative and antioxidant activities of Turkish pomegranate (*Punica granatum* L.) accessions. International Journal of Food Science and Technology 49 (1), 82–90.
Kevers, C., Falkowski, M., Tabart, J., Defraigne, J.O., Dommes, J., Pincemail, J., 2007. Evolution of antioxidant capacity during storage of selected fruits and vegetables. Journal of Agricultural and Food Chemistry 55, 8596–8603.
Kim, H., Moon, J.Y., Kim, H., Lee, D.-S., Cho, M., Choi, H.-K., Kim, Y.S., Mosaddik, A., Cho, S.K., 2010. Antioxidant and antiproliferative activities of mango (*Mangifera indica* L.) flesh and peel. Food Chemistry 121, 429–436.
Kumar Reddy, C.V., Sreeramula, D., Raghunath, M., 2010. Antioxidant activity of fresh and dry fruits commonly consumed in India. Food Research International 43, 285–288.
Lako, J., Trenerry, V.C., Wahlqvist, M., Wattanapenpaiboon, N., Sotheeswaran, S., Premier, R., 2007. Phytochemical flavonols, carotenoids and the antioxidant properties of a wide selection of Fijian fruit, vegetables and other readily available foods. Food Chemistry 101, 1727–1741.
Leong, L.P., Shui, G., 2002. An investigation of antioxidant capacity of fruits in Singapore markets. Food Chemistry 76, 69–75.
Leskovar, D.I., Bang, H.J., Crosby, K., Maness, N., Franco, J.A., Perkins-Veazie, P., 2004. Lycopene, carbohydrates, ascorbic acid, and yield components of diploid and triploid watermelon cultivars are affected by deficit irrigation. Journal of Horticultural Science and Biotechnology 79 (3), 75–81.
Low, J.H., Rahman, W.A.W.A., Jamaluddin, J., 2015. Structural elucidation of tannins of spent coffee grounds by CP-MAS ^{13}C NMR and MALDI-TOF MS. Industrial Crops and Products 69, 456–461.

MacLean, D., Martino, K., Scherm, H., Horton, D., January 4, 2014. Pomegranate Production. Retrieved August 1st, 2015, from: http://extension.uga.edu/publications/files/pdf/C%20997_4.PDF.

Madrigal-Carballo, S., Rodriguez, G., Krueger, C., Dreher, M., Reed, J., 2009. Pomegranate (*Punica granatum*) supplements: authenticity, antioxidant and polyphenol composition. Journal of Functional Foods 1 (3), 324–329.

Matkowski, A., Kus, P., Goralska, E., Wozniak, D., 2013. Mangiferin—a bioactive xanthonoid, not only from mango and not just antioxidant. Mini-Reviews in Medicinal Chemistry 13 (3), 439–455.

Mélo, E.A., Lima, V.L.A.G., Mciel, M.I.S., Caetano, A.C.S., Leal, F.L.L., 2006. Polyphenol, ascorbic acid and total carotenoid contents in common fruits and vegetables. Brazilian Journal of Food Technology 9 (2), 89–94.

Mena, P., Calani, L., Dall'Asta, C., Galaverna, G., Garcia-Viguera, C., Bruni, R., Crozier, A., Del Rio, D., 2012. Rapid and comprehensive evaluation of (poly)phenolic compounds in pomegranate (*Punica granatum* L.) juice by UHPLC-MSn. Molecules 17 (12), 14821–14840.

Mercado Silva, E., Mondragón Jacobo, C., Rocha Peralta, L., Álvarez Mayorga, B., 2011. Efectos de condición del fruto y temperatura de almacenamiento en la calidad de granada roja. Revista mexicana de ciencias agrícolas 2 (3), 449–459.

Murillo, E., 2013. Combinación de tecnologías de membranas para la purificación de L (+) – ácido láctico a partir de mosto de banano (Musa AAA, variedad Cavendish cultivar Gram naine) obtenido de un desecho agroindustrial. Project Graduation. Food Engineering. Universidad de Costa Rica.

Murphy, M.M., Barraj, L.M., Rampersaud, G.C., 2014. Consumption of grapefruit is associated with higher nutrient intakes and diet quality among adults, and more favorable anthropometrics in women, NHANES 2003–2008. Food and Nutrition Research 58, 22179.

Mussatto, S.I., Carneiro, L.M., Silva, J.P., Roberto, I.C., Teixeira, J.A., 2011a. A study on chemical constituents and sugars extraction from spent coffee grounds. Carbohydrate Polymers 83 (2), 368–374.

Mussatto, S.I., Machado, E.M., Martins, S., Teixeira, J.A., 2011b. Production, composition, and application of coffee and its industrial residues. Food and Bioprocess Technology 4 (5), 661–672.

Ng, L.-K., Hup, M., 1998. Analysis of sterols: a novel approach for detecting juices of pineapple, passionfruit, orange and grapefruit in compounded beverages. Journal of the Science of Food and Agriculture 76, 617–627.

Pappu, A., Patil, V., Jain, S., Mahindrakar, A., Hague, R., Thakur, V.K., 2015. Advances in industrial prospective of cellulosic macromolecules enriched banana biofibre resources: a review. International Journal of Biological Macromolecules 79, 449–458.

Patthamakanokporn, O., Prapasri, P., Anadi, N., Prapaisri, P.S., 2008. Changes of antioxidant activity and total phenolic compounds during storage of selected fruits. Journal of Food Composition and Analysis 21, 241–248.

Pavan, R., Jain, S., Kumar, A., 2012. Properties and therapeutic application of bromelain: a review. Biotechnology Research International Article ID 976203, 6 p.

PEN—Programa Estado de la Nación, 2009. Estado de la Región. San José, Costa Rica. In: http://www.estadonacion.or.cr/files/biblioteca_virtual/centroamerica/004/cap04_economico-REGION004.pdf.

PEN, 2014. Programa Estado de la Nación en Desarrollo Humano Sostenible Estadísticas de Centroamérica. San José, Costa Rica. 96 p. In: http://www.estadonacion.or.cr/images/stories/biblioteca_virtual/otras_publicaciones/ECA-2014.pdf.

Percival, S.S., Talcott, S.T., Chin, S.T., Mallak, A.C., Lounds-Singleton, A., Pettit-Moore, J., 2006. Neoplastic transformation of BALB/3T3 cells and cell cycle of HL-60 cells are inhibited by mango (*Mangifera indica* L.) juice and mango juice extracts. Journal of Nutrition 136, 1300–1304.

Pérez, A.M., 2005. Caracterización físico-química de frutas de exportación: piña, mango, melón y banano. Estudio sobre su contenido en pigmentos y compuestos volátiles. Informe final de proyecto. Centro Nacional de Ciencia y Tecnología de Alimentos, Universidad de Costa Rica. Proyecto MICITT-CONICIT 147-01-FI, 26 p.

Pérez, E.G., Hernández, P.P., Díaz, M.J.J.G., 2010. Necesidades de Investigación y Transferencia de Tecnología de la Cadena Agroalimentaria Toronja en el estado de Veracruz. Tepetates, Veracruz Colegio de Postgraduados, FUNAPROVER, SAGARPA Veracruz, México. http://www.cofupro.org.mx/cofupro/Publicacion/Archivos/penit148.pdf.

Perkins-Veazie, P., Beaulieu, J.C., Siddiq, M., 2012. Watermelon, Cantaloupe and Honeydew. In: Sidiqq, M. (Ed.), Tropical and Subtropical Fruits: Postharvest Physiology, Processing and Packaging. John Wiley & Sons, Inc., New Jersey, pp. 549–568.

Procomer, 2015. Estadísticas del sector exportador. http://www.procomer.com/contenido/sector-alimentario.html.

Proteggente, A.R., Pannala, A.S., Paganga, G., Van Buren, L., Wagner, E., Wiseman, S., Van De Put, F., Dacombe, C., Rice-Evans, C.A., 2002. The antioxidant activity of regularly consumed fruit and vegetables reflects their phenolic and vitamin C composition. Free Radical Research 36 (2), 217–233.

Quesada, C., 2013. Determinación de las condiciones óptimas del tratamiento enzimático acoplado a un proceso de microfiltración tangencial para la obtención de jugo clarificado de piña en una empresa procesadora de frutas y vegetales. Project Graduation. Food Engineering. Universidad de Costa Rica.

Quesada, K., Alvarado, P., Sibaja, R., Vega, J., 2005. Utilización de las fibras del rastrojo de piña *(Ananas comosus, variedad champaka)* como material de refuerzo en resinas de poliéster. Revista Iberoamericana de Polímeros 6 (2), 158–179.

Ramsaroop, R.E.S., Saulo, A.A., 2007. Comparative consumer and physicochemical analysis of Del Monte Hawaii gold and smooth Cayenne pineapple cultivars. Journal of Food Quality 30, 135–159.

Rani, D.S., Nand, K., 2004. Ensilage of pineapple processing waste for methane generation. Waste Management 24, 523–528.

Reddy, G.V., Ravindra Babu, P., Komaraiah, P., Roy, K.R.R.M., Kothari, I.L., 2003. Utilization of banana waste for the production of lignolytic and cellulolytic enzymes by solid substrate fermentation using two Pleurotus species (*P. ostreatus* and *P. sajor-caju*). Process Biochemistry 38, 1457–1462.

Robledo, A., Aguilera-Carbo, A., Rodriguez, R., Martinez, J.L., Garza, Y., Aguilar, C.N., 2008. Ellagic acid production by *Aspergillus niger* in solid state fermentation of pomegranate residues. Journal of Industrial Microbiology and Biotechnology 35 (6), 507–513.

Roache, S., 2007. Central America's Regional Trends and U.S. Cycles. Economic and Financial Linkages in the Western Hemisphere. Western Hemisphere Department. Intenational Monetary Fund. In: https://www.imf.org/external/np/seminars/eng/2007/whd/pdf/session2-1.pdf.

Rocha Ribeiro, S.M., Queiroz, J.H., Ribeiro, L., de Queiroz, M.E., Campos, F.M., Pinheiro Sant'ana, H.M., 2007. Antioxidant in mango (*Mangifera indica* L.) pulp. Plant Foods for Human Nutrition 62, 13–17.

Sauma, J., March 03, 2014. Costa Rican Bananas Competing in a Saturated World Market. Fresh Plaza.com. In: http://www.freshplaza.com/article/119001/Costa-Rican-bananas-competing-in-a-saturated-world-market.

Seeram, N., Lee, R., Hardy, M., Heber, D., 2005. Rapid large scale purification of ellagitannins from pomegranate husk, a by-product of the commercial juice industry. Separation and Purification Technology 41 (1), 49–55.

Sepúlveda, L., Buenrostro-Figueroa, J., Ascacio-Valdés, J., Aguilera-Carbó, A., Rodríguez-Herrera, R., Contreras-Esquivel, J., Aguilar, C., 2014. Submerged culture for production of ellagic acid from pomegranate husk by *Aspergillus niger* GH1. Micología Aplicada International 26 (2), 27–35.

Shofian, N.M., Hamid, A.A., Osman, A., Saari, N., Anwar, F., Pak Dek, M.S., Hairuddin, M.R., 2011. Effect of freeze-drying on the antioxidant compounds and antioxidant activity of selected tropical fruits. International Journal of Molecular Sciences 12, 4678–4692.

SIAP, 2014. Cierre de la producción agrícola por cultivo. Retrieved 25 July, 2015, from: http://www.siap.gob.mx/cierre-de-la-produccion-agricola-por-cultivo/.

Sureda, A., Córdoba, A., Ferrer, M., Pérez, G., Tur, J., Pons, A., 2010. L-Citrulline-malate influence over branched chain amino acid utilization during exercise. European Journal of Applied Physiology–Springer 110, 341–351.

Tarazona-Díaz, M.P., Alacid, F., Carrasco, M., Martínez, I., Aguayo, E., 2013. Watermelon juice: potential functional drink for sore muscle relief in athletes. Journal of Agricultural and Food Chemistry 61, 7522–7528.

Ti, T.C., 2000. FAO Bangkok Thailand. In: Subhadrabandhu, S., Chairidehai, P. (Eds.), The Global Pineapple Economy. Procedings of III International Pineapple Symposium. Acta Horticuturae. 529. In: http://www.actahort.org/books/529/529_3.htm.

Tlili, I., Hdider, C., Lenucci, M.S., Ilahy, R., Jebari, H., Dalessandro, G., 2011. Bioactive compounds and antioxidant activities during fruit ripening of watermelon cultivars. Journal of Food Composition and Analysis 24, 923–928.

United Nations Program for Development–PNUD, 2014. Human Development Report 2014. In: http://hdr.undp.org/en/content/human-development-index-hdi.

United States Department of Agriculture (USDA), 2015a. National Nutrient Database for Standard Reference. Agricultural Research Service. Release 27 http://ndb.nal.usda.gov/ndb/nutrients/report/nutrientsfrm?max=25&offset=0&totCount=0&nutrient1=203&nutrient2=&nutrient3=&subset=0&fg=&sort=f&measureby=g.

USDA, 2015b. Full Report (All Nutrients): 09116, Grapefruit, Raw, White, All Areas. Retrieved July 29th, 2015, from: http://ndb.nal.usda.gov/ndb/foods/show/2274?fgcd=&manu=&lfacet=&format=Full&count=&max=35&offset=&sort=&qlookup=09116.

Vagneron, I., Faure, G., Loeillet, D., 2009. Is there a pilot in the chain? Identifying the key drivers of change in the fresh pineapple sector. Food Policy 34, 437–446.

Valente, A., Albuquerque, T.G., Sanches-Silva, A., Costa, H.S., 2011. Ascorbic acid content in exotic fruits: a contribution to produce quality data for food composition databases. Food Research International 44, 2237–2242.

Wang, L., Xu, H., Yuan, F., Pan, Q., Fan, R., Gao, Y., 2015. Physicochemical characterization of five types of citrus dietary fibers. Biocatalysis and Agricultural Biotechnology 4 (2), 250–258.

Wang, Y., Zhang, H., Liang, H., Yuan, Q., 2013. Purification, antioxidant activity and protein-precipitating capacity of punicalin from pomegranate husk. Food Chemistry 138 (1), 437–443.

Wolfe, K.L., Kang, X., He, X., Dong, M., Zhang, Q., Liu, R.H., 2008. Cellular antioxidant activity of common fruits. Journal of Agricultural and Food Chemistry 56 (18), 8418–8426.

World Economic Forum (WEF), 2015. The Global Competitiveness Report 2014–2015. In: http://reports.weforum.org/global-competitiveness-report-2014-2015/.

Wu, X., Beecher, G.R., Holden, J.M., Haytowitz, D.B., Gebhardt, S.E., Prior, R.L., 2004. Lipophilic and hydrophilic antioxidant capacities of common foods in the United States. Journal of Agricultural and Food Chemistry 52, 4026–4037.

Wu, G., Collins, J.F., Perkins-Veazie, P., Siddiq, M., Dolan, K.D., Kelly, K.A., Heaps, C.L., Meininger, C.J., 2007. Dietary supplementation with watermelon pomace juice enhances arginine availability and ameliorates the metabolic syndrome in Zucker diabetic fatty rats. Journal of Nutrition 137, 2680–2685.

CHAPTER 3

Plants By-Products and Fibers' Industrial Exploitation

S. Narasimhan
Asthagiri Herbal Research Foundation, Chennai,Tamil Nadu, India

B.S. Srikanth
Intrim Business Associates, Chennai, Tamil Nadu, India

P. Poltronieri
CNR-ISPA, Lecce, Italy

3.1 Introduction

Some of the basic feedstocks for many industrial biotechnology processes are carbohydrates (different sugars ranging from sugar syrup, to crystalline sugar, starch, glucose, and molasses) and cellulosic and hemicellulosic fibers. Non-European industries have access to cheap raw materials. Therefore, the access to world market prices for sugar is a condition for industrial biotechnology companies to support their activities. One alternative represents the use of price competitive waste materials from the fruit and vegetable transforming industry, which could provide a cheap alternative carbohydrate resource for bioderived chemicals. Plants use the light energy from the sun to convert water and carbon dioxide to sugars that can be stored through the process of photosynthesis. Some plants, like sugar cane and sugar beets, store the energy as simple sugars. These are mostly used for food. Other plants store the energy as more complex sugars, such as starches, and pectin, formed by galacturonic acid and neutral sugars, such as rhamnose, arabinose, galactose, and xylose. Another type of plant matter called cellulosic biomass is made up of very complex sugar polymers and is not generally used as a food source. This type of biomass is under consideration as a feedstock for bioderived chemical production. The main components of these types of biomass are polysaccharides. Cellulose, accounting for 40–60% by weight of the biomass, a complex polysaccharide made from the six-carbon sugar, glucose. Its crystalline structure makes it resistant to hydrolysis, the chemical reaction that releases simple, fermentable sugars from a polysaccharide. Hemicellulose is also a major source of carbon in biomass at levels of between 20% and 40% by weight. It is a complex polysaccharide made from a variety of five- and six-carbon sugars (glucose, xylose, arabinose, etc.). Lignin is a complex polymer, which provides structural integrity in plants. It makes up 10–24% by weight of biomass.

Biotransformation of Agricultural Waste and By-Products. http://dx.doi.org/10.1016/B978-0-12-803622-8.00003-3

Food processors are becoming increasingly aware that waste and by-products offer new market opportunities. Research addresses the issue of plant (fruit and vegetable) wastes and by-products by taking a comprehensive approach looking for sustainable processes to turn food processing by-products and nonfood plant fibers into valuable industrial products (fabrics and textiles, vegetal leather, elements for the automotive and building industry, cases for electronics products or household appliances, composites, bioresins for plastic materials, lightweight structural composites based on biobased polymers, reinforced with cellulose pulp fibers). There are many additional applications for plant wastes, including the use of wood fibers to produce lower carbon, energy efficient construction materials with antiseismic properties, or the use of bio-based and biodegradable agricultural mulch films for improving agricultural productivity.

To be successful, the bioeconomy must take into account the balance of industrial use of biomass resources for multiple uses including food, feed, chemicals, fibers, and fuels. To achieve this, the concepts of smart and sustainable use of biomass should be applied.

3.2 India Food Industry Wastes: Mango Case Study

Mango is cultivated over an area of 2,500,000 hectares with an annual production of 13 million metric tons, accounting for about 45% of global production. The exports data: mango fruit 69,096 tons (1.28 Billion Rs) Pulp 134,613 tons (3.64 Billion Rs) (Source: APEDA).

Ripe mangoes are usually eaten as a fruit. Further, items like jams, jellies, pickle, smoothies, ice cream, sweet chili sauce, mango drinks, milkshakes, etc., are made from mango.

3.2.1 Processed Mango

The processing of mango is mainly in the form of canning, producing considerable quantities of pulp. India is a major exporter of mango pulp in the world. However, it should be noted that canneries generate waste to the extent of 40–50 percent of total fruit weight of processed mangoes.

Thus, a large quantity of mango by-products such as seed, kernel, peel, and stone are unutilized and disposed as waste (Table 3.1). These contains a variety of interesting molecules, which when utilized, can radiate new revenue and economic growth with sustainable mango ecosystem.

3.2.2 Barriers

1. Lack of a critical ecosystem of value-added processors of fruits and vegetables looking beyond the core product value.
2. Production areas in many cases are far removed from existing processing hubs of India.
3. Needed national mission beyond biofuels pitch to address possibilities in feed, food, and fiber dimensions for perceived waste from fruits and vegetables.

Table 3.1: Potential of Mango Waste Components.

Mango Waste (By-products)	Chemicals Present (Reported)	Bioactivity (Reported)	Reported Uses
Mango seed	Starch Fat Protein		
Mango seed kernel	Saturated fatty acid—44–48% Unsaturated fatty acid—52–56% Essential and non-essential amino acids—with Leucine, valine and lysine—high contents Polyphenols—tannin, gallic acid, coumarin, caffic acid, vanillin, mangiferin, ferulic and cinnamic acid. Phytosterols—campesterol, sitosterol and tocopherols Dietary fiber K, Mg, Ca, P, Na, Fe, Zn, and Mn Vits. C, E, A, K, and B1, B2, B6, B12	DPPH radical scavenging activity of seed is 24.2%	Functional food ingredient Antimicrobial activity (Ayala-Zavala et al., 2011). Antioxidants (Imran et al., 2013; Larrauri et al., 1997) Cosmetics Mango kernel starch solubility range from 0.141 to 0.149 g/g, the DSC analysis show good thermal property. The amylase content in the starch is low and it is oval to elliptical in shape, with 1.5–28 μm in size. These values are less than potato starch but closer to rice and corn (Kaur et al., 2004)
Mango seed kernel oil	Free fatty acid—5.35% Palmitic acid, stearic, oleic, linoleic, linolenic and arachidic acid are present. Saturated fatty acid—44.5% Unsaturated fatty acid—55.5% Iodine value—39.5 Saponification value—207.5		Good blend for vegetable oil Mango seed fat Substitute for cocoa butter Substituted as natural antioxidant in food Cosmetics
Cake after oil extraction	Two essential amino acid—methionine and lysine (Berardini et al., 2004)		Chicken feed (Joseph and Abolaji, 1997)
Mango seed kernel flour			Good composite blend for wheat flour bakery products
Mango stone—after decortications of seed			Low cost adsorbent with up to 60% dye removal.

Continued

Table 3.1: Potential of Mango Waste Components.—cont'd

Mango Waste (By-products)	Chemicals Present (Reported)	Bioactivity (Reported)	Reported Uses
Mango peel	Polyphenols, phytochemicals, carotenoids, dietary fiber, Vitamins C and E Rich source of dietary fiber, cellulose, lipids, protein, enzymes and pectin (Ajila et al., 2007). Minerals—Ca, Mg, K=Na, Fe, Cr, Cu, and Mn. 334 unique protein species in peel (Fasoli and Righetti, 2013)	DPPH radical scavenging activity of peel is 53.3%. Gallotannin derivatives—show anti-tumor, antioxidant, anti-cardiovascular, and hepatoprotective effects	Mango peel powder and kernel powder are used as composite flour in biscuit, adding nutrient value with enhanced antioxidant and storage property of the bakery product (Yatnatti et al., 2014; Aslam et al., 2014; Ashoush and Gadallah, 2011). Naturally dried peel has potential use as functional food or value added ingredient in diet because of its rich source of bioactive components (Ajila et al., 2007).
Mango peel and kernel	Different drying methods, viz., freeze drying, hot air, vacuum, and IR radiation.		The phenolics, carotenoids, and antioxidant properties were lower whereas solubility, water, and oil absorption values are adequate for utilization (Sogi et al., 2013). Mango waste or by-products (Kittiphoom, 2012; Ravani and Joshi, 2013; Jahurul et al., 2015)

4. Policy needs for stringent norms on waste disposal for peels, seeds, etc. (besides use in organic manure) and of policies to sustain the extraction of phytochemicals having implications as alternative sources of proteins, etc.
5. Lack of availability of technical infrastructure, competent human pool, and indigenization of machinery for extraction of various phenolic- and phytochemical-based by-products having implications for animal feed, human food, etc.

3.2.3 Opportunities

Creating a range of table consumable mango variants and food supplements and preparatives provides immense scope for growth if the government, private sector, Research and Development bodies, and farmers communities can get their acts correct by minimizing postfarm wastage of mangoes to international benchmarks of 3–5% and creating a downstream value chain from food→feed→fuel→fiber.

Extensive research and development will be required to innovate and refine the new products and processes having impacts on the dynamics of food processing, waste treatment sustainability and economics, waste supply, transport, end product delivery, and costs–benefits of the mango ecosystem.

Also, creating new markets and applicative uses of active compounds as cosmeceuticals and nutraceuticals, evaluating the applicative uses of active compound mixes as animal feeds and animal health-promoting nutrients in respect to greenhouse gas reduction (less land for animal-feeding crops) and assessing the sustainability of exploitation of wastes as fermentation substrates for biobased biotechnology (biochemical production of ethanol, glycerol, itaconic acid, and organic acids), and fermentation substrates for microbe mixes productions suitable for bioremediation adds extra dimensions to the use of mango by-products.

In summary, mango by-products are rich sources of natural food ingredients. The important challenge is the recovery and utilization of these by-products, ensuring a sustainable production in the food industry; utilization of by-products is an important aspect of solid waste management. An extensive study is needed to explore and exploit specific compounds for industrial applications.

3.3 Market Making in the Bioeconomy of Plants' By-Products in Europe and in the World

Europe's fruit and vegetable industries generate about 30 million tons of waste a year. With the increase in production of processed fruit products, the amount of fruit and plant wastes generated is increasing enormously. A large amount of these wastes poses the problem of disposal without causing environmental pollution.

A recent agricultural outlook provided projections to 2024 for major agricultural commodities, biofuels, and fish. The 2015 report focused on prospects and challenges for Brazilian agriculture. The market projections not only cover OECD member countries (European Union as a region), but also FAO member countries, notably Brazil, the Russian Federation, India, People's Republic of China, and South Africa (OECD/Food and Agriculture Organization of the United Nations, 2015).

The bioeconomy and sustainable growth are top topics, both in Europe and around the world. There is a need to exploit all the added value in the present productions and to unravel new markets and technologies for the future. An important topic in plant and tree varieties showing good prospective and economy potential is the exploitation of plants with high content of fibers, combined with improved processes and industries, to be applied to cellulose industry pulp recycling, novel fabrics- and fiber-based insulating material, automotive industry furniture, and biofuel production.

As a world leader in advancements in bioeconomy research and innovation, the EU is well placed to improve the management of its resources and to open up new and diverse markets in food, feed, fiber, energy fuel, and biobased products. As the EU bioeconomy strategy identified, there is a need to establish a more competitive bioeconomy. Europe holds great potential for creating sustainable economic growth and jobs in rural, coastal, and industrial areas. At the same time, development of bioeconomy markets will help address societal challenges such as food and energy security, natural resource scarcity, the need for sustainable economic recovery, and mitigation of the impacts of climate change.

The Strategic Working Group "Sustainable Bio-Resources for a Growing Bioeconomy" (SWG SBGB) is a thematic working group under the Standing Committee on Agricultural Research (SCAR). The SCAR was established in 1974 by a regulation of the council of the European Union. The Strategic Working Group on Bio-resources discusses how renewable bioresources can be sustainably produced for the bioeconomy. This is a matter of producing more renewable biological resources from fewer inputs (such as agricultural land area, water, fertilizer, use of machinery). But it is also about logistical questions, the biomass potential of different European regions, how much biomass is available, and how much of that can be used. It is furthermore about how farmers can profit from the bioeconomy and how they can get involved as stakeholders. Other questions the SWG addresses are how the implementation of the bioeconomy strategy impacts upon agriculture, forestry, fisheries, and aquacultures; how the bioeconomy fosters new connections between well-established sectors; and how this in turn might change these sectors and their sustainable and profitable operations in the future. The SWG has developed a work plan on these different issues and will give strategic advice from the SCAR perspective on the bioeconomy. A Global Bioeconomy Summit will take place in Berlin, in November 2015 (see https://www.scar-swg-sbgb.eu/).

As for EU strategy, policies and economy studies highlighted that the bioeconomy is currently worth €2 trillion in annual turnover and accounts for more than 22 million jobs and approximately 9% of the workforce. In the coming years, significant growth is expected to arise from sustainable primary production and food processing. In addition, the development of biorefinery will enable many sectors within the bioeconomy to convert biomass into higher-value everyday products, such as food, feed, chemicals, plastics, textiles, pharmaceuticals, and cosmetics, providing an alternative to fossil carbon sources.

3.3.1 Feedstock Biomasses

Sustainable biomass feedstock is the key to sustainable bioeconomy. The alternative energies may have a positive or negative impact on social and environmental aspects, depending on local conditions and the design and implementation of specific projects. That means we cannot say a specific feedstock is always sustainable, or unsustainable, but that it depends on how and where the feedstock is produced. Sugar beet, sorghum, corn, and sugarcane are feedstocks for first-generation fuel bioethanol. Perennial grasses like switchgrass, miscanthus (Felten and Emmerling, 2012; Felten et al., 2013), and wood trees such as eucalyptus, willow, and poplar are source of lignocellulosic biomass, feedstock for second-generation cellulosic ethanol.

There have been numerous attempts in recent years to understand the cellulose and hemicellulose biosynthetic pathways. Noteworthy, feedstock biomass can potentially be increased by changing the environmental growth conditions such as carbon allocation in the form of carbon dioxide, oxygen, water supply, and soil nutrient content.

Agronomic traits such as biomass yield and lignocellulosic composition are targeted for plant improvement to attain industrial and economic advantage. The most important among these selection criteria are the agronomic traits of a lignocellulosic biomass crop. These traits include (1) photosynthetic efficiency; (2) canopy duration: longer canopy duration with a short perennial life span; and (3) translocation of photosynthesized carbohydrates into structural lignocellulose. Such traits are common in perennial species such as miscanthus and switchgrass, plants able to grow on marginal soils. The recalcitrance of perennial crops to *Agrobacterium*-mediated transformation for plant genetic improvement is not well understood and needs additional research.

3.3.2 Cellulose

Cellulose is a well-known and widely exploited material. Besides its traditional uses, recent technological developments are opening up the opportunity for the use of cellulose in new and higher added-value applications. This does not only enhance the competitiveness of the concerned industrial sectors, but also significantly improves their environmental performance. While demonstration activities are being pursued to this end, the challenge lies in demonstrating at an industrial scale first-of-a-kind cost-effective biorefinery concepts leading to the production of economically competitive cellulose-based products for bulk materials and volume applications. The techno-economic viability of biorefinery concepts leading to new cellulose-based products with tailored functionalities has been envisaged. Microfibrillar cellulose (MFC)-based products have a potential for exploitation in various applications and market segments (food and/or industrial applications). MFC-based demonstrated products match or outperform properties of competing conventional products (e.g. synthetic polymers like acrylics, gums, hydrocolloids, thickeners, and high molecular weight polymer stabilizers), and show a CO_2 footprint 25% lower than competing technologies. MFC-based additives provide enhanced properties such as control of rheology; barrier and strength properties for stabilizers of emulsions and dispersions; biodegradable viscosifiers; barrier and strength enhancers; industrial thickeners in glues, paint, pastes, and slurries; and many more new applications. Industrial-scale processes have been realized and are underway to produce MFC in suitable quantities, such as 1000 ton/y of MFC-based products, with an optimized efficiency in terms of performance, throughput, and energy requirements so as to allow a suitable price for a significant industrial production.

Lightweight structural composites, based on biobased polymers that are reinforced with cellulose pulp fibers, are another sector of application of fibers from plants and their by-products. There are still technological challenges represented by the different polarities and viscosities between the cellulose and the matrix. Applications of these composites are represented by end products in markets such as structural elements for the automotive and building industry, cases for electronics products, and household appliances. Structural composites on fiber-based biopolymers replacing pure fossil-based materials may compete with pure polymer solutions, mineral filled, or short-fiber-glass reinforced composites in terms of mechanical properties, while exhibiting a significantly lower weight, reaching at least 25% weight reduction and a 100 times reduced tool wear compared to existing materials.

There is a need of industrial symbiosis and integration of actors along the whole value chain and, wherever possible, to make use of existing facilities and to realize a cascading use of forest biomass while providing access to sufficient raw material to set up the new value chain.

Life cycle analysis has demonstrated good environmental and socioeconomic performances of these bioproducts. The development of entrepreneurial activities throughout the entire value chain is a favorable requisite that may give advantages for the primary sector (i.e. forest), the secondary sector (e.g. logistics, bioproduct transformation industry), and the tertiary sector. The biobased alternatives may, in the future, reach price requirements from the industry while functionally matching similar products obtained from fossil fuels.

3.3.3 EU Textile Industries and Vegetal Fibers

Combining wool with vegetal fibers, it is possible to manufacture insulating panels for thermal and acoustic isolation. Using rice straw, similarly to wheat and maize straw, it is possible to produce bioresins and plastic materials. Combining bioresins with other vegetal fibers, such as hemp or flax, it is possible to produce a vegetal leather applied to the industry of sport shoes.

The potential exploitation of plant by-products is huge: agricultural by-products accounted to 135,000 tons.

3.3.4 Innovative Companies Involved in Recycling of Plant Residues, Transforming Them in Textiles, Biobased Biochemicals, and Fiber Applications

As well as highlighting these different aspects of making markets in the bioeconomy, case studies have been analyzed on industrial activities in the bioeconomy field (EU Bioeconomy panel, 2014). The report, "Where next to EU bioeconomy?" has presented the latest thinking on the bioeconomy from the European Bioeconomy Panel and the Standing Committee on Agricultural Research. It contains two issues papers, one on biomass supply and one on market making in the bioeconomy. It also contains a report from the Standing Committee on Agricultural Research, which is made up of representatives of EU governments, on sustainable bioresources for a growing bioeconomy. The panel's work on market making in the bioeconomy presented a collection of 13 case studies, which illustrate many of the issues raised in this chapter. Selected case studies are reported in the next paragraphs. See http://ec.europa.eu/research/bioeconomy/news-events/news/20141009-08_en.htm.

The **Frumat** Company in Bolzano, Italy, in 2009 launched the production of leather-like fabrics made from apple peels and apple waste, designed by engineer Alberto Volcan. In 5 years, the amount of fruit wastes used to realize the sustainable products, handkerchiefs, paper rolls, packaging boxes, and leather, similar for shoes and furniture dressing, reached 30 tons/month, of which apple waste contributes 25%, up to 30%. The results were presented in the contest of BioEnergy Italy, Green Chemistry Conference and Exhibition, and Food Waste Management Conference, on February 5–27, 2015.

The **Orange Fiber** Company, Italy, produces innovative threads for fabrics, which are made starting from oranges, approximately 700,000 tons of wastes that processing industry in Italy produces every year. Through nanotech solutions, wastes are transformed into ecology-friendly fabrics, made by silk-like cellulose acetate fiber, that release vitamins A, C, and E upon contact with the skin. Orange Fiber received the Award Fondazione Marisa Bellisario 2015 "Germoglio d'Oro" to creativity, innovation and sustainability. See http://www.orangefiber.it/.

Istituto Italiano di Tecnologia in Genova is the Italian counterpart of MIT, US; researchers there designed and tested a method to produce bioproducts from vegetal waste natural fibers, while other groups developed Smart Materials, such as antibacterial bioproducts from parsley and cinnamon.

Renewcell. Sweden reveals world's first garment made entirely from recycled cotton. The new technology that recycles cotton could reach scale and offer hope for fast fashion. See http://renewcell.se.

Renewcell, by sourcing 2000–3000 tons of high quality raw material for a first pilot plant, produces a product similar to conventional dissolving pulp. Dissolving pulp is the raw material for a number of different products, but the main use for the pulp is the production of textile fibers such as Lyocell and viscose. In 2010, the world consumed a record 69.7 m tons of clothes. That's up from 47.4 m tons just 10 years earlier, according to statistics from the United Nations Food and Agriculture Organization (FAO). The textiles are ripped and dissolved, which allows for the separation of cellulosic fibers form noncellulosic fibers and nontextile components. After dissolving the textiles, Renewcell's patented and unique process is able to create pure and homogenous dissolving pulp. The Renewcell pulp can now be used in the traditional viscose or Lyocell processes, leading to a textile fiber made from 100% recycled fibers.

The **Textifood** exhibition, originating from the project Futurotextiles, was dedicated to technical and innovative fabrics, presenting fibers obtained from plants' nonedible tissues. Created for the France Pavilion at Expo Milano 2015, the exhibition treated the world of fabric biodiversity (http://www.futurotextiles.com/fr/expo/milan-2015-milan-italie-2015-05-01) to show synergies between agro-food production systems and the textile industry.

At the Institute Français in Milano, the presentation of Lille3000 was held on the techniques and manufacturing of highly valued fabrics in the centuries and around the world, sponsored by the Métropole Européenne de Lille and Nord-Pas-de-Calais region. In this context, highly valued fabrics have been given the center stage.

In Japan, from XIII century, banana wastes were uses to produce "jusi," a silky, highly lightly cotton-like fabric, nowadays used in kimonos. It is derived from the stems that bring the attached bananas. The stylist Dita Sandico has transformed the interior fibers from leaves of banana trees into abacà, "banana tree silk," while in the US, industries make use of "abaca," a typical banana fiber traditional fabric in Philippines, to create textiles and carpets.

Coffee: Dresses made from coffee powder wastes, such as a jacket with hydrophobic power, are made especially for the internal textiles and have odor binding properties; this property has enabled the production of bioproducts adsorbing odors and heavy metals, made from coffee residues.

Soya: In China, dresses made from soya fiber are common. Henry Ford patented a fiber named Azlon to use in the textile industry, but for the production, chemicals of high toxicity were required. In 1999, Li Guanqi, industry owner and scientist in Shanghai, set up Soybean Protein Fiber: a very soft fabric adapted for underwear, elegant evening dresses, called "vegetal cashmere," which has properties such as antibacterial, transparent, able to block UV radiation, and three times more resistant to breaks with respect to wool.

Bamboo: In ideal environment bamboo grows rapidly, 1 meter in few hours. The textiles are transparent and have capacity to ventilate and adsorb humidity, with antibacterial and deodorant action.

Nettle: Since the Second World War, Germans used nettle fibers as an alternative to cotton to manufacture uniforms. Nowadays, it has entered the fashion industry, such as by the Grado Zero Company that created a jacket in 100% nettle fiber. The advantages are: no need to fertilize fields, no need for watering, making nettle one of the most sustainable plants. Using innovative technologies, it is possible to obtain very thin threads, flexible and soft, transparent as linen and brilliant like silk.

Hemp: A fiber is produced from hemp that is used by the fashion industry. The fabrics are fresh and transparent in summer, and warm and enveloping in winter, with antiallergy properties, while they have low environmental impact. *Cannabis sativa* L. is the scientific name for the hemp plant, which has virtually zero drug content. Hemp is a tall, woody plant that will grow in practically any climate. Hemp produces mainly fiber, shivs (woody part of the stem), and seeds. After processing, these parts of the hemp plant are used in a number of industrial products.

Algae: From algae is possible to obtain a resistant fiber that is transparent and with a low percentage of shrinkage, adapted to washing machines. Minerals and bioactives present in algae facilitate an exchange between the plants and the body, acting antiinflammatory.

SeaCell: http://www.smartfiber.info/seacell.

Textiles can be made from seaweed to produce skin care products, protecting the body while also feeling incredibly soft. The brand SeaCell™ of the fiber producer Smartfiber AG has, for many years, been a recognizable term on the international textile market. SeaCell™ means wellness for the skin.

The cellulose fiber is made from renewable resources, including seaweed and the bioactives contained in algae. The fibers can be ideally used in all kinds of textiles and can already be found in numerous collections of renowned international providers. It is produced through the natural Lyocell process, which is an environmentally friendly production method. The Lyocell process is virtually closed and integrated into the nature cycle process and meets the expectations of an industry of the future. This is also why the European Union awarded the procedure the European Environmental Award 2000 in the category "technology for sustainable developments." In 2010, Smartfiber was honored with the German Innovation Prize (see http://www.smartfiber.info/seacell).

3.3.5 EU Business Analysis of Case Studies

A collection of case studies is described in a publication on market making in the bioeconomy released by the European Bioeconomy Panel in October 2014 (EU bioeconomy panel, 2014).

HempFlax Company, The Netherlands: In addition to the necessary adaptation of the existing production and processing machinery, certain infrastructure developments were needed. As HempFlax's goal was to limit the growing area to a radius of 80 km from the processing point, dedicated factory storage spaces for both raw material (hemp straw) and finished products were built. The construction of a decortication factory and fiber refinery as well as storage and packaging were also required.

Europe allows industrial hemp to be grown in most of its member states. In addition, a fiber subsidy was granted to producers for several years. This was especially important in the startup period and enabled producers to decrease feedstock costs. Today, however, the dedicated subsidies have been discontinued, prompting fierce competition for cultivation areas with heavily supported energy crops. The HempFlax initiative came into play at the right time because there was a need for a new cash crop that could also fit well in the crop rotation.

Hemp has a deep rooting system, has a favorable influence on the soil structure, and curtails the presence of nematodes and fungi. After cultivation, the soil is left in optimum condition due to the complete weed suppression following the high shading capacity of hemp. A study by Bócsa and Karus (1998) reports 10–20% higher wheat yields after the cultivation of hemp. At the time of the startup, the region experienced low costs for main crops like potatoes, sugar, beet, and wheat.

There were frequent discussions about modifying the EU directives in favor of green materials, but to date, this has not been accomplished. While a German market introduction program for natural insulation materials has existed in the past, some current members' state regulations pose significant barriers in terms of market access.

Nevertheless, HempFlax has grown to be a professional hemp processing company, currently employing 22 people. The company has enabled the development of state of the art hemp (bast fiber) decortication and fiber separation technology (breaking the woody core, separation of bast fiber and shivs, cleaning of fibers), which is now available on the market. In addition, contracting 1200 ha of hemp in The Netherlands, Germany, and now Romania provides an important source of income to farmers and benefits the rural economy in general.

Today, hemp fiber composites are used in many middle- and high-class automotives. While this market has the potential to expand further, other sectors, such as the furniture industry, could also make good use of this new material group. Unfortunately, these sectors are currently unaware of the existence of such a sustainable, renewable biobased alternative, making communication on the subject all the more necessary.

Standardization and authorization: As an emerging sector, another significant challenge that the hemp industry faced was that tests and standards in both the automotive and the construction industries were originally set up for conventional materials, but they were not in place for natural materials. This lack of evaluation criteria initially posed a barrier to market entry for hemp products, as these are necessary for industry to have assurances of the quality and safety of products and materials.

Today, in the case of hemp fiber-based products, there are several technical hurdles in the field of standards and regulations:

- Lack of specifications, which results in prejudices against biobased properties;
- Irrelevant or misleading information for the consumer;
- A lack of compliance with less important specifications due to the intrinsic physical properties of biobased products; and
- A lack of incorporation of biobased alternatives, due to a low level acceptance or awareness, help provide extremely low-weight biocomposites for automotive interiors, such as door panels, made using natural fibers, which have only 40% of the area weight of glass fiber-reinforced plastic alternatives.

Adapting fibers to market needs so that they can be used in conventional applications and technologies has increased their market potential tremendously. In addition, the whole supply chain benefits from the developments in the natural fiber sector because it encompasses a large number of actors, from farmers to processors, to Tier 1 companies and original equipment manufacturers. Today, every European car contains an average of 2 kg of natural fiber, but the technical potential reaches up to 5 times more, amounting to 150,000 tons of natural fibers per year.

Tackling regulatory barriers: The increased use of hemp as an insulation material is currently hampered by individual member state building regulations and a lack of coherence in policy. For example, current legislation does not recognize some of the intrinsic benefits of biobased insulation and imposes unnecessary limits to its use. Legislation could instead play an enabling role in creating new markets for biobased products.

Cost barriers leveling the playing field: With regards to the price of inputs, the cost of raw material is one of the most important price components of biobased products. The heavy focus on subsidizing bioenergy has had a hugely increasing effect on the cost of growing. Farmers indeed benefit from a higher income if they grow biomass crops for bioenergy purposes rather than for biobased products. Unfortunately, the great potential of biobased products, including those produced from hemp, cannot be realized due to this existing political framework.

On the other hand, European hemp fibers are completely unprotected by the competition from imported exotic fibers like jute, kenaf, or sisal, which may not even show a sustainability certification like imported biofuels. The social and environmental standards of these tropical fibers can be low. The European Industrial Hemp Association (EIHA) is calling for a binding sustainability certification for all imported (and domestic) natural fibers to guarantee fair competition.

- The need to level the playing field between energy and material uses of biomass; existing incentives for biofuels and bioenergy should be opened for hemp-based products too.
- The absence of incentives for member states for the development of new biobased markets (directives and bans in construction and insulation due to GHG reductions and for mulch).

The main barrier for the scale-up of investment in integrated biorefineries is the missing market opportunities for by-products like hemp shivs (woody core, pith) and dust. In the absence of a viable commercial and local market for hemp by-products, the business case cannot be profitable. Subsequently, incentives for the development of new markets for these by-products should be developed. A prime example is the development and use of shiv in the construction market. The marketing and development of products like Hemcrete as a construction material should be stimulated in other EU countries, like it has been done in France and the UK.

In conclusion, hemp processing is based on one-third technology and two-thirds operator skills. Investment decisions are, however, also based on the transportation distances between the growing region and processing installation. Subsequently, putting in place the necessary infrastructure for processing is required in feasible growing regions.

The main barrier for the scale-up of investment in integrated biorefineries is the missing market opportunities for by-products like hemp shivs (woody core, pith) and dust. In the absence of a viable commercial and local market for hemp by-products, the business case cannot be profitable. Subsequently, incentives for the development of new markets for these by-products should be developed. A prime example is the development and use of shiv in the construction market.

A healthy growth of the EU hemp industry with reliable suppliers of hemp fibers and other hemp products to the various markets will enable the successful integration of natural hemp fibers in the value chain. For instance, a continual and reliable supply is of utmost importance for the automotive industry. The insulation and construction industry, as well as furniture and the whole reinforced (biobased) plastic market, also hold great potential.

3.3.6 Cargill Starches

Cargill has developed high-performing and innovative solutions for paper coating applications, which successfully replace petroleum-based synthetic binders and chemicals, thus enabling cost and energy optimization of paper making processes.

Cargill's coating starches are produced from cereals, and they can replace the traditional fossil fuel-based binders. The products are 100% biodegradable, chlorine free, and support more sustainable papermaking processes.

Cargill has developed some coating starch series that are specially designed for high solids coating. These high powder solids captured the imaginations of coating technologists, who have known about the advantages of high solids coating for a long time. In the coating area, there is excellent rheology or fluidity (which is a physical test method indicating the flow properties of the coating during the application process, which has a high influence on its processing at the normally high speeds) of the coating starches, which perfectly meets the requirements of this demanding application.

Advanced instrumentation techniques have helped in the understanding of rheological behavior of starch, its interactions with other coating color components, and its interactions between the coating colors and the base paper.

3.3.7 Tomatoes Plant Waste Recycling Industrial Group

The different wastes considered include fibres from the plant stems, rich in antibacterial agents, and the tomatoes themselves which are discarded when market price falls too low or when tomatoes do not meet retailer/consumer requirements in terms of quality and appearance.

Once the tomato crop is finished in summer, there is a large amount of plant waste remaining in the form of tomato plants and stems. Disposing of this waste has historically represented a cost burden to vegetable growers who have to pay for its removal, either for incineration or composting.

Numerous sectors and companies were involved in the industrial development of the tomato-based by-products, including: Smurfit Kappa, Provalor, Koppert Biological Systems, TNO, Wageningen University, Biobase Westland, The Greenery, various tomato growers, the food industry, the packaging industry, and recycling companies. The cooperative focused on developing integrated approaches centered on development of strong value chains and creation of "biomass hubs" to minimize the cost of transportation of biomass. To measure and evaluate progress, different objectives were set. These included the development of feasibility studies to analyze technical development and economic viability, market research, business modeling of biobased products in new supply chains, the development of communication strategies, trials of new products on the market, and the codevelopment of strategies between growers and retailers. It was also necessary to invest in infrastructures for the collection, storage, and refining of the organic waste streams.

3.3.8 Novamont Case Study on Bioplastics

The need to tackle the overconsumption of plastic bags was addressed by a ban on nonbiodegradable single-use plastic bags in January 2011. The ban encouraged the use of durable, thicker, reusable plastic bags and compostable single-use plastic, which can be re-used as "multi-purpose" waste bags and are suitable for collecting residual waste (any waste that cannot be separated before collection) and for biowaste (e.g. kitchen waste), and was in line with European standards on compostable packaging. This approach is improving the quality and quantity of biowaste collection and recycling. Fewer nonbiodegradable plastics are contaminating compost. The risk of a nonbiodegradable bag being improperly used to collect biowaste is eliminated if the householder only receives biodegradable compostable bags. This, in turn, improves the quality of organic recycling and brings important environmental benefits.

Bioplastics are a broad family of materials with widely varying properties. Ultimately, bioplastics can find a place in market segments where conventional plastics are used and where biodegradability represents added value for the entire system. In many of these market segments, bioplastic alternatives are already available today. However, they face market entry barriers when compared with fossil-based plastics, which can be cheaply made and are not required to demonstrate sustainability.

Biodegradable plastics are materials whose properties and characteristics of use are very similar to those of traditional plastics, but at the same time, they are biodegradable and compostable, according to the European standard UNI EN 13432. As a result of these characteristics of biodegradability and compostability, bioplastics can play a valuable role in optimizing biowaste collection and management, reducing environmental impact, and contributing to the development of a closed loop of recycling with significant advantages along all the production–consumption–disposal cycle.

According to the European Bioplastics Association, global bioplastics' production capacity is set to grow 500% by 2016.

Since 2010, there has been an increasing level of controversy regarding the use of food-based feedstocks for the production of industrial materials such as bioplastics. In reality, at current production levels, bioplastics rely on only 0.0063% of a global agricultural area of 5 billion hectares. Even if bioplastics' capacity grows as predicted by 500% by 2016, this percentage would only increase to 0.022%. Therefore, the production of bioplastics poses no competition to the production of food.

Among the benefits and enabling factors, there are the creation of jobs across the value chain from agriculture, R and D, green chemistry, and waste management (60 jobs per 1000 tons of bioplastics), the improvements in biowaste collection and recycling with an 8% decrease of compost impurities and 30% decrease of greenhouse gas (GHG) emission achieved since 2012, and new investments in research and innovation (new crops, reuse of food and agro waste, new biotechnological processes, etc.) are envisaged and put into action.

3.3.9 Wheatoleo

The French company Wheatoleo, owned 100% by ARD, a leading company in biorefinery research and development activities, is involved in the project through industrialization and commercialization of new biosurfactants based on pentoses extracted from plant cell walls. The growing need for the replacement of fossil resources with renewable resources has created a new market for biobased surfactants in the applications such as detergents, home care, and personal care products.

At ARD (located in Bazancourt–Pomacle, in the Champagne–Ardennes region of France), research and development in the production of biosurfactants from coproducts of the farming industry first began in 1997. Cutting-edge technologies in the fields of plant cracking, industrial biotechnology, and green chemistry have been employed with the aim of utilizing plant biomass to produce high value products. These research projects involved several European partners and were co-financed by the EU (FAIR-CT 97–3130). ARD is now considered as a leader in the field of pentose chemistry, which ultimately led to the formation of Wheatoleo, a new biosurfactant company.

Since its creation in 2009, Wheatoleo has produced and commercialized a range of new bio (based) surfactants. The first marketed products are known as alkyl-poly-pentosides (APPs) (also known under their trade name: APPYCLEAN). These compounds are made from pentoses (biobased chemicals extracted from plant cell walls) and fatty alcohols. These "green" surfactants have applications in cosmetics and detergents, cleaning, oil fields, and agrochemicals sectors. They are made from a combination of extracts from sugar beet, wheat, and vegetable oil. The most innovative technological advancement was the conversion of pentoses (xylose) into the targeted surfactants molecules without compromising efficiency on the applications. This singular technology is paving the way toward unlocking the potential of added value, lignocellulosic-based chemistry.

3.4 Conclusion

While Europe plays a leading role in much of the cutting-edge science and technology enabling the bioeconomy, for a number of reasons, it has been much less successful in converting this into commercial and socially valuable innovations. While some emerging, green-technology renewable industries have received significant legislative support for their development, many sectors within the bioeconomy have not. The absence of long-term framework support and policy predictability continues to make these sectors unattractive for investment in the EU. At the same time, significant incentives and offers of longer term commitments continue to draw successful industries away from the EU bioeconomy overseas to the US, Brazil, China, and Southeast Asia. Without putting specific market making measures in place, the EU will not meet its target of raising the contribution of industry to GDP to 20% by 2020.

The prospective presented here touched upon the positive and negative impacts of the costs of inputs necessary in the EU, as well as in India and around the world, including raw materials, land, energy, infrastructure, logistics, and skills. They also highlight the role of financing, collaboration, technology transfer, demand-side measures, and communications in tackling fragmentation in the EU, and in rest of the world, which will be necessary in order to develop markets within the bioeconomy.

References

Ajila, C.M., Bhat, S.G., Prasada Rao, U.J.S., 2007. Valuable components of raw and ripe peels from two Indian mango varieties. Food Chemistry 102, 1006–1011.

Ashoush, I.S., Gadallah, M.G.E., 2011. Utilization of mango peels and seed kernels powders as sources of phytochemicals in biscuit. World Journal of Dairy and Food Science 6 (1), 35–42.

Aslam, H.K.W., Raheem, M.I.U., Ramzan, R., Shakeel, A., Shoaib, M., Sakandar, H.A., 2014. Utilization of mango waste material content and antioxidant properties of biscuit. Journal of Global Innovations in Agriculture and Social Sciences 2 (2), 76–81.

Ayala-Zavala, J.F., Vega-Vega, V., Rosas-Domínguez, C., Palafox-Carlos, H., Villa-Rodriguez, J.A., Wasim Siddiqui, Md, Dávila-Aviña, J.E., González-Aguilar, G.A., 2011. Agro-industrial potential of exotic fruit byproducts as a source of food additives. Food Research International 44, 1866–1874.

Berardini, N., Carle, R., Schieber, A., 2004. Characterization of gallotannins and benzophenone derivatives from mango (*Mangifera indica* L. cv. 'Tommy Atkins') peels, pulp and kernels by high-performance liquid chromatography/electrospray ionization mass spectrometry. Rapid Communications in Mass Spectrometry 18, 2208–2216.

Bócsa, I., Karus, M., 1998. The Cultivation of Hemp: Botany, Varieties, Cultivation and Harvesting. Hemstitch. ISBN 13: 9781886874039.

EU Bioeconomy panel, 2014. Where Next for the European Bioeconomy? Case Studies of Market Making in the Bioeconomy. European Commission Directorates, Research and Innovation. Bioeconomy, Brussels. https://www.scar-swg-sbgb.eu/documents.

Fasoli, E., Righetti, P.G., 2013. The peel and pulp of mango fruit: a proteomic samba. Biochimica Biophysica Acta 1834, 2539–2545.

Felten, D., Emmerling, C., 2012. Accumulation of *Miscanthus*-derived carbon in soils in relation to soil depth and duration of land-use under commercial farming conditions. Journal of Plant Nutrition Soil Science 175, 661–670.

Felten, D., Fröba, N., Fries, J., Emmerling, C., 2013. Energy balances and greenhouse gas-mitigation potentials of bioenergy cropping systems (Miscanthus, rapeseed, and maize) based on farming conditions in Western Germany. Renewable Energy 55, 160–174.

Imran, M., Butt, M.S., Anjum, F.M., Sultan, J.I., 2013. Chemical profiling of different mango peel varieties. Pakistan Journal of Nutrition 12 (10), 934–942.

Jahurul, M.H., Zaidul, I.S., Ghafoor, K., Al-Juhaimi, F.Y., Nyam, K.L., Norulaini, N.A., Sahena, F., Mohd Omar, A.K., 2015. Mango (*Mangifera indica* L.) byproducts and their valuable components: a review. Food Chemistry 183, 173–180.

Joseph, J.K., Abolaji, J., 1997. Effects of replacing maize with graded levels of cooked Nigerian mango-seed kernels *(Mangifera indica)* on the performance, carcass yield and meat quality of broiler chicken. Bioresource Technology 61, 99–102.

Kaur, M., Singh, N., Singh Sandhu, K., Sindh Guraya, H., 2004. Physicochemical, morphological, thermal and rheological properties of starches separated from kernels of some Indian mango cultivars (*Mangifera indica* L.). Food Chemistry 85, 131–140.

Kittiphoom, S., 2012. Utilization of mango seed—mini review. International Food Research Journal 19 (4), 1325–1335.

Larrauri, J.A., Ruperez, P., Saura-Callixto, F., 1997. Mango peel fibres with antioxidant activity. Zeitschrift für Lebensmitteluntersuchung und Forschung A 205, 39–42.

OECD/Food and Agriculture Organisation of the United Nations, 2015. OECD-FAO Agricultural Outlook 2015. OECD Publishing, Paris. ISBN: 978-92-64-23190-0. http://www.oecd-ilibrary.org/agriculture-and-food/oecd-fao-agricultural-outlook-2015/biofuels_agr_outlook-2015-13-en. http://dx.doi.org/10.1787/agr_outlook-2015-en.

Ravani, A., Joshi, D.C., 2013. Mango and its by product utilization—a review. Trends in Post Harvest Technology 1 (1), 55–67.

Sogi, D.S., Siddiq, M., Greiby, I., Dolan, K.D., 2013. Total phenolics, antioxidant activity, and functional properties of 'Tommy Atkins' mango peel and kernel as affected by drying methods. Food Chemistry 141, 2649–2655.

Yatnatti, S., Vijayalakshmi, D., Chandru, R., 2014. Processing and nutritive value of mango seed kernel flour. Current Research in Nutrition and Food Science Journal 2 (3), 170–175.

CHAPTER 4

Organic Recycling of Agricultural Waste Today: Composting and Anaerobic Digestion

N. Mortier, F. Velghe, S. Verstichel
Organic Waste Systems NV, Gent, Belgium

Abbreviations

AD anaerobic digestion
BVS mass of biogas produced per unit of organic matter
GC-TCD gas chromatography – thermal conductivity detector
LCA life cycle assessment
PDMS polydimethylsiloxane
PLA polylactic acid
ppb parts per billion
SWOT strengths, weaknesses, opportunities and threats
TS dry matter or total solids
VS organic matter
WWTP wastewater treatment plant

4.1 Introduction

Organic waste has been around for a long time and many treatment options have been developed and installed to deal with this waste. This chapter focusses on two commonly applied technologies: composting and anaerobic digestion. Landfilling and incineration are not discussed as they are less sustainable than composting and anaerobic digestion and are therefore being faded out in waste management. First the policy in Europe and America on organic waste treatement is discussed, after which the different composting and anaerobic digestion technologies are explained.

4.2 EU & American Policy on Organic Waste

Organic waste is the oldest type of waste known to mankind, and throughout history, different means to deal with this waste have been applied. Whereas for the longest part of history, these means were primarily driven by economic incentives (e.g. reuse or dumping, depending on

Biotransformation of Agricultural Waste and By-Products. http://dx.doi.org/10.1016/B978-0-12-803622-8.00004-5

the specific needs of the society); in more recent times, the fate of organic waste is mainly determined by the legislative framework that is installed in a country or region. As in the second half of the 20th century waste has become more and more problematic (sanitary, environmentally), legislation was developed by many countries to install a more sustainable approach to waste collection and treatment. In this subchapter, an overview will be provided on the legislative path that has led to today's waste management in both Europe and the Americas, together with an outlook on future evolutions.

4.2.1 Europe

The starting point of the European waste policy dates back to the 1970s and 1980s when a number of problems and scandals related to the handling of waste surfaced. This arose awareness in many European member states and led to the Waste Frame Directive in 1975 (European Commission, 2005). This framework has since then been changed and updated on a regular basis, and many other directives have been put in place. The most important directives influencing the fate of organic waste will be shortly discussed below.

Waste Framework Directive (2008/98/EC)

This directive was first introduced in 1975 and has been updated for the last time in 2008. The new Waste Framework Directive gives specific attention to prevention, reuse, and recycling, and also contains new provisions regarding the distinction between waste and nonwaste and clearly states what waste activities can be classified as recovery. The member states should have adopted the new Waste Framework Directive by the end of 2010. This directive primarily focuses on municipal and industrial waste and excludes manure, straw, and other natural nonhazardous agricultural material used in farming, forestry, or the production of energy from such biomass. Nevertheless, Article 22 of the 2008 revision refers, for the first time at the European level, specifically to biowaste, defined as biodegradable garden and park waste, food, and kitchen waste from households, restaurants, caterers, and retail premises, and comparable waste from processing plants. The directive states that the member states shall take appropriate measures to promote:

- Separate collection of biowaste with the objective to compost and/or anaerobically digest the waste;
- The processing of biowaste in such a way that a high level of environmental protection can be guaranteed; and
- The use of environmentally safe products produced from biowaste.

Article 11 of the Waste Framework Directive introduces reuse and recycling targets. Biowaste, however, is not included in the waste types that are to be collected separately or for which recycling targets have been established. Nevertheless, member states are allowed and encouraged to include more waste streams to promote high-quality recycling. To this end,

Table 4.1: Recycling Targets for Biodegradable Waste as Prescribed by the Landfill Directive.

Timeframe[a]	% of Biodegradable Waste Going to Landfill
After 5 years	75
After 8 years	50
After 15 years	35

[a]The year 1999 has been taken as reference year.

they introduce additional separate collection schemes of waste where this is technically, environmentally, and economically feasible and appropriate to meet the necessary quality standards for the relevant recycling sectors.

The 2008 revision of the Waste Framework Directive also introduces an important new element on energy recovery through anaerobic digestion (AD) of biowaste. Unlike composting plants, AD plants do no longer require an environmental permit for recycling activities.

Landfill Directive (1999/31/EC)

The Landfill Directive was introduced in April 1999 and must be implemented by its member states by July 16, 2001. Article 5 of this directive states that the member states should set up a national strategy for the implementation of the reduction of biodegradable waste going to landfills by means of organic recycling, material recycling, and/or energy recovery. The targets as set forth by this Landfill Directive, which need to be reached after implementation on a national level, are shown in Table 4.1. Member states that landfilled more than 80% of their collected municipal waste in 1995 may postpone the attainment of the targets by a period not exceeding 4 years.

Municipal waste is in the Landfill Directive defined as waste from households, as well as other waste, which, because of its nature or composition, is similar to the household waste. However, the precise definition of biodegradable municipal waste varies from member state to member state.

The main motivation for these targets and measures was to reduce the production of methane gas from landfills, and as such, reduce global warming. Furthermore, the targets should also aim at encouraging the separate collection of biodegradable waste, sorting in general, recovery, and recycling.

Renewables Directive (2008/28/EC) (Directive, 2009)

The Renewables Directive entered into force in 2008 and stipulates that national measures must be taken by the member states aiming at replacing 20% (calculated on the basis of energy content) of all fossil fuels with renewables by 2020, although specific targets were set for each member state depending on their geographical and economic situation. According to the directive, biomass means the biodegradable fraction of products, waste, and residues from agriculture (including vegetable and animal substances), forestry, and related industries, as well as the biodegradable fraction of industrial and municipal waste. Biogas from agricultural byproducts is specifically mentioned in this directive to provide a significant contribution to these targets in rural or remote areas.

Biowaste Directive

Following the clause included in the Waste Framework Directive asking the European Commission to carry out an assessment on the management of biowaste, the Biowaste Coalition was established in 2006. This coalition, composed of Austria, Belgium, Cyprus, Czech Republic, Estonia, Germany, Hungary, Italy, Portugal, Slovakia, and Spain, launched the demand for legislative measures on the EU level on biowaste under the form of a biowaste directive, which includes binding targets and a mandatory separate collection of biowaste.

In 2008, the European Commission published a Green Paper questioning whether such a biowaste directive is indeed needed and whether progress in the biowaste management, according to the waste hierarchy and the potential economic, social, and environmental benefits, is, after all, possible. According to this paper, there is currently no preferred option for processing biowaste because this is highly dependent on national or local circumstances (even though making separate collection mandatory would also contribute to achieving the European landfill reduction targets and the targets for recycling household waste set down in the European Waste Framework Directive). Finally, in 2010, the European Commission stated that, due to the presence of the Waste Framework Directive, no policy gaps are present preventing member states from taking appropriate actions with regard to biowaste management. Therefore, there is no need to promote biowaste recycling through one European legal instrument like a biowaste directive. Nevertheless, reality is learning that this is not completely correct, resulting in the fact that the Biowaste Coalition will proceed with their work and try to have the European Commission setup a biowaste directive.

Apparently, not all member states are convinced of the need for a biowaste directive, as some believe that the separate collection should indeed be primarily the responsibility of the local authorities:

- Belgium: Both the industry and government have a positive opinion
- France: Negative, as they believe this is not needed
- Germany: Positive
- Italy: Positive, although there is a difference between the North (+) and the South (−)
- Luxembourg: Unclear, but more positive than negative
- The Netherlands: The sector is in favor, but the government is not
- Spain: Positive
- United Kingdom: Also here the sector is positive, but the government is not

On the other hand, such a biowaste directive would also provide the necessary "legal push" for member states where biowaste is still (partially) landfilled.

4.2.2 South and Central America

Latin America and the Caribbean (LAC) is the most urbanized region in the world, with the urban waste generation of 1.1 t/habitant on average (Hoornweg, 2012). For a comparison an European citizen (EU-27) produces 0.5 t of waste (Eurostat). Unfortunately, the development regarding municipal solid waste does not meet European standards. In Latin America waste sector is responsible for 5% of total CO_2 emissions. This is a particularly large share once compared with the global average of 3.3%. Methane originating from landfills, controlled dump sites or open dumps – the most popular waste disposal method in LAC – are mainly responsible for this large contribution to CO_2 emissions (Center for Clean Policy, 2014). On average collected waste are very humid (45–50%) and have a high organics content (40–50%) (UNEP International Environmental Technology Centre, 2006).

The Latin America waste sector presents many barriers to implement Integrated Solid Waste Management (ISWM), what could significantly improve the situation. First of all, the alternatives to landfills are not supported in regulatory frameworks. In Colombia for example the definition of public waste management is limited to street sweeping, collection, transport, and disposal. It does not include alternative treatment technologies, which does not allow operators to receive compensation for diverting waste from landfills. Moreover, nonhazardous MSW is not considered as potential fuel (e.g., under a Colombian's regulation). Second important problem of the national and regional legislation is that it does not require a development of 3R (Reduce, Reuse, Recycle) waste management strategy in the Southern American countries (Center for Clean Policy, 2014). Very often public sector of large cities commercialize the solid waste management and privatize waste collection. Unfortunately, due to the limited experience in monitoring of compliance with contract conditions, the performance standards are hardly satisfactory and efficiencies are rather low (UNEP International Environmental Technology Centre, 2006).

Traditionally solid waste management has been a responsibility of local governments, which usually do not take enforcement measures due to the lack of resources. The establishment of policies, the development of management plans and the assignment of budgets continue to be national and regional responsibilities. Various municipalities have a solid waste management plan (SWMP). However, it does not guarantee the quality, applicability and execution of these plans. On the one hand, because there is no national SWMP to guide the municipalities and on the other hand, due to the lack of resources, personnel training or simply insufficient specification of tasks, time frames, stakeholders and source of financing (PAHO/WHO/IBN, 2010).

Seven countries of the region have proclaimed a national legal framework. All these laws aim towards valorization of waste, sorting-at-the-source programs, minimization of waste, energy generation from biogas, recycling programs and financial sustainability of services. The majority of these laws includes also producer responsibility obligations (PAHO/WHO/IBN, 2010).

- Brazil – Brazil's National Law on Solid Waste (2010) – after many years of negotiation, the Brazilian Senate approved Legislative Bill 384 of 1989, which establishes the National Policy on Solid Waste. The law requires development of solid waste treatment plans at federal, state, and municipal level and establishment of recycling goals and programs. It prohibits unofficial dumpsites and requires correct disposal of waste. The municipalities are responsible for ensuring that illegal dump sites no longer exist and for implementing systems to separate recyclable materials. The law enhances recycling efforts by strengthening the role of cooperatives that collect recyclable materials as waste management agents with access to financial support. The law also requires that packaging should be manufactured with reusable or recyclable materials.
- Argentina – Integral Management of Solid Urban Waste (2005) – a Zero waste law for the region of Buenos Aires
- Peru – General Law on Solid Wastes (2000 modified by legislative decree of 2008)
- Paraguay – Solid Waste Management Law (2009)
- Mexico – General Law for Prevention and Integral Management of Wastes (2004)
- Venezuela – Waste Management Law (2010)
- Costa Rica – Integrated Waste Management Law (July 2010)

4.2.3 North America

In the USA, the Environmental Protection Agency (EPA) is the independent government entity in charge of enforcing the rules and allocates aid in relation to waste management programs throughout the US.

At national level, the Resource Conservation and Recovery Act (RCRA) of 1976 provides, in broad terms, the general guidelines for waste management. It amends the Waste Management Act of 1965 by introducing national targets on risks, preservation of national resources and energy, and waste (impact) reduction.

The RCRA regulations are contained in Title 40 of the Code of Federal Regulations (CFR) Parts 239 through 299. The CFR is a collection of all federal regulations codified and enforced by all federal agencies. Title 40 (Protection of the Environment) contains all regulations governing US EPA programs. Part 246 (Source Separation for Materials Recovery) outlines requirements for recyclable materials but does so far not address organics. Diversion of food and yard waste is included in the EPA Resource Conservation programs, which encourage—but do not mandate—diversion from landfills. Only open dump sites (unmanaged landfills) are being banned. Until recently, the program has nearly exclusively been seen in the context of composting (Title 40 – Protection of the Environment).

In the absence of a federal mandate for organic waste management, it can be concluded that the US legislative and regulatory context for organic waste treatment via composting and AD is complex. The different facilities and pathways for processing organic waste are covered under different agencies and rules at federal, state, and local levels. The relevant rules will depend upon where the facility is located and the materials it accepts. In the US, current and projected organic waste diversion to AD facilities falls into the following five categories.

Municipal wastewater treatment plant facilities

Garbage disposal units, installed under the kitchen sink and shredding food waste into pieces small enough to pass plumbing, are widely used in the US. In 2009, an estimated share of almost 75% of the newly built houses had a garbage disposal unit installed. As such, a vast majority of the food waste produced at household level enters the wastewater stream and is treated in municipal wastewater treatment plants (WWTP). Furthermore, the garbage disposal units are increasingly gaining in popularity. In New York City, there was a specific ban on garbage disposable units, but the ban was lifted at the end of the 1970s (US Census Bureau, 2011).

In the US, 1054 anaerobic WWTPs use the produced biogas for energy, with 270 producing electricity. These facilities are regulated by the Clean Water Act, and their primary responsibility is municipal sewage and other wastewater management to achieve a final water quality target. The digestate from the AD, referred to as biosolids, is regulated both at the federal and state level and cannot be certified as organic since it originates from human waste, thereby limiting its market value. Some of these facilities have begun accepting high-strength liquid organics and slurried organic waste, predominantly from commercial venues as additives for the AD to increase biogas and self-generated energy production. One notable exception is the East Bay Municipal Utilities District Facility in San Francisco, where up to 120 tons of residential organic waste per day are digested (www.biogasdata.org; US EPA).

Agricultural facilities

As of November 2013, there are 220 AD plants operating at commercial livestock farms. Nearly all AD plants are located on dairy and swine facilities and are currently designed to treat low solids (~10% total solids) manures and other on- and off-farm organic residuals (AgSTAR database).

Federal laws do not require solid waste permits for manure; however, the acceptance of municipal organic waste may designate the AD plant as a waste processing facility in some states. These waste processing facilities are required to meet federal regulations under the Resource Conservation and Recovery Act (RCRA) Subtitle D (which covers nonhazardous solid wastes) and 40 CFR Part 258 (which covers landfills). These regulations include specifications for the management of these wastes.

The Clean Water Act regulates Concentrated Animal Feeding Operations (CAFOs) that discharge or propose to discharge to US waters and requires these facilities to obtain a National Pollutant Discharge Elimination System (NPDES) permit. Discharges include the result of inappropriate land application of manure. Large CAFOs that discharge must be permitted and develop and maintain Nutrient Management Plans (NMPs) to ensure appropriate land application of manure as well as digestate. Inclusion of biowaste in the digester may trigger a review of the farm's NPDES permit.

Integration with existing compost facilities

Compost facility management is regulated (i.e., siting, permitting, and management) at state level, except for biosolids and animal manures. The insertion of an AD process with an existing compost facility is increasingly seen as the most expeditious regulatory pathway if the facility is already permitted at state level to accept food waste. The regulations vary considerably by state, but in most cases, the additional permitting activity is limited to demonstrating that inclusion of AD will have no or only limited impact on the existing facility's environmental footprint. If the composting facility only accepts yard waste, the inclusion of other organics, whether or not the facility also adds an AD plant, will require compliance with applicable regulations for the broader set of organic feedstocks.

Integration with landfills or refuse-derived fuel facilities

The integration of an AD process at landfill sites is being considered and implemented at several sites in the US. In most cases, the permitting requirements are limited to any changes that would result from the addition of the AD plant. Normally, the facilities are able to demonstrate a "no significant impact" or "mitigated minimal impact." US EPA regulates landfills, requiring those exceeding certain size specifications to capture, collect, and destruct landfill gas as well as control and treat leachate. The regulations for alternative daily cover also vary by state, with many still allowing green or other organic waste. Neither the federal or state regulations currently require waste stabilization prior to landfilling.

Stand-alone facilities

Stand-alone or greenfield facilities encounter the highest regulatory permitting hurdle. Developers can choose to construct stand-alone or greenfield AD plants that process a variety of organic residuals, with a composition that is driven predominantly by economics. The project opportunities arise when some combination of the following exists:

- Control of feedstocks is established, tipping fees are high, and transport to cheaper landfills is cost prohibitive;
- Feedstocks are sufficient and energy revenues are high; and
- Captive supply of feedstocks and use of produced energy.

In Mexico, in 1988, the General Law for Ecological Equilibrium and Environment Protection was passed. This law touched aspects related to waste reduction, although this did not lead to concrete actions. In 1996, this law was reformed following the North American Free Trade Agreement between Canada, the USA, and Mexico, which demanded environmental regulations (Law for Ecological Equilibrium and Environment Protection, 1988; Study of Waste Collection Systems in Mexico, 2005).

In 1992, Mexico signed the Rio Declaration of Sustainable Development. In the same year, the National Ecology Institute (INE) and the Federal Attorney for Environmental Protection (PROFEPA) were created. PROFEPA is still in charge of controlling law fulfillment. In 1994, the environmental agency SEMARNAP (SEMARNAT from 2000 on) became officiated, focusing mainly on sustainable development and reduction of environmental damage (Study of Waste Collection Systems in Mexico, 2005).

In 2003, the General Law for Waste Prevention and Waste Integral Management was published. This law considers waste as (1) a possible contaminant that must be avoided, reduced, and managed in an environmentally adequate manner; and (2) a possible material or energy resource (Law for Waste Prevention, 2003).

The reformed standard NOM-083-SEMARNAT-2003 specifies the protection of the environment in the selection of the location, design, construction, operation, and monitoring of municipal and industrial waste. This norm provides guidelines for the construction of landfills on an environmentally friendly way. There is no law in force prohibiting landfilling of certain types of waste (including organic waste) (Study of Waste Collection Systems in Mexico, 2005).

In 2009, the National Program for the Prevention and Comprehensive Management of Waste 2009–2012 was presented. The main objectives are to:

- Reduce waste generation;
- Encourage waste segregation;
- Reuse and recycle;
- Reduce the amount of waste destined for landfilling; and
- Develop legal and economic instruments to minimize waste production.

Despite its good intentions, this program contains no binding targets for effectively reducing waste from being landfilled (National Program for Waste Prevention and Waste Integral Management, 2008; National Waste Management Program, 2009).

In 2013, a new national waste management plan was published for the period 2013–2018 to strengthen municipal solid waste management by increasing the operational and structural abilities of federal and local authorities for collection, separation, and recycling of waste.

The 2008 Law for the Use of Renewable Energy and Finance of Energy Transition forms the basis for the support of AD in Mexico. This law created a fund for the energy transition and sustainable use of energy, assuring the financing of projects. The yearly budget was fixed at \$200 million (€145 million) for the period 2009–2011. In 2012, the Mexican Senate approved the integration of long-term climate targets into national legislation. Some of these targets are (Law for the Use of Renewable Energy and Finance of Energy Transition, 2008):

- 35% of energy must come from renewable resources by 2024;
- Government agencies are obliged to use renewables; and
- 30% reduction in emissions by 2020 and 50% by 2050 (compared to the business-as-usual scenario).

The climate change law also established a Climate Change Fund with the objective of raising funds from public and private entities for the implementation of adaptation plans (General Law on Climate Change, 2012).

4.3 Industrial Composting

4.3.1 Principle

Composting is the controlled biological transformation of organic waste under aerobic conditions, i.e. in the presence of oxygen, by indigenous microorganisms such as bacteria, actinomycetes, yeasts, and fungi. The degradation of the organic matter leads to the release of CO_2, H_2O, and heat and to the production of plant-available minerals, biomass, and humus (Fig. 4.1) (Ministry of the Environment, 1992; FFTC, 1994; ORCA, 1992).

In average, 65% of the organic matter is transformed during the 12 weeks of active composting under industrial conditions, resulting in a decrease in volume of approximately 50%. The remaining organic matter will degrade at a much slower rate, thereby slowly releasing nutrients for plant life, and the humus present in the compost improves the soil structure by binding to the soil particles and as such, providing a stable structure and good water retention in the soil.

Compost can be considered as a long-term biofertilizer on the one hand, and as a soil conditioner on the other hand. The high temperatures, which are reached during the industrial composting process, guarantee the kill-off of pathogens and weed seeds. Consequently, environmental risks when applying compost in soil are eliminated (ORCA, 1992; Biological Waste Management, 1995; Temesi).

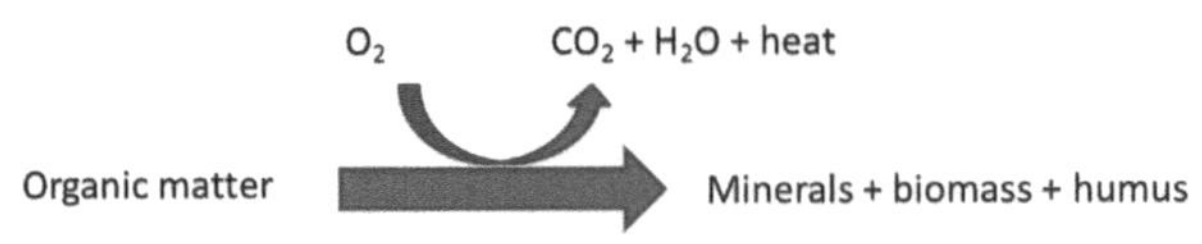

Figure 4.1
Organic matter transformation into humus, biomass, and minerals.

4.3.2 Input

Organic fractions, characterized by a good or considerable biodegradability, can be used as input material for an industrial composting process. Some examples of suitable input materials are vegetable, garden and fruit waste, yard trimmings, clippings, shredded wood, digestate, etc. (EMIS).

The quality of the input material is mainly determined by three parameters: (a) carbon/nitrogen (C/N) ratio; (b) moisture content; and (c) structure. Some typical values for these parameters are given in Table 4.2 for some biodegradable materials.

Carbon/Nitrogen ratio

During composting, microorganisms require C for growth and energy and N for protein synthesis. Optimum composting conditions are obtained at C/N ratios of 20–30. At higher C/N ratios, certainly above 40, the microbial population will lack nitrogen, and the composting process will slow down. At lower C/N ratios, excess nitrogen will be released as ammonia (NH_3), causing odor and environmental problems (FFTC, 1994; ORCA, 1992).

As the C/N ratio entirely depends on the composition of the incoming waste and as such cannot be controlled, the C/N ratio can be adjusted in the pretreatment phase by mixing different types of waste. Tree cuttings can be added in order to compensate for the rather low C/N ratio of grass or vegetable and fruit waste (see Table 4.2). During the composting process, the C/N ratio will gradually decrease to an average value of approximately 10 (Ministry of the Environment, 1992; ORCA, 1992).

Moisture content

The optimum moisture content of organic wastes for composting ranges from 45% to 55%. If the moisture content drops below 30%, all microbial activity stops. On the other hand, if the moisture content exceeds 70%, aeration of the organic waste becomes very difficult, and anaerobic conditions may prevail (FFTC, 1994; ORCA, 1992).

The moisture content can be adjusted in the pretreatment phase by mixing different types of waste. During the composting phase, evaporation will naturally occur as a result of the elevated temperatures and aeration. The moisture content is then adjusted by water supply systems or via aeration. Approximately 100 L/ton organic waste needs to be added during the

Table 4.2: Main Characteristics of Some Biodegradable Materials.

Optimum Value	20–30	45–55	Loose for Air Access
Grass	12–20	80–90	Poor
Food, vegetable waste	12–25	70–90	Poor
Leaves	30–60	40–50	Average
Tree and bush clippings	100–150	Moist to dry	Good

composting process in a tunnel composting system. The humidity at the end ideally lies between 30% and 35% (Ministry of the Environment, 1992; EMIS).

Structure

A small particle size is advantageous for the composting process, as it provides a higher surface area on which the microorganisms can feed. At the same time, too fine material will result in compaction and loss of porosity, and therefore, a higher chance of anaerobic conditions. Hence, the feedstock ideally contains a good mixture of fine materials, such as fruit and vegetable waste, and coarse or more porous materials, such as garden waste, mulch, or wood chips (FFTC, 1994; ORCA, 1992).

When the structure is too fine, a sufficient amount of coarse material should thus be added. The different types of materials are thoroughly mixed in the pretreatment step by mixing drums (ORCA, 1992).

4.3.3 Pretreatment Steps

The pretreatment phase is used to optimize the material characteristics of the waste, such as C/N ratio, moisture content and structure, as well as to remove contaminants and reduce obstructive materials that could hinder the composting process. Equipment that is used include tipping floors, front-end loaders, bag openers, shredders, screw mills, mixing drums, screens, magnetic separators, grinding equipment, centrifuges for dewatering, etc. (ORCA, 1992).

An example of a composting process, including some commonly used pretreatment steps (trommel screen, manual sorting, shredding, metal removal and mixing), is given in Fig. 4.2.

4.3.4 Composting Process

The composting phase can be divided into two subphases: a first, high-rate decomposition phase or preliminary phase, and a second, slow-rate decomposition phase, also known as maturation or stabilization phase. During the preliminary phase, a significant temperature increase is observed, caused by the decomposition of easily degradable material, like sugars and starch, at a high rate during a period of one or several weeks. During this period, effluents and noxious odors are generated and need to be strictly controlled and treated prior to the release in the environment. Temperature will initially rise to about 45°C in the so-called mesophilic incubation phase, after which it will further increase to 60–70°C in the so-called thermophilic phase. At temperatures around 58°C, the decomposition rate finds itself at its optimum. Hygienization of the material occurs when pathogens are killed and any seeds present are sterilized. The first phase yields a decontaminated but unstable intermediate called fresh compost. Once the majority of the easily degradable materials starts to vanish, the bacterial activity slows down, and temperatures will gradually drop off to a lower, stable level (approximately 40°C). It is the start of the much slower degradation of more complex

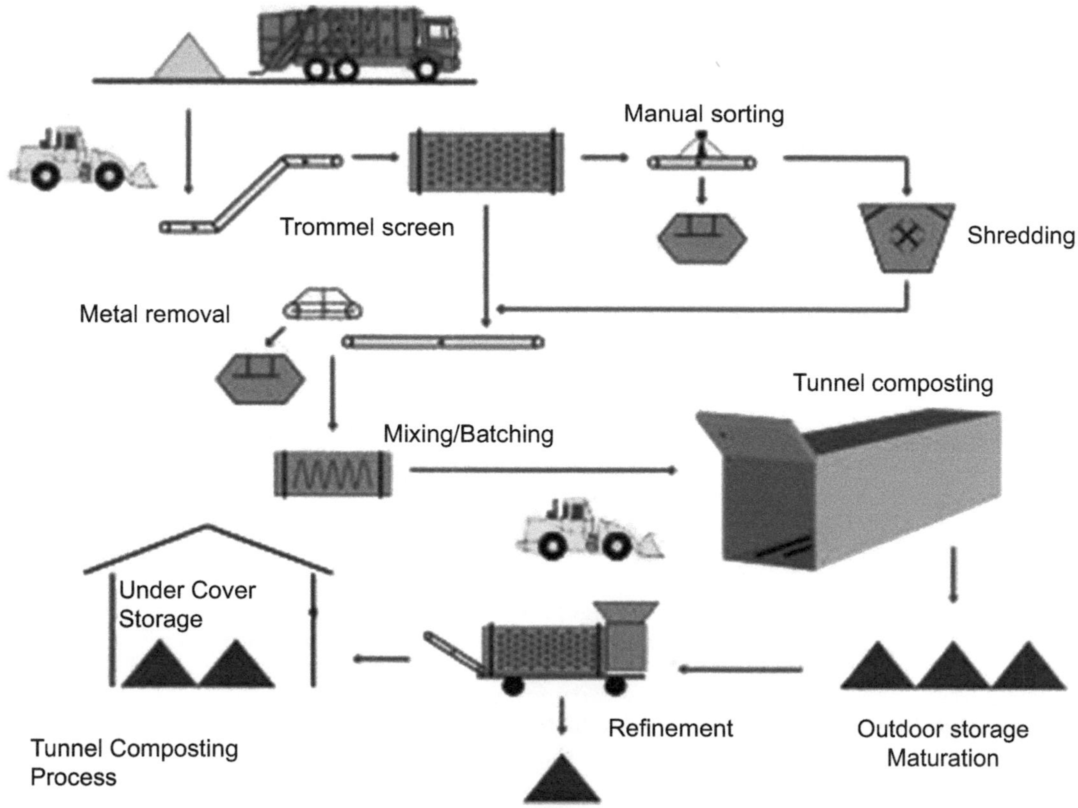

Figure 4.2
Example of composting process (AP Business Technology Consultancy).

materials, e.g. lignin (one of the basic substances of wood), by bacteria, and in particular, by fungi. At the end of the stabilization phase, ready-to-use matured compost is obtained (FFTC, 1994; ORCA, 1992; Periurban Vegetable Project).

A typical example of a temperature profile and a pH profile during the composting phase is given in Fig. 4.3.

The duration of the preliminary phase and the stabilization phase varies in function of the composting system. Windrow and static pile systems are comparable in process time, while more complex composting systems (in-vessel composting systems, etc.) or composting systems with higher turning frequencies are characterized by a shorter duration. Some examples are given in Table 4.3.

The degree of maturation depends on the amount of readily biodegradable organic matter and can be expressed by the so-called "Rottegrad," which is based on the potential heating ability of aerobic compost as shown in Table 4.4 (Ministry of the Environment, 1992; Biological Waste Management, 1995).

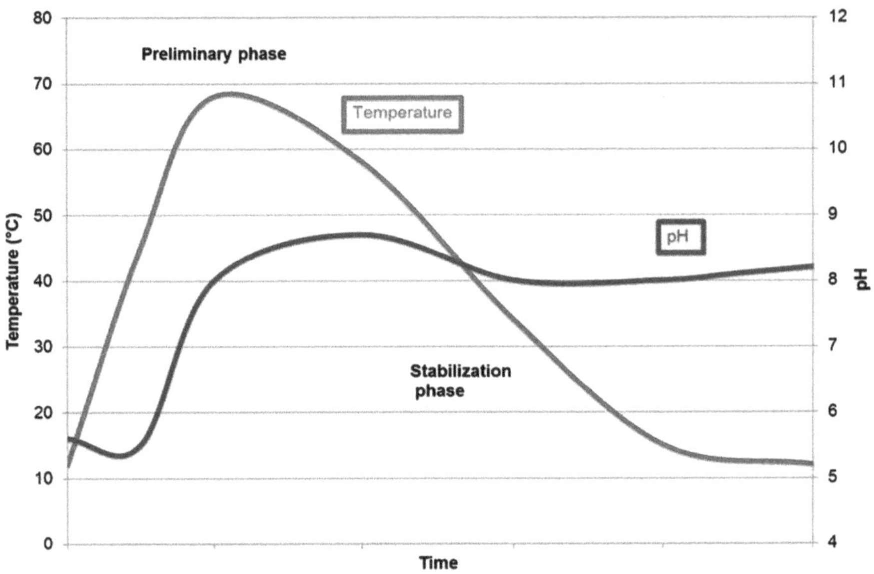

Figure 4.3
Typical temperature and acidity profile during composting.

Table 4.3: Duration of Some Industrial Composting Processes (Large-scale Compost; Deyerling & Fuchs; Productie Van Compost; Offaly).

Example	Preliminary Phase	Stabilization Phase
Windrow composting system	3–9 weeks, depending on type of material and frequency of turning	
Aerated static pile composting system	3–5 weeks	
Windrow composting system covered with GORE™ Cover System (Central Europe)	8 weeks	
Windrow composting system covered with GORE™ Cover System (South Europe)	6 weeks	
Closed hall windrow composting system (WIPS, Belgium)	8 weeks	Some additional weeks
Tunnel composting	7–21 days	4–8 weeks (composting in piles)
Automated in-vessel composting system (Laois Council Canteen Composting Program, Ireland)	2 weeks	4–6 weeks

Fresh compost is a hygienized product that still contains a significant amount of readily biodegradable organic matter. The resulting microbial activity can negatively affect plant growth, since microorganisms will compete for the use of oxygen and nitrogen, and moreover, they may partially attack plant roots. Therefore, fresh composts should be only readily applied in absence of plants, foreseeing a sufficient in-soil stabilization period between

Table 4.4: Levels of Rottegrad Based on the Potential Heating Ability of Compost.

Rottegrad	Maximum Temperature (°C)	Product Description
I	>60	Raw
II	50–60	Fresh
III	40–50	Fresh/mature
IV	30–40	Mature
V	<30	Mature

compost application and seeding (Biological Waste Management, 1995; Handbook on Composting and Anaerobic Digestion on VGF Waste, 1991).

Mature or stable compost has reached an advanced stage of decomposition and will thus contain only small amounts of readily biodegradable organic matter, resulting in a low and constant level of microbial activity. Full stability can only be obtained after a sufficiently long composting period (Biological Waste Management, 1995; Handbook on Composting and Anaerobic Digestion on VGF Waste, 1991; Compost Stability, 1996).

A generally accepted, stricter definition of "stability" and "maturity" is not available, and apart from the Rottegrad, several methods and corresponding parameters exist to describe the stability of compost, such as age, C/N ratio, cation exchange capacity, lipid extraction, carbon dioxide evolution, and mechanical or electronic oxygen uptake (Periurban Vegetable Project).

4.3.5 Composting Methods

Composting methods are manifold. Generally, the differences only concern the first phase of the composting process, i.e. the preliminary phase. The subsequent stabilization phase generally takes place in piles. Basically, a distinction can be made according to the handling or moving of the material, which can take place in batch (static methods, e.g. windrow composting, tunnel composting), semicontinuously (semistatic methods, e.g. table composting), or continuously (dynamic methods, e.g. drum reactors) (Ministry of the Environment, 1992; Biological Waste Management, 1995).

In Germany, the composting systems were divided in different categories called "Baumuster." An overview of these categories is given in Table 4.5.

In the following, the most important methods for industrial scale composting will be discussed.

Composting in piles

Windrow composting

Windrow composting is a simple method with low investment costs and a low level of process control. In this method, the waste is piled in triangular or rectangular heaps of 2–3 m high,

Table 4.5: Overview Baumuster Composting Categories (Activity report, 2013).

				Windrows	Open Windrows		
Boxes/Container	**Briquets**	**Tunnel/Line**	**Trommel**	**Enclosed**	**Open**	**Roofed Over**	**Windrow Enclosed (Membrane Cover)**
1.1 Herhofboxen	**2.1** Brikollare (42 days)	**3.1** Gicom-tunnel	**4.1** Envital	**5.1** Horstmann/ Kompo Plus/ SutcoKompoflex (7 weeks)	**6.1** Dreiecksmiete belüftet	**6.7** Tafelmiete, belüftet	**7.1** GORE™ cover (6 weeks)
	2.1A Brikollare (21 days)			**5.1A** Horstmann/ Kompo Plus/ SutcoKompoflex (10 days)			
1.2 Biodegma		**3.2** Bioferm-tunnel		**5.2** BühlerWendelin	**6.2** Dreiecksmiete unbelüftet	**6.8** Dreiecksmi-eteunbelüftet	**7.2** GORE™ cover (14 days)
1.3 ML-container		**3.3** Geotec-tunnel (14 days)		**5.3** AE und Koch	**6.3** Tafelmiete (I), unbelüftet		**7.3** Humivit/Plus
1.4 BRV-boxen		**3.3A** Geotec-tunnel (7 days)		**5.4** Thyssen-Dynacomp	**6.5** Tafelmiete, belüftet		
		3.4 Linde KCA-tunnel		**5.5** Stratmann	**6.6** Tafelmiete (II), unbelüftet		
		3.5 Sutco-BiofixZeile		**5.6** KNO Bremen	**6.9** WURM Komp-Aktiv		
		3.6 Horstmann WTT-tunnel					

3–5 m wide, and up to 100 m long. The heaps are turned mechanically to allow for heat release and to enhance aeration and homogenization (see Fig. 4.3). The need of turning can be reduced by installing air piping under the pile, providing passive aeration (chimney effect) or aeration by over or under pressure. The windrows can be placed in open air (Fig. 4.4) or in a closed hall (Ministry of the Environment, 1992; Biological Waste Management, 1995; American Bio Tech).

When windrow composting is used for the treatment of fresh organic waste, it should be executed in closed halls in order to control the effluents and noxious odors, which are released during the preliminary decomposition phase. For the stabilization phase, however, open air and roofed windrow systems are widely used, as its requirements for aeration, moistening, and air control are less stringent (Biological Waste Management, 1995).

Table composting

Table composting takes place in a hermetically sealed composting hall in which the biowaste is piled up into a stack and constantly aerated through perforated floor panels. A fully automatic turning machine moves the organic matter from one end, where new shredded organic matter is added, to the other end, where the finished compost can be removed (see Figs. 4.5 and 4.6). The resulting outlet gas can be cleaned using a biofilter. It takes 8–12 weeks for the compost to mature (Temesi; Bekon).

Compared to windrow systems, compost tables have the advantage that they allow a semicontinuous composting process, while windrow composting is a batch process. Moreover, composting tables need less space, the aeration conditions are more homogeneous, and aeration and watering can be adjusted according to the decomposition phase (Temesi).

Figure 4.4
Open air windrow system.

Figure 4.5
A fully automated turning machine.

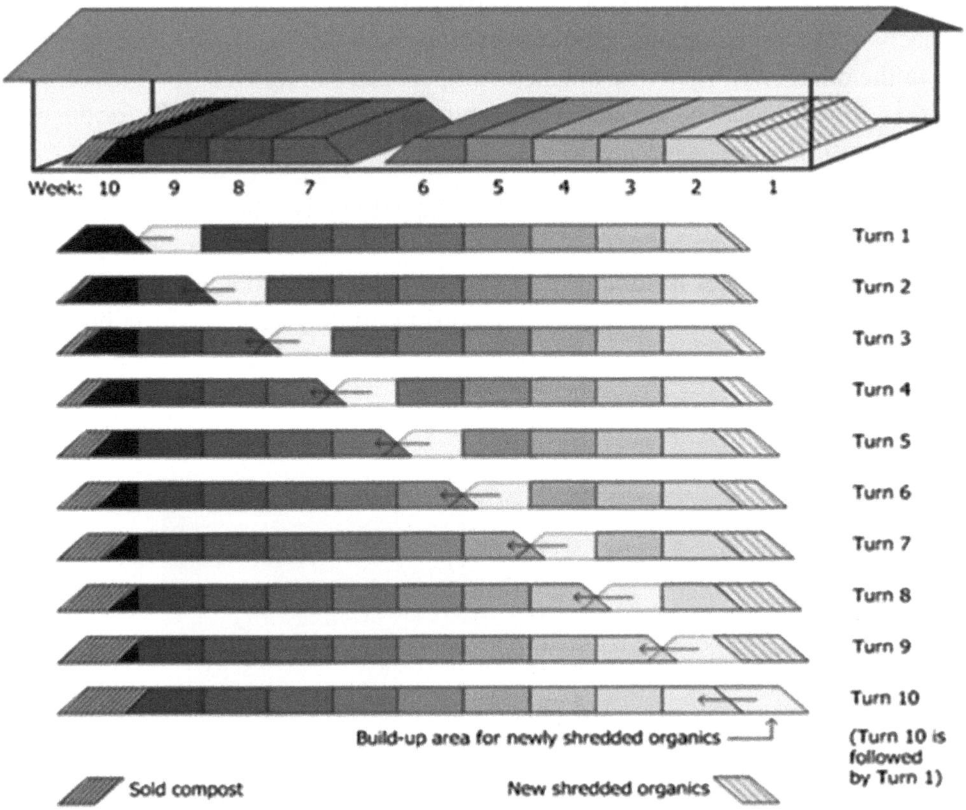

Figure 4.6
Schematic overview of turning of table composting piles.

Row composting

Composting in rows is similar to table composting, differing only in the fact that the piles are subdivided and sheltered, creating several smaller closed tables next to each other. Compared to the table system, the system is less straightforward but has the main advantage that the overall construction does not need protection against aggressive composting gases, and that each table can be monitored independently (Biological Waste Management, 1995).

Tunnel composting

In this method, preliminary decomposition takes place in closed tunnels or containers (also called biocell reactors). The tunnels can be loaded by wheel loaders or automatically, and are provided with aeration, moistening, and temperature control systems. Because of their small volume, the method allows for a precise monitoring of process parameters and control of emissions. In some cases, the material is led through several tunnels and is sieved in between. Preliminary decomposition in tunnels takes between 7 and 21 days and produces a fresh compost, which can be turned into mature compost in about 4–8 weeks by composting in piles (Ministry of the Environment, 1992; Biological Waste Management, 1995; Biowaste Producers Forum, 1995).

Drum reactors

One of the dynamic preliminary decomposition methods are drum reactors (Fig. 4.7). The organic waste follows a helical course from entry to exit and is thoroughly and continuously rotated and mixed by the drum's revolutions. This is an advantage, since it allows the processing of composting feedstocks from all sources, with constant variations in type and composition. In the drum, the stock is further comminuted in a selective and controlled manner, according to residence time (Ministry of the Environment, 1992).

Figure 4.7
Drum composting reactor (DTEnvironmental).

4.3.6 Posttreatment Steps

Sieving and recirculation

Whereas easily biodegradable substances are largely degraded during the preliminary composting phase, some other organic materials, e.g. woody substances, branches, shells, fruit peels, and also corncobs and hard plant leaves, will need a longer degradation period. Therefore, recirculation of these materials after the preliminary phase or even after the stabilization phase is necessary and is implemented by adding a sieving step, whereby the oversize fraction is sent back to the start of the process. Although the overall retention time is slightly increased by recirculation, it is an advantageous methodology: the recirculated wood and hard plant material improve the structure of the incoming waste, and consequently, the need for addition of extra structural material, which may be not readily available on site, is diminished.

Refining

The oversize fraction is removed and recirculated, and the resulting compost is prepared for use. The compost is cleaned to remove all remaining impurities like plastics, ceramics, stones, and glass, and is sieved according to the various commercial requirements. Bagging equipment may be used to pack products destined for the retail market. Bulk material for agricultural use is stored in sheltered areas awaiting seasonal application (ORCA, 1992).

4.3.7 SWOT Analysis

Strengths

Production of compost

The produced compost is a soil improver. The application of compost improves the physical, biological, and chemical soil properties, and carbon is incorporated in the soil. Moreover, compost is considered an organic fertilizer as it contains nitrogen, phosphate, potassium, magnesium, and calcium. Finally compost can also be used in order to replace peat and bark in growing media. In growing media for hobby gardening, 40–50 vol% compost can be used, while in growing media for professional use 20–30 vol% compost can be used (JRC, 2008).

Simplicity

The industrial composting process is characterized by a higher simplicity when compared to AD. Therefore, composting technologies can be considered to be reliable systems in order to convert organic waste to compost.

Low investment cost

Investment cost for industrial composting is lower when compared to AD.

Weaknesses

No energy production

Whereas AD yields biogas (= energy), no energy is produced during the composting process. In contrast energy is required for the turning and the aeration of the compost.

Air emissions during the process

During the composting process gaseous emissions (e.g. carbon dioxide, ammonia, volatile organic compounds, bio aerosols, particulates, methane, nitrous oxide, etc.) can be formed, which are often associated with odor emissions. The emissions can be reduced in closed systems using biofilters. In open systems, however, industrial composting can lead to emissions and odor.

Difficulties in marketing

Compost producers have often difficulties in the marketing of their product due to a lack of understanding of potential use sectors. When compared to competing products, declaration, advertisement, and marketing are often less developed (JRC, 2008).

Opportunities

Reduction of biowaste going to landfills

Composting is a valid recovery route for biowaste in order to reduce the amount of landfilled biowaste.

Threats

Replacement by end-of-life systems with energy recuperation

The energy of the biowaste is lost during the composting process. From an energetic point of view, alternative techniques (e.g. AD) will be more suitable.

Price of compost

The price of bulk compost is still lower than the production costs (i.e. the costs of treating biological wastes in a composting plant). In Central Europe, the price for compost for agricultural use is rarely higher than 5 €/ton of compost, while the cost for composting is estimated at 35 to 60 €/ton of waste for best practice plants in closed systems and 20 €/ton of waste for low-tech windrow composting. The sale of compost becomes more difficult in countries where there exists a competition with manure (e.g. Belgium, the Netherlands). The low price of compost forms a limitation to the distances over which the transport of compost for applications in agriculture is still possible from an economic point of view (JRC, 2008).

4.3.8 Composting Within the TRANSBIO Project

Agro–industrial byproducts from the production of vegetables like sweet corn, green beans, peas, potatoes, endive, lettuce, broccoli, bananas, etc., are suitable sources in order to compost. The C/N ratio of most of the products lay in the optimum range between 20 and 30. For those products that this is not the case (e.g. broccoli and lettuce), the byproducts should be mixed with an appropriate material in order to increase the ratio to the optimum range. The disadvantage of composting the byproducts is that no energy is recovered. However, mature compost will be produced that can be used as soil fertilizer in agriculture.

4.4 Anaerobic Digestion

4.4.1 Principle

AD is a microbiological process in which organic material is degraded in the absence of oxygen. During this degradation, the organic matter is converted to biogas (mainly methane (CH_4) and carbon dioxide (CO_2)), and a digestate, which contains the undegraded organics (e.g. lignin), biomass, and nutrients. Unlike the composting process, degradation in an anaerobic environment is carried out exclusively by bacteria, and not by fungi, actinomycetes, or other microorganisms. As a consequence, organic matter containing high amounts of lignin will only be poorly degraded. The main advantage of AD over composting is the production of methane (50–70% in the biogas), which can be valorized as biofuel (ORCA, 1992). The digestate can be transformed into mature compost in an aerobic postcomposting step (usually 2–3 weeks).

AD is usually performed in a (semi)industrial setup, treating volumes of 10 ktons to more than 100 ktons of organic substrates per year. In developing countries like India, China, and many African countries, domestic AD units have been introduced as a means for many households to convert their organic waste to cooking fuel (biogas). Also in some European countries with high manure surpluses (e.g. Flanders, the Netherlands and some regions in Germany), small-scale farm digesters have gained popularity for treating manure onsite without any cosubstrates. The technology used in these domestic and small farm digesters is very basic compared to the industrial plants treating organic waste.

4.4.2 Input

Many organic feedstocks can be treated by AD, woody biomass being the exception, as anaerobic bacteria are not able to degrade lignin. Anaerobic digesting facilities can be roughly divided in the following categories depending on their main input:

- **Wastewater**. Apart from the classical aerobic WWTPs, anaerobic wastewater treatment is a valuable alternative for wastewaters containing high concentrations of degradable matter. Many municipal WWTPs and also industrial WWTPs use this technology to treat the wastewater and simultaneously produce energy.
- **Sludge**. During aerobic wastewater treatment, a sludge is produced, which can be further valorized in a digester for the production of biogas. Another advantage is the further reduction of the amount of sludge to be disposed of.

- **Manure**. In many countries, farm-scale AD facilities are implemented to treat manure. The advantages of manure digestion are odor control, elimination of uncontrolled CH_4-production during storage, energy recovery, and production of digestate with improved fertilizer value. Because biogas potential of manure is rather low, codigestion of manure with other substrates is the most commonly used technique.
- **Agricultural waste**. In most agricultural crops, only part of the plant is used as a marketable product (e.g. grain, fruits, and most vegetables). The remaining part of the plant can be further valorized in AD. This can be done as a co-feed with manure in farm digesters or in industrial digesters as only input feet or in combination with other agricultural or industrial organic waste products. Due to the increased demand for renewable energy and the food versus fuel discussion in recent years, agricultural waste has gained a lot of interest as input material for biogas production.
- **Energy crops**. Maize (silage) is the most used energy crop in AD. In this case, crops are grown specifically for use as input for biogas production. Mainly in Germany, a lot of energy crop digesters have been built in the last decade, although recently, not many new plants are being constructed due to the feed versus fuel discussion. Energy crops are also used in codigestion plants with manure.
- **Industrial waste**. During industrial food processing, many organic waste and byproducts are generated, which can be further valorized with AD. Depending on the available amounts, these products are treated as the main feedstock in industrial digesters, or as co-feed in manure digesters.
- **Municipal organic waste**. AD of municipal organic waste has emerged as an alternative for landfilling and composting since the 1990s and has been a growing sector ever since. Two main categories can be discerned: AD of source separated VGF (vegetable, garden, and fruit waste), which usually results in a high-quality compost, and AD of mixed (unsorted) household waste (or the organic fraction after pretreatment), which has as a main goal the production of biogas and the reduction of material going to landfill.

4.4.3 Pretreatment Steps

Similar to the composting process, the objective of a pretreatment step is to improve the material characteristics (removal of contaminants, size reduction, etc.). As most organic input materials (manure, sludge, agricultural and industrial waste, energy crops) contain little contaminants, a pretreatment step is usually not necessary or is limited to a removal of stones (protection of equipment) or a size reduction (to increase the degradation rate and improve miscibility). Municipal organic waste, on the other hand, typically contains a significant amount of undesirable material, hence the need for an elaborate pretreatment phase. This phase typically includes tipping floors, front-end loaders, bag openers, shredders, screw mills, mixing drums, screens, magnetic separators, and grinding equipment.

The type of equipment not only depends on the characteristics of the incoming waste, but also on the digester type (e.g. solubilization and separation of sinking/floating material for wet digestion).

4.4.4 Anaerobic Digestion Process

Microbial communities

AD is a chain of processes, which can roughly be subdivided into four steps, as shown in Fig. 4.8. As the output of one step is being used as the input for the next step, a good functioning of all steps and equilibrium between these different steps is very important. If one step fails, this will have significant consequences on the next step(s) and could lead to a breakdown of the biogas production altogether.

The first step in the AD process is the **hydrolysis**. In this step, complex, insoluble, high-molecular structures like proteins, carbohydrates, and fats are enzymatically degraded to their building blocks: proteins to amino acids, carbohydrates to monosaccharides (sugars), and fats to fatty acids. These building blocks are water soluble and can be further degraded by incorporation by bacteria. The hydrolysis step is generally regarded as the speed limiting step in the process.

Once soluble products are formed as a result of hydrolysis, the acidogenic bacteria will further degrade these products to less complex products (organic acids, alcohols, H_2, and CO_2)

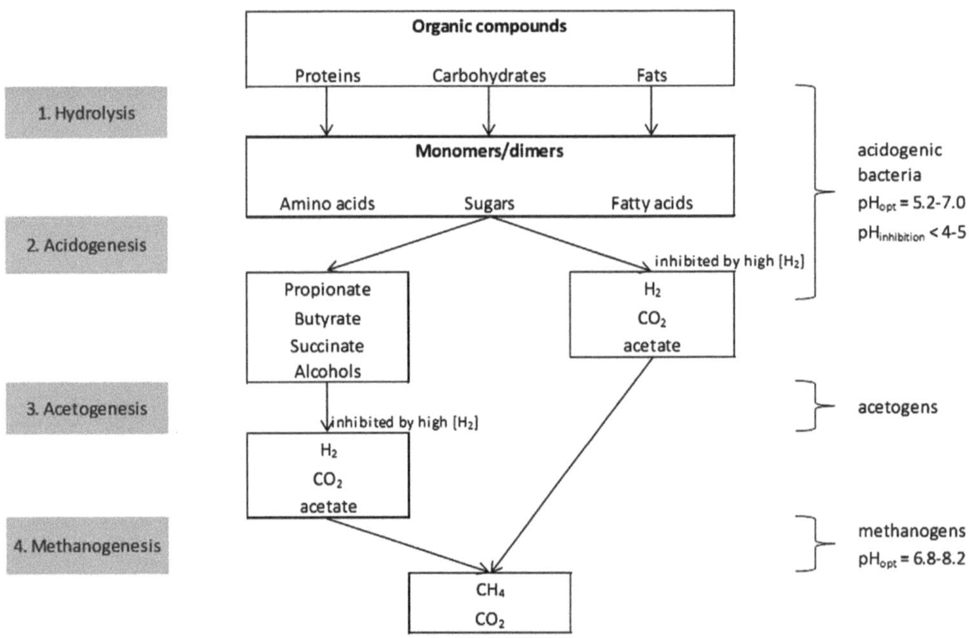

Figure 4.8
The four steps in the anaerobic digestion process (Biogashandbuch Bayern, 2004; Weiland, 2010).

in the **acidogenesis**. The product composition depends on the type of bacteria present and on process conditions. In this step, also ammonium is produced during the degradation of amino acids. When process conditions become too acidic (this is the case when the produced acids are not further degraded in the following process steps) and pH drops below a value of 4, the acidogenic bacteria become inhibited, and the process comes to a halt. The optimal pH varies around a value of 7 (Zhang et al., 2005).

Acetogens transform fatty acids and alcohols to acetate, H_2, and CO_2 in the **acetogenesis**. If H_2 concentration should become too high, then acetogenesis becomes energetically unfavorable for the microorganisms, and the process will stop. Therefore, H_2 should be converted to methane (step 4) at the same rate as it is produced. This results in a high level of symbiosis between acetogens and methanogens (Ernährung und Landwirtschaft, 2006).

In a last step, methane is produced in the **methanogenesis**. Two bacteria groups can be discerned in this step: *acetoclastic methanogens* degrade acetate to methane and CO_2, and *hydrogenotrophic methanogens* produce methane and water from hydrogen and CO_2. In a typical AD process, the acetoclastic bacteria are responsible for 70–80% of methane production, the hydrogenotrophics for 20–30%. The methanogens are very sensitive to fluctuations in process conditions (pH, temperature, concentration of organic acids, and ammonium). The optimal pH varies between 6.8 and 8.2 (Zhang et al., 2005).

Factors affecting the digestion process

Similar to composting, the characteristics of the input material in AD will have a significant impact on the overall process conditions.

The C/N ratio should be convenient for the microorganisms. For AD, this typically lies around 20–40, although the nitrogen release rate will also play a very important role. A too low C/N ratio can lead to nitrogen toxicity (ammonia), whereas a too high C/N ratio will lead to growth limitation due to lack of nitrogen for protein formation. The C/N ratio is inherent of the organic waste and cannot be controlled, although it is possible to improve the C/N ratio of the input by mixing several different substrates to obtain an optimum input recipe. During the AD process, the C/N ratio will decrease as C is leaving the system as CH_4 and CO_2.

The composition of the organic waste has an important impact on the composition of the biogas, as shown in Table 4.6. Furthermore, the presence of lignin and, to a lesser extent,

Table 4.6: Theoretical Amounts of Methane in Biogas.

Starting Material	Gas Yield (L/kg Organic Matter)	Theoretic Amount of CH_4 in Biogas (%)
Carbohydrates	790	50
Fats	1250	68
Proteins	700	71

hemicellulose, will limit the biodegradability of the organic waste because these components are not or difficult to be degraded under anaerobic conditions (as mentioned above). Degradable matter (e.g. cellulose) can be enclosed inside a lignin capsule, making it inaccessible for degradation. A high lignin content usually also comes with a more fibrous structure of the waste. This, in turn, will have an impact on the miscibility of the digestate (high fiber content leads to higher viscosity). Also, the presence of unwanted material will have an impact on the process (depending on the technology used). For example, the presence of heavies (stones, sand, etc.) will lead to sedimentation in the reactor, decreasing the active reactor content. Plastics can form a floating layer or damage the mixing equipment.

The optimal pH for the hydrolyzing/acidogenic bacteria lies around 5.2–6.3; for the methanogenic bacteria, this is 6.8–7.5. In a two-phase process, both bacteria groups can be operated at their optimum pH, whereas in a single-phase process, a compromise has to be found. The hydrolyzing/acidogenic bacteria can operate above optimum pH values with only a minor reduction of their activity, whereas the methanogens are bound to a neutral pH. Therefore, in single stage AD, the optimum pH is set at a value of 7–8. Nevertheless, in a normal process, the pH will be buffered, mainly by CO_2 and NH_4. In case the methanogens are inhibited, acid concentration will increase above the capacity of this buffer and pH will subsequently decrease, further inhibiting the methanogens. A good process control around a stable pH is therefore of paramount importance (Ernährung und Landwirtschaft, 2006).

During the anaerobic fermentation process, proteins are degraded and ammonium is produced. The concentration of ammonium to be found in the digestate depends on the total nitrogen concentration of the input and of the degradation rate of the proteins. Too low ammonium content will lead to nitrogen shortage and hence limit the growth of the bacteria, thus leading to a suboptimal process. Too much ammonium will shift the ammonium$\leftrightarrows$ammonia equilibrium to the right side, causing ammonia inhibition and reactor failure in extreme cases.

In contrast to aerobic composting, AD is a thermodynamically less favorable process for microorganisms. Part of the energy stored in the organic matter is converted to chemical energy (in the form of CH_4) and cannot be used for microbial metabolism. As a consequence, less energy will be set free, as heat and temperature will not rise automatically. This does not imply that AD is an energy consuming process. On the contrary, energy is still available in the form of biogas. Burning this biogas in a cogeneration unit will not only yield electricity, but also process heat, which largely exceeds the heating requirements to keep the process at the desired temperature.

AD is mainly performed in two temperature ranges: mesophilic (35–42°C) and thermophilic (47–55°C). Both temperature intervals are characterized by different bacteria groups, hence it is difficult to switch from mesophilic to thermophilic or vice versa. Mesophilic digestion is considered more robust and thus more stable than thermophilic digestion, and also needs less heat input to maintain its temperature. Thermophilic digestion, on the other hand, is known to yield more biogas from the same input material and at a faster rate. Moreover, thermophilic temperatures also guarantee a form of hygienization. The downside is that at higher temperature the bacteria are more prone to ammonia toxicity.

The bacteria responsible for the production of methane (methanogens) are strictly anaerobic, which means they cannot function in the presence of oxygen. Therefore, a hermetically closed reactor setup is an absolute necessity. Hydrolyzing bacteria and acidogens are facultative anaerobic, which means they can still function in the presence of oxygen. Oxygen entering the system with the feedstock material will first be consumed by these bacteria, thus preventing the buildup of O_2 to lethal concentrations for the methanogens.

4.4.5 Digestion Methods

AD plants can be subdivided on the basis of several operational and/or technical parameters.

Operating temperature (Converti et al., 1999; Remade Scotland, 2003; Ziemba et al., 2010; Bayr et al., 2012)

AD can be operated at two different temperature ranges: mesophilic (35–42°C) and thermophilic (47–55°C). Mesophilic digestion is more stable (mainly with regard to ammonium toxicity) and has less need for heating compared to thermophilic digestion. Thermophilic digestion, on the other hand, is characterized by its much higher efficiency (hydrolysis rate at thermophilic temperature is five times higher than at mesophilic temperature). Another advantage of thermophilic digestion over mesophilic digestion is the higher destruction of pathogens and weed seeds. The choice for the optimal temperature will be mainly determined by the input characteristics. Due to the higher sensitivity for ammonium toxicity in thermophilic digestion, mesophilic digestion is the most obvious choice for treating organic material rich in nitrogen. For low N-containing substrates, thermophilic digestion is recommended for its higher efficiency.

Moisture content (Remade Scotland, 2003; CSCE/EWRI, 2002)

Although several definitions are circulating, dry digestion can be defined as digestion with more than 15% dry solids inside the reactor, while wet digestion usually has less than 10% dry solids inside the reactor. Wet and dry digesters both have their advantages and disadvantages, and it is important to realize that the choice of technique also has an influence on the pre- and posttreatment of the substrates.

Dry digestion seems the logical choice when dealing with dry materials like (the organic fraction of) municipal solid waste. Because of the lower water content in the reactor, higher loading rates can be achieved. This has, in turn, the advantage that a smaller reactor is needed, reducing the investment costs. Another advantage is that no phase separation will occur, and there will be no crust formation or floating layer on top of the reactor or sedimentation at the bottom, both decreasing the effective reactor volume. The resulting digestate after dry digestion will be more suited for postcomposting, as no or less water must be extracted. A disadvantage of dry digestion is the need for more robust (and more expensive) equipment for handling the drier material. Two orientation types can be discerned: a horizontal design and a vertical design. The advantage of the latter is that it needs little space and that no mixing inside the digester is needed. Plug flow is obtained by gravity, and mixing can be done outside the reactor. Disadvantage is that heavy pumps are needed to transport the material to the top of the reactor.

An advantage of wet digestion over dry digestion is that cheaper equipment can be used for mixing and transporting the digestate and/or waste because of the lower viscosity. This must, of course, be balanced by the extra costs for a larger reactor and the energy needed for dewatering the residue. As a lot of agricultural and industrial byproducts are characterized by a high moisture content, wet digestion is usually the preferred technology for these substrates.

Complexity

AD is a common denominator for the different processes that occur simultaneously and sequentially in the biomass, resulting in the production of biogas. These processes all require different optimal conditions. Therefore, some constructors developed a technology using different vessels with the aim of optimizing the process and maximizing biogas production. In that case, usually two reactors are built: in the first reactor hydrolysis and acidogenesis take place, while in the second reactor, the acetogenesis and methanogenesis take place. A disadvantage of a two-phase process is the capital costs resulting from the high complexity.

Homogeneity of feedstock (Remade Scotland, 2003; Garcia-Peña et al., 2011)

Codigestion is a process where the AD unit treats a mixture of various input substrates. Many farm-scale digesters are codigesting manure together with energy crops, agricultural, and/or industrial (organic) waste streams. The advantage of codigestion lies in the fact that a well-balanced input mixture improves the digestibility (optimum dry matter content, good ratio of important nutrients, good ratio of carbohydrates, proteins, and fats) and thus can enhance the biogas productivity.

In the case of municipal waste, already a heterogeneous material is present, and the need for other cosubstrates will be small, although in some cases, an improvement of the anaerobic process can be obtained. In case digestion of a single feedstock is only suboptimal due to the lack of one or more nutrients, nutrients in the form of a concentrated additive can also be added as such.

Operating mode (Ernährung und Landwirtschaft, 2006)

The feeding modus of an AD system has a large impact on the availability of fresh substrate to the bacteria and thus on the biogas productivity.

In batch mode, at full-scale, usually applied only in dry box digesters, the anaerobic reactor is filled with fresh substrate and inoculum, and closed so no air can enter the system. The substrate/inoculum mixture is left to digest without adding or removing anything from the system (except biogas). During digestion, biogas production starts slowly and at very low CH_4 levels, and both speed and methane content increase over time. At the end of the process, biogas production will decrease again, but CH_4 levels will have reached a maximum value (depending on the feedstock). In other words, during digestion, no constant biogas flow and quality is obtained, which makes it difficult to find the optimal treatment for the biogas. A way of obtaining a more constant biogas flow and quality is to install several batch fermenters, each at a different stage of fermentation, thus averaging the differences in biogas composition. The main advantage of this system is the low investment costs, and the main disadvantage is the lower process control and hence less optimal digestion conditions and lower biogas production.

In a (semi)continuous feeding mode, the anaerobic reactor is fed on a regular basis (at least once per day) with fresh input substrate. In the case of wet digestion, this is usually done in a continuously stirred tank reactor system in which fresh substrate is added to the reactor and intensively mixed to assure a good contact between the anaerobic bacteria and substrate. Digestate is extracted on a regular basis to guarantee a constant reactor volume. The advantage of this system is that a good mixing can be expected, resulting in a good interaction between the different bacteria. The disadvantage is that relatively fresh input substrate can be extracted from the system without being completely converted to biogas (short-circuiting of biomass).

In dry digestion, a (semi)continuous feeding mode can be found in plug flow designs. In this setup, fresh input substrate is mixed with digestate and passes through the system until complete digestion (first in–first out). The advantage of this system is that the residence time is guaranteed (no shortcuts of fresh input substrate), but mixing is more difficult, and different layers can be discerned (acidification at beginning, methanogenesis at end), which could lead to suboptimal process conditions.

Also, a combination of plug flow and continuously mixing is possible in dry digestion. In that case, the flow-through time in the reactor does not last the complete retention time but only a few days, after which the majority of the digestate is again mixed with fresh substrate and passes through the reactor again. This system has the advantage that a minimum residence time of a few days is still guaranteed, while also obtaining a good mixing, thus avoiding different zones and suboptimal process conditions.

Mixing methods (Ernährung und Landwirtschaft, 2006)

Mixing of a digester is very important for several reasons. The first reason is to ascertain a good homogenization and contact between fresh input material and the bacteria inside the digester. The better the input is dispersed in the reactor, the faster the bacteria can degrade the material to biogas. A constant mixing also eliminates the buildup of "toxic" intermediary products, which could slow down the degradation process (e.g. acetogenesis will come to a halt when the end products of this stage are not quickly enough converted to methane by other bacteria). In wet digestion, mixing is also needed to prevent the formation of floating layers or sedimentation layers. The first can lead to a blockage of gas outlet from the digestate; the second will decrease the active reactor volume. Finally, mixing will allow for a good distribution of heat inside the reactor.

Several mixing types are available, as described below. In many digestion systems, more than one type of mixing is used to obtain the most optimal conditions.

- **Hydraulic mixing**. Hydraulic mixing is the continuously pumping of digestate through the digester with centrifugal or lobe pumps (wet digestion). By continuously extracting digestate from the reactor at one or several points and reintroducing the digestate at one or several others points back into the digester, mixing is obtained. Although there are no moving parts inside the digester, this type of mixing can be considered vulnerable to impurities blocking or damaging the pumps. Hydraulic mixing can also be used with (vertical) dry digestion, using heavy-duty pumps. In that case, it is mostly used together with (external) paddle mixers to improve mixing.
- **Paddle mixers**. Paddle mixers are usually mounted on an axis with an external engine. They have a relatively large propeller blade and are slow turning, which means that the energy consumption is considerably lower compared to propeller mixers. They are also less prone to damage from impurities. When installed properly, they can avoid the buildup of floating or sedimentation layers, although they are less suited to remove these layers when they are already present. They are most suited for use in wet digestion with high viscosity of the digestate. Repair or replacing paddle mixers usually requires complete removal of the digestate from the reactor.
- **Propeller mixers**. Propeller mixers are either completely submerged (including the engine) in the digestate or are mounted on an axis with the engine outside the digester. A common characteristic is that these propeller mixers are all fast-turning mixers with relatively small propeller blades. Because of the high speeds that are reached, a thorough mixing is assured, making it possible to remove floating or sedimentation layers by altering the angle or altitude of the mixers inside the reactor. The disadvantage is that they have high energy consumption and are prone to damage from impurities in the digestate (stones can damage propeller blades, plastics can entwine around the blades or axis, causing it to break). They are also less effective in case of digestate with a higher viscosity. Repairing or replacing propelling mixers needs partial removal of digestate from the reactor above a certain level.
- **Pneumatic mixing**. Pneumatic mixing is realized by injecting biogas inside the digester. The upward flow of biogas through the digestate leads to a vertical movement and thus,

mixing of the digestate. The main advantage of this system is that no moving parts are present in the reactor, decreasing the risk for damage through impurities. Sedimentation layers can be prevented, although they cannot prevent floating layers from forming. In case the biogas injection nozzles become clogged or need to be replaced, the reactor has to be emptied completely.

- **Percolation**. Percolating is a mixing system used mostly in batch fermentation. In this setup, liquid digestate is percolated on top of the input/inoculum and extracted again from the bottom. This limited form of "mixing" avoids that the reactor content dries out on top, and provides a better dispersion of bacteria and degradation products throughout the reactor. This form of mixing also provides the operator with a limited process control. The downside of this technology is that this system can easily become clogged and that preferential pathways can be formed, resulting in partial mixing.

Residence time (Velghe and Wierinck, 2013)

The residence time or retention time inside an AD reactor determines the time available to the bacteria to degrade the input substrates to biogas. The shorter the residence time, the more efficient the reactor volume is used, although there is a limitation where a too-low residence time will lead to negative effects. Two parameters will determine the limiting residence time for a given process. First of all, the residence time cannot be shorter than the replication rate of the bacteria, which is approximately 12 days (theoretical minimum residence time). If this would become the case, bacteria would leave the system at a faster rate than they are renewed, eventually leading to reactor failure. Secondly, the time needed by the bacteria to degrade a given substrate to biogas plays an important role. This time depends largely on the type of input substrate. Easily converted substrates like manure and vegetable waste are readily transformed to biogas and therefore need a shorter residence time than more recalcitrant substrates like fats or products rich in fibers. Also, the particle size plays an important role: the smaller the particles, the higher the surface area to volume ratio and the faster the product is degraded. In AD of organic waste, the degradation rate of the input will determine the minimum retention time, which is usually around 20–30 days. Because thermophilic digestion has a faster degradation rate, the minimum retention time will be closer to 20 days, whereas mesophilic digestion will need minimum 30 days.

Loading rate (Velghe and Wierinck, 2013)

The loading rate of a digester is a way of expressing the amount of work that is carried out by the anaerobic bacteria and can be expressed in several ways. A first way is the organic loading rate, which shows how much organic matter is fed to the reactor on a daily basis $\left(\text{kg OM/m}^3_{\text{reactor}}\text{/day}\right)$. The higher the organic loading rate, the higher the workloads for the bacteria to convert the organic matter to biogas. This means that the reactor volume is used at a more efficient level but also leads to a higher sensitivity for suboptimal process conditions (high S- or N-content, low nutrient level, toxic compounds).

Biogas productivity is another way of expressing the loading rate of an anaerobic digester. This parameter gives the amount of biogas produced per reactor volume and per day $\left(m^3\ \text{biogas}/m^3_{\text{reactor}}/\text{day}\right)$. Generally, a higher organic loading rate will also yield higher biogas productivity. When this is not the case, the bacteria could be inhibited by suboptimal process conditions. As with the organic loading rate, the higher the biogas productivity, the more efficient the reactor volume is used, but also the more sensitive the process will become for inhibitive conditions.

4.4.6 Posttreatment Steps

At the end of the AD process, digestate remains. In the case of agricultural and industrial waste digestion (depending on the input characteristics and fertilizer regulations), this digestate can be applied directly on land as an organic fertilizer (replacing manure). To improve the application of the digestate however, the digestate is often separated in a liquid and solid phase using either a centrifuge or press. Depending on the application, the press or centrifuge cake could be further composted, dried, and/or pelletized.

With municipal waste digestion, posttreatment of the digestate is a standard practice. Several posttreatment steps can be applied, depending on its characteristics and the end use of the digestate. Dewatering of the digestate is applied in many installations by either a press or a centrifuge. The resulting process water can be recirculated to inoculate the incoming waste and to maintain a good moisture content inside the digester. Excess water can be further treated in a wastewater treatment facility. The press or centrifuge cake can be transformed to mature compost in an aerobic postcomposting step. In case of municipal waste, this cake is usually first sieved to remove contaminants (e.g. plastics). It is also possible to postcompost the digestate obtained after dry AD without dewatering. In that case, the digestate is mixed with fresh incoming organic waste (mainly garden waste with high dry matter content) to obtain an optimal moisture content for composting. At the end of the composting process, the mature compost can be further refined (cleaning, sieving, packing).

In case of AD of the organic fraction of mixed (unsorted) household waste, the digestate will still contain high amounts of impurities (plastics, textiles, glass, stones, etc.), which makes this less suitable to be used as a high-quality compost. In that case, valuables can be recovered from the digestate by means of a wet separation step. Fractions that can be recuperated with this treatment step are sand, fibers, glass, metal (ferrous and nonferrous), refuse-derived fuel (RDF), and organic sludge. This sludge is then usually aerated to obtain a mature compost containing little contaminants.

4.4.7 Biogas Treatment

The most valuable product obtained in AD is biogas, which consists of methane (CH_4) and CO_2, together with some other minor components (H_2O, H_2S, NH_3, etc.). Depending on the final application of the biogas (see section 4.4.8), the biogas needs to be treated to improve its characteristics.

Removal of water (SGC, 2001; Schomaker et al., 2000)

When the biogas is leaving the AD reactor, it is saturated with water vapor. This corresponds to ±4% moisture content in mesophilic digestion, or almost 10% moisture content in thermophilic digestion. For all possible valorization options, this water must be removed. During the dewatering step, usually NH_3 is also (partly) removed from the biogas. Several options are available.

- **Condensation**. The easiest way to remove the water from the raw biogas is by cooling the gas. This water removal step is used with the majority of existing biogas plants. In many small- and/or farm-scale plants, this is done passively by leading the gas through long pipes in which the gas is cooled by the outside temperature or by ground temperature in the case of underground pipes. These pipes are sloping toward a condensation separator, which must be kept frost-free and emptied at regular intervals. This type of passive cooling has varying efficiencies, depending on the outside or ground temperature. To have a more reliable cooling, the gas can be compressed to below the dew point and then decompressed again to the desired pressure. The condensed water is removed from the gas by either a demister, a cyclone filter, or a moisture trap compressor.
- **Adsorption**. Biogas can be dewatered by pumping it over a silica filter. Usually two silica columns are used simultaneously: one is used for adsorption, while the other is being regenerated (e.g. evaporation by heating). Besides silica, also active carbon, aluminum oxide, and magnesium oxide can be used for this purpose. Compared to condensation, a higher removal efficiency of water can be obtained. Because of the higher operational costs (adsorption material), this option is less frequently used.

Removal of H_2S (SGC, 2001; Schomaker et al., 2000; Krich et al., 2005; Wellinger and Lindberg, 2000)

H_2S is a toxic gas with an unpleasant odor, which can form sulfuric acid in combination with water vapor and cause corrosion to pipes and machinery. Therefore, this component should always be removed from the biogas before valorization. To avoid H_2S toxicity on the microbial processes, H_2S can be removed in the reactor. To safeguard the biogas pipes and boiler, engine or upgrading equipment, an extra H_2S removal step can be placed after the AD reactor.

Biological H_2S removal in the AD reactor

By adding a limited amount of oxygen in the raw biogas, H_2S can be oxidized to elemental S or to liquid H_2SO_4 by a specialized group of aerobic bacteria (*Thiobacillus* sp.). A support material for these bacteria to grow on must be present in the reactor (this can be the reactor walls, a floating layer, or a specially designed structure). The advantage of this technology is its simplicity and low cost. A disadvantage is that H_2S removal is difficult to regulate.

Because it is a biological process, the bacteria cannot always cope with sudden changes in H_2S levels, e.g. when feeding composition changes. Another disadvantage is that because of the O_2 dosage, an explosive gas mixture can be created locally. The air injection pipe(s) should be positioned on the opposite side of the biogas output pipes. When the elemental sulfur (S) accidentally falls back into the reactor, this can anaerobically be reconverted into H_2S, leading to unexpected peak concentrations and possible negative effects in the biology.

Chemical H_2S removal in the AD reactor

H_2S can also be removed by adding iron (Fe)-containing chemicals to the input substrates or directly to the digestate in the digester. The added Fe reacts with the sulfur to form FeS. The most reliable way of dosing Fe is in the form of $FeCl_2$, although also Fe-containing sludge or water from physicochemical water treatment plants can be used for this purpose. The disadvantage of the latter is that the Fe concentration in these products varies more, and its functionality is difficult to assess, making it more difficult to maintain an optimal H_2S removal. The advantage is that it is a cheaper way of adding Fe to the reactor. Compared to the biological H_2S removal, adding Fe is more reliable, as sudden increases in H_2S production can be countered quickly by increasing the Fe dosage.

Biological H_2S removal outside the AD reactor

The biological process that is responsible for converting H_2S to elemental S inside the digester can also be applied in a specifically designed desulphurization reactor. The advantage compared to H_2S removal in the digester is an optimization of the biological process for S removal (optimized O_2 injection, addition of nutrients and moisture). There is also no risk that the produced S enters the digestate, again leading to sudden peaks of H_2S. Instead, the S can be reused as a fertilizer. Mostly scrubbers or biotrickling filters are used for this purpose. The bacteria grow on a porous carrier material (usually plastic elements), water is sprayed on top of the filter to provide moisture and nutrients for the bacteria, and the biogas is flowing in countercurrent.

Chemical H_2S removal outside the AD reactor

H_2S can be removed from the biogas using different products:

- Adsorption using $Fe(OH)_2$ or Fe_2O_3. This reaction results in the formation of Fe_2S_3. Usually wood flakes are impregnated with iron oxide to obtain a high surface/volume ratio. The reaction is slightly endothermic and requires a temperature of minimum 12°C but has an optimum between 25 and 50°C. Dewatering is necessary to reduce the risk of clogging of the filter. The filter can be regenerated with air or oxygen, resulting in the formation of elemental sulfur. This reaction is exothermic and usually two columns are combined, in which the heat released during regeneration can be used for maintaining an optimum temperature in the active column. After several cycles, the packing material must be replaced because the elemental sulfur after each regeneration step builds up, covering the surface of the Fe_2O_3, decreasing the efficiency.

- Adsorption using liquid NaOH, $FeCl_2$, or $Fe(OH)_3$ solutions. The liquid and the biogas are mixed in a countercurrent system.
- Adsorption using activated carbon. For this purpose, the activated carbon is usually treated with KI or H_2SO_4. The adsorption reaction is optimal at 7–8 bar and 50–70°C. The activated carbon can be regenerated using warm N_2 gas or steam.
- Membrane separation. By leading the biogas alongside a semipermeable membrane, H_2S can be separated from CH_4 and CO_2.

Removal of CO_2 (SGC, 2001; Schomaker et al., 2000; Rutledge, 2005; Gomes and Hassan, 2001)

When the biogas is to be used for injection into the natural gas grid or as vehicle fuel, all or most of the CO_2 must be removed from the biogas. Because much CO_2 removal equipment is very sensitive, water and H_2S must be removed prior to CO_2 removal. Several technologies are available to remove CO_2.

Adsorption

CO_2 is separated from biogas by washing at high pressure. In this technology, biogas is fed at the bottom of a washing tower at 10–20 bar. Water enters the tower from the top. A packed bed is used to increase the contact between water and biogas. Because CO_2 is more soluble than CH_4, the biogas leaving the column is enriched in methane. Also, H_2S can be partly removed using this technique. Because CH_4 is also soluble in water, the water leaving the tower will also contain CH_4. The water leaving the tower is therefore decreased in pressure in a flash tank. The CH_4 and CO_2 that is released in this step is recycled to the tower. A further pressure decrease is releasing the remaining CO_2. Instead of water, polyethylene glycol can be used. This product has a low vapor pressure, so there is little loss of chemicals during the process. Compared to water, CO_2 has a higher solubility, so the volume of solvent is smaller. The biogas can also be upgraded using alkanolamines. Contrary to water or polyethylene glycol where CO_2 is only dissolved, a chemical reaction occurs between CO_2 and the alkanolamines. This reaction occurs at atmospheric pressure and can be regenerated. To avoid efficiency loss, H_2S should be removed beforehand.

Pressure swing adsorption and vacuum pressure swing adsorption

In pressure swing adsorption (PSA) and vacuum pressure swing adsorption (VPSA), a molecular sieve (usually activated carbon) is used to adsorb CO_2 and let CH_4 pass through. It is a cyclic process in which the adsorption takes place at high pressure and desorption (regeneration) at low pressure. H_2S binds irreversibly on the molecular sieve and must therefore be removed beforehand to avoid poisoning of the activated carbon. PSA usually consists of four beds, each operating at four different stadia: adsorption, pressure release, desorption, and pressurization. In the first step, biogas flows through the filter until the filter is saturated with CO_2, whereas CH_4 leaves the filter unaltered. After saturation, the pressure in the bed is released gradually, resulting in a release of CH_4/CO_2 gas mixture rich in CH_4, which is recycled to the PSA unit. In a next step, the filter is regenerated to apply vacuum (VPSA) or further reduce pressure (PSA). This step releases CO_2 and small amounts of CH_4. As a final step, the bed is pressurized, and the cycle can start again.

Membrane separation

Membrane separation uses the selective permeability of membranes to separate CO_2 and CH_4. To obtain a CH_4-rich gas with little CH_4 loss, the difference in permeability of CH_4 and CO_2 must be high. Two principles can be discerned: high pressure membrane separation with gas phase at both sides of the membrane, and low pressure membrane separation, in which the permeated molecules are adsorbed in a liquid phase.

Cryogenic separation

Because CH_4 and CO_2 condense at a different temperature, biomethane can be produced by cooling and compressing the biogas to obtain liquid CO_2 that can easily be separated from the remaining CH_4 gas. The raw biogas is compressed to 80 bar in several stadia with intermittent cooling. Drying of the biogas beforehand is necessary to avoid ice formation. The biogas is cooled to −45°C, after which the condensed CO_2 is removed. This CO_2 is then treated to release any CH_4 that was dissolved in the liquid CO_2. The gas is cooled further to −55°C and expanded to 8–10 bar. This results in a temperature of −110°C, at which the CO_2 is in a solid state, and CH_4 still in a gas phase. With this separation technology, a very pure CO_2 gas is obtained, which can be used in other processes.

Removal of other components (SGC, 2001; Wellinger and Lindberg, 2000; Hagman et al., 2001; Ahjar and Melin, 2006)

Oxygen (O_2) and nitrogen (N_2) can be present in the biogas, although both are not natural components of the biogas. When present, this is usually a sign that air can leak into the gas storage or transport system, and the cause of this leak must be remediated to prevent the buildup of an explosive gas mixture. The occurrence of air in the biogas is more common in landfill gas than in AD reactors. Both O_2 and N_2 can be removed with membranes or PSA, although this is very expensive. A cheaper and more reliable alternative is to prevent both gases from entering the system.

Ammonia (NH_3) is usually present in the biogas due to the equilibrium with NH_4^+ in the digestate. Part of the ammonia will be removed together with the water in a cooling step. If a more thorough cleaning is necessary, a washing tower using HNO_3, H_2SO_4, or an activated carbon filter can be used.

Siloxanes are mostly found in landfill gas or in biogas originating from residual household waste, but seldom in biogas originating from agricultural or industrial waste. Siloxanes can cause serious damage to engines and must therefore be removed prior to any further application. Activated carbon can be used for this purpose, but because the siloxanes bind irreversibly, the activated carbon needs regular replacing making this an expensive solution. Cryogenic separation (at temperatures of −70°C) or polydimethylsiloxane (PMDS) membranes can also be used for this purpose.

Depending on the input composition, a wide array of volatile organic components can be present as trace amounts (ppb (parts per billion) level) in the biogas. For most applications, these trace components must not be removed. If a high quality methane is to be obtained, an activated carbon filter can be used to remove these components.

4.4.8 Biogas Valorization

Production of heat

The easiest way to use the produced biogas is burning it in a boiler for warm water or steam production. This option is the most recommended if there exists a large heat demand near the biogas production site. The biogas can also be transported to be burnt near a location with high heat demand. Because a boiler is a relatively simple technology, little gas cleaning is required (only water and particle removal), and capital and operational costs are kept to a minimum. Because biogas has a lower calorific value than natural gas (due to the presence of CO_2), boiler efficiencies are usually lower than can be obtained in a natural gas boiler. An efficiency of 80–85% is typical for biogas boilers.

Production of electricity (and heat)

The most common option to valorize the biogas is burning it in a CHP unit (combined heat and power) in which both electricity and heat are produced. This technology is more complicated than a boiler, so quality restrictions to the biogas are higher. As with burning the biogas in a boiler, water and particles must be removed. Specifically H_2S should be as low as possible to avoid corrosion of the engines (mainly the cylinders and pistons), but most engines also have restrictions on halogenated hydrocarbons and siloxanes. The most used type of CHP units are combustion engines coupled with a generator (typically $>100\,kW_e$). Alternatives are a stirling motor (typically $>50\,KW_e$, lower efficiency), micro gas turbines (typically $<200\,kW_e$), and fuel cells.

A typical efficiency of 90% is possible (±40% to electricity and ±50% to heat), provided all the heat can be used (preferably on site to avoid heat losses during transport, or by transporting the biogas to a site with a high heat demand). Unfortunately, this is not the case with many existing biogas plants. Especially early generation biogas plants have been built only for electricity production without considering heat valorization. This is now changing, partly because in many countries, the governments are imposing stricter rules for total energetic efficiency, partly because the sale of electricity alone is not sufficient to have an economically sustainable plant.

Production of a natural gas substitute

A final treatment option for biogas is upgrading the biogas to natural gas quality by removing CO_2 and other impurities. This upgraded gas can then be injected into the existing natural gas grid or can be compressed for use as vehicle fuel. Therefore, this valorization option has the highest quality restrictions. The upgrading technology is the latest development in biogas valorization, and is gaining more and more interest. This upgrading technique has the highest efficiency: up to >90%.

This treatment option is the most interesting when there is no large heat demand onsite or nearby. By upgrading the biogas to natural gas quality, the "biomethane" can be injected into the natural gas grid and transported to locations with a specific energy demand. This way, little energy is lost (e.g. as waste heat with CHP), and existing infrastructure is used for transporting the gas. In many countries, access is guaranteed to the gas grid for upgraded biomethane, and incentives are offered to choose this type of valorization (e.g. in the Netherlands, Sweden, Germany, etc.). Standards have been developed to assure the safety of the biomethane injection.

When there is no access to a natural gas grid near the biogas production site, the upgraded biogas or biomethane can be compressed and stored locally for use as biofuel. In countries like Sweden, Germany, and Italy, CNG (compressed natural gas) or CBG (compressed biomethane gas) is already widely used.

4.4.9 SWOT Analysis

General SWOT analysis for anaerobic digestion

Strengths

- Combined production of **energy and soil improver**: AD yields two valuable products. On the one hand, there is the production of biogas that can be transformed into several final uses (electricity, heat, natural gas). On the other hand, a stable soil improver and fertilizer is obtained, which can be used in agriculture. This results in a considerable advantage over many other technologies. Composting also yields a valuable soil improver, but is a net energy consumer. Incineration of (organic) waste is usually done with energy recovery, although this is less optimal (many conversion losses and organic waste is not an optimal fuel source, due to its high moisture content), and the remaining ashes have little value as fertilizer. An additional strength of biogas production as a renewable energy source is the reliability. Whereas wind and solar energy depend on climatological conditions, biogas from organic waste can be regarded as a stable and continuous energy producer.
- Suitable for **heterogeneous waste** products like organic waste: Unlike many specialized technologies (production of bioethanol, biochemicals, bioproducts), AD can use a wide variety of input products to produce biogas. This makes organic waste, which is

characterized by a great heterogeneity, a very suitable feedstock for AD. Whereas specialized technologies still need rather "pure" feedstocks to produce biofuels or chemicals, AD can produce high-quality products (biogas and compost) from "low-grade" organic waste products. Therefore, this technology can also be used to valorize the remaining organic matter after high-value bioproducts have been extracted.

- **Proven technology** with optimization ongoing: AD has been implemented since the early 1990s. Since then, it has continued to increase its popularity for treating organic waste. Still, AD is more and more recognized by operators and decision makers as a sustainable and proven technology to treat heterogeneous waste. Furthermore, optimization of AD has been going on, both on technical issues and on biological factors, both on lab- and full-scale. This makes AD a reliable and advanced technology for treating organic waste products.
- **Odor control**: Organic waste is typically characterized by an unpleasant smell. Treating this waste with AD will drastically reduce these odors due to the conversion of odorous organic compounds in the biogas. Because AD needs closed vessels, there is no risk of odor problems during the process. Any odorous compounds escaping the anaerobic reactor with the biogas will not enter the environment, but are either burnt in a combined heat and power (CHP) plant or removed in a biogas upgrading facility.

Weaknesses

- Process stability **depends on input characteristics**: AD is a biological treatment technology, and therefore, process conditions inside the reactor are very important. Although there are ways to influence and control the process conditions, the most important factor in this matter remains the feedstock composition. If the feedstock is not optimal, then conversion to biogas will also be suboptimal. Ammonium toxicity is one of the most important limitations for the use of some waste products in AD (e.g. chicken litter). Codigestion can provide an answer for this problem, combining several feedstocks to obtain a suitable mixture.
- **Lower biodegradation**: Because fungi, which are needed for the biological degradation of lignin, cannot survive or are deactivated in the absence of oxygen, the biodegradation during the AD process is lower compared to a composting plant. This also implies that the digestate is not as stable as compost and could have negative effects when applying directly on land (oxygen consumption in the soil caused by further degradation could kill off part of the soil microbiology). By integrating AD and composting, this problem can be remediated, resulting in compost with a similar economic as well as agricultural value.
- **Complex process**: The biological process is very complex, as this is an intricate equilibrium between different bacteria groups interacting with each other. This implies that good process knowledge is paramount for sustaining a stable process. If plants fail, in many cases, this could be traced back to poor process knowledge.

Opportunities

- Easily **integrated with bioproduction plants**: Production of biobased materials gains more and more interest all around to world to shift from a fossil fuel-based economy to a biomass-based economy. Most of the bioproduct facilities only use a small fraction of the biomass, leaving large amounts of "low-value" residual biomass. This biomass can still be used as input to AD, and through the production of biogas even supply the bioproduct facility of the necessary energy, making this an energy neutral alternative for fossil fuel-based materials.
- Easily **integrated with existing composting plants** (for municipal waste): All around the world, composting of organic waste is common practice, and many plants have been installed, exceeding the number of AD plants on household waste. As the need for renewable energy sources grows, emphasis shifts more and more to AD of organic waste. AD has the advantage that the area needed for a reactor is relatively small compared to the place needed for a composting plant, although this also depends on the technology used. This makes it relatively easy to expand an existing composting plant with AD, thus improving the energy balance of the waste management, while still making optimal use of already existing capacity and equipment.
- Increased **value of compost and nutrients**: Although the intrinsic value of the nutrients and stable organic matter (even more so after postcomposting of the digestate) in the digestate or compost is known and appreciated by many scientists and policymakers, so far this does not yet translate into a competitive market price. Especially in the South of Europe, there is an increasing demand for nutrients due to the decreasing amount of organic matter in many agricultural soils (desertification). Therefore, it is likely that this will change in the future. Proof of this is the increasing number of projects (technical, LCA (life cycle assessment)) that focus on closing the nutrient cycle and improving soil fertility. Awareness has also been raised with many policymakers to change legislation and make the market accessible for fertilizers and soil improvers originating from waste material.
- Growing market for **renewable energy**: The growing world population and growing economies like India and China have led to a continuously growing demand for energy. Fossil fuel reserves are limited, and nuclear energy brings high risks, so renewable energy is gaining more and more success. Since no renewable energy source alone will suffice to deliver enough energy, and since wind and solar energy are less suited to deliver a base load of energy (e.g. on a windless cloudy day), renewable energy from biomass is needed to guarantee an around the clock energy production. From a sustainable point of view, energy from waste biomass is preferred over cultivated feedstocks (e.g. energy crops), which brings new opportunities for AD of organic waste.

Threats

- Growing **competition for biomass**: biomass is gaining more and more attention from different technologies to be used as a resource in their processes. In reality, there is, of course, only a limited amount of biomass, leading to a growing competition between technologies for the same biomass. Also, organic waste materials that have since long been used in AD have attracted attention from other markets (e.g. bioethanol, enzymes, pharmaceuticals, etc.). Because AD is a rather robust technology that is able to cope also with more "difficult" substrates (e.g. low sugar and/or starch content), AD can be combined with these new technologies, treating their waste products.

- More **costly** than composting alone: Despite the many positive effects of a combined AD and composting plant on the energy balance and environmental aspects (see above), this also brings higher costs with it. Depending on the energy prices and financial incentives of a specific country or region, AD might not be economically feasible compared with other waste treatment technologies.
- **Different regulations**: Legislation on waste management, renewable energy, fertilizers, and soil improvers varies from country to country, and even within one country, differences can exist between regions. This could slow down the implementation of AD on a wider scale.
- No need for **biogas as an alternative fuel**: Although the need for renewable energy is a worldwide phenomenon, biogas production is not always the most efficient way. In countries with a lot of hydro energy like Canada, Russia, and the Scandinavian countries, biogas production is much more costly compared to hydro energy. In those cases, the role of AD as an economically feasible source of renewable energy from waste largely depends on the targets set by the responsible authorities (national, regional, local, etc.). Apart from other sources of renewable energy, also fossil fuels could limit the need for biogas. The discovery and exploitation of shale gas in the US is an example that could largely reduce the fossil fuel prices and therefore make biogas production unfeasible.

SWOT analysis of mesophilic digestion (±37°C)

Strengths

Mesophilic digestion is the most frequently used temperature range worldwide, and hence, most experience is available in this temperature range. Because of the lower temperature compared to thermophilic digestion, less energy is needed for heating the reactor. Mesophilic bacteria come in a wider variety than thermophilic bacteria and are therefore known to be more resistant to suboptimal process conditions, making this the most stable temperature range.

Weaknesses

As with all biochemical processes, the lower the temperature, the slower the process rate. This means that mesophilic digestion is less efficient in converting organic matter to biogas compared to thermophilic digestion. In case of waste treatment, hygienization can be an important issue in selecting the most suited treatment. Because of the lower temperature, mesophilic digestion is not contributing to pathogen kill, requiring as such an additional hygienization step (if necessary).

Opportunities

As the ammonia (NH_3)/ammonium (NH_4^+) equilibrium is greatly affected by temperature (the higher the temperature, the more the equilibrium tends to the toxic ammonia), mesophilic digestion can be considered as the preferred temperature range when dealing with organic matter with high nitrogen content, like vegetable, fruit, and/or restaurant waste.

Threats

If legislation in a certain country or area demands a certain degree of hygienization of the waste in order to obtain a marketable end product, then mesophilic digestion might be less suited unless hygienization is guaranteed in a pre- or posttreatment step (making it more expensive).

SWOT analysis of thermophilic digestion (±52 °C)

Strengths

The main strength of thermophilic digestion over mesophilic digestion is the higher conversion rate of organic matter, leading to shorter retention times. Thermophilic digestion also provides better pathogen kill (hygienization), due to the higher process temperature.

Weaknesses

Thermophilic digestion is more prone to process instability. This is especially the case for ammonium inhibition. Due to the higher process temperature, more heat is required.

Opportunities

If waste is to be converted to a qualitative product (compost in this case), hygienization is an important factor to improve public acceptance. Because thermophilic digestion assures a higher reduction of pathogens and weed seeds, thermophilic digestion might be best suited for treating organic waste that contains animal byproducts.

Threats

Since many organic waste streams contain higher amounts of nitrogen, thermophilic digestion might be less favorable, unless a pretreatment to reduce nitrogen toxicity is included (which would then makes this more expensive).

SWOT analysis of wet digestion

Strengths

Wet AD is the most suitable technique for waste products with high moisture content like manure, sewage sludge, and many waste streams from food industry. Because the digestate inside a wet digester has a low viscosity, it is easily mixed and pumped. Investment costs are therefore generally lower than a dry digestion system. The low viscosity also allows for a good mixing of the digestate, guaranteeing a good contact between the bacteria and the input substrates.

Weaknesses

Waste products with high total solids content can only be treated in a wet system if diluted with (process) water, which makes it more costly. Wet AD is also more sensitive to the occurrence of a floating layer (fibers, plastics, etc.) and a sedimentation layer (sand, stones, etc.). Especially in the case of organic waste treatment, these impurities are likely to be present in the input, making wet digestion less suitable. Also, to protect mixing and pumping equipment from damage by larger contaminants, a more extensive pretreatment is necessary, increasing the chances that part of the organic material is lost before it can be converted to biogas. Finally, because of the low total solids content in wet digestion, larger reactors are needed to produce the same amount of biogas compared to dry digestion.

Opportunities

The best opportunities for wet digestion lay in farm-scale digesters (high amounts of manure) and in sludge digestion. Also, pure industrial waste products with high moisture content are best treated in a wet digestion system. All of these waste streams (agricultural, sludge, pure industrial) are usually characterized by a low contamination factor (physical contamination), so most disadvantages of wet digestion do not come into view, and the advantages or strengths can be used to the maximum.

Threats

Because organic waste from household collection usually has rather high total solids content and contains many impurities, wet digestion can only be applied with the installment of an extensive pretreatment step to remove physical impurities or by using extra (process) water. In case of using (process) water, dewatering will be a must prior to postcomposting the digestate.

SWOT analysis of dry digestion

Strengths

Dry AD is the most suitable technique for waste products with high total solids contents (e.g. organic waste from households, straw). Because the digestate inside the reactor has a high viscosity, there is no risk of floating or sedimentation layers. The higher total solids content inside the digester also implies that higher loading rates are possible, reducing the need for high reactor volumes. As less water is present, less energy is needed for heating compared with wet digestion. The digestate obtained from dry AD has a high dry matter (TS) content, which means that dewatering is not always needed before postcomposting.

Weaknesses

Due to the high viscosity of the digestate inside the digester, mixing is more difficult compared to wet digestion and is therefore usually performed in a plug flow design. A form of mixing can be simulated by creating a fast pass-through rate in the plug flow, with a high recirculation rate. The higher viscosity also calls for more heavy-duty pumps.

Opportunities

(Source separated) organic waste typically has a high total solids content, making dry AD the most suitable technique for this waste, especially if the digestate is postcomposted (less need for dewatering). Furthermore, a lot of agricultural waste products like straw and stover have very high TS content and are usually fibrous, making dry AD the best option to energetically valorize these waste products while also obtaining digestate with good soil improving properties (compared to burning this dry waste).

Threats

Because of the higher dry matter content in dry digestion, ammonium that is formed during fermentation is less diluted compared to a wet digestion system. This implies that dry digestion is not the preferred anaerobic fermentation technique for organic waste with high nitrogen content.

SWOT analysis of single-stage digestion

Strengths

As all processes take place inside one reactor, the process design is simpler and therefore cheaper, compared to multistage digestion.

Weaknesses

A limitation of single-stage digestion is that process conditions will not be optimal for all groups of bacteria.

Opportunities

The simple process design makes single-stage digestion the most frequently used technology worldwide. The knowledge built up with single-stage digestion and the optimization that has been going on has eliminated or reduced most of the disadvantages compared to the multistage systems. The lower cost makes single-stage digestion the most likely technique to be implemented in new plants or in composting plants that are expanded with AD.

Threats

Although most types of organic waste products can be treated via single-stage digestion, difficult waste streams would still better be treated in a multistage system.

SWOT analysis of multistage digestion

Strengths

The main advantage of multistage digestion is that optimum conditions can be created for all bacteria groups, leading to better process stability and higher throughput rates.

Weaknesses

The higher complexity of this design and the higher cost that comes with it should be considered as the biggest disadvantage.

Opportunities

AD of organic waste that is not easily degradable can be the biggest opportunity for a multistage system. A first example is organic waste with very high nitrogen content. By first hydrolyzing the waste and removing (part of the) ammonium before treatment in the second reactor, methanogenesis can be much improved. Another example is the digestion of some bioplastics that require a thermal pretreatment step, which is not possible in a single stage reactor design (e.g. PLA (polylactic acid)).

Threats

The biggest threat to the implementation of multistage AD is the complexity (both in design and control) and the higher cost as a result of this.

SWOT analysis of mono-digestion

Strengths

The main advantage of mono-digestion is that no extra substrates are needed. In most cases, a plant manager starts from a certain waste stream that needs to be treated. If this waste stream can be treated without the need for other products (that need to be searched for in the market), this is a big advantage (no need for elaborate storage capacity, one pretreatment technology, no need for several types of dosage equipment, etc.).

Weaknesses

AD is a biological process, and the bacteria, as all living creatures, need a well-balanced diet (carbohydrates, proteins, fats, micronutrients) in order to function optimally. In the case of mono-digestion of one product, chances are high that not all necessary components are available, leading to suboptimal process conditions. In some cases, this can be remedied by specific additions (e.g. micronutrients), but it will be hard to mono-digest a product that, for example, consists of almost 100% fats.

Opportunities

In organic waste treatment, the input is determined by the region in which it is collected, and it is usually undesirable to treat other products with it (e.g. agricultural or industrial byproducts), since this falls outside the scope of a waste treatment plant. Mono-digestion of the collected organic waste is therefore the most suited technology.

Threats

In case an organic waste stream has a very undesirable composition, mono-digestion might be impossible to treat such a waste stream (e.g. 100% fats).

SWOT analysis of codigestion

Strengths

By codigesting several different products, an ideal mix can be selected to reach optimal process conditions.

Weaknesses

By digesting several types of products, the need for different storage capacity increases (e.g. storage of wet and dry products, products that cannot be stored together for safety reasons, etc.). If both liquid and solid substrates are to be treated in one digester, then there is need for different dosing equipment (pumps, screws, etc.). This could make codigestion more expensive than mono-fermentation.

Opportunities

In the case of organic waste streams that by themselves are not suitable for AD, codigestion could create a new mixture that has better qualities.

Threats

In some cases (e.g. municipal waste treatment facilities), the search for different waste products on the market is not an option from a managerial viewpoint. Suitable cosubstrates should preferably be found in the vicinity of the plant, which is not necessarily the case for all locations, or at a reasonable price.

SWOT analysis of batch digestion

Strengths

The simplicity in design of the batch digestion (no mixing, no pumps, simple construction) is the main advantage, which leads to very low investment costs.

Weaknesses

A batch digestion system is usually characterized by a poor process control. Once a reactor is filled with inoculum and fresh input material and hermetically sealed, the operator has no or very limited influence on the process. This often leads to suboptimal conversion rates of organic matter to biogas. As usually no mixing goes on inside a batch digester, the different stages of AD are occurring more or less sequentially. This means that biogas production and composition change significantly over the course of one retention time, negatively influencing the continuous use of the biogas (conversion to electricity in a CHP/upgrading). Batch digestion shows a higher explosion risk, as both at start and at the end of a run, the atmosphere inside the reactor will pass through a stage in which an explosive mixture of air and methane is present. Because of the design, this technology is also less suited for treating wet waste products.

Opportunities

In cases where biogas production is less important, batch digestion could be an interesting alternative, due to the very low investment costs. Filling and emptying the reactor is done with a front-end loader, making pumps redundant.

Threats

The limited process control makes this technology one of the least efficient in converting organic waste to valuable end products. Examples from the past have also shown that batch conversion could lead to unhygienic situations. For instance, if the percolation system becomes clogged, slurry can be formed, which will flow out of the reactor when opened. This could lead to odor problems and contamination.

SWOT analysis of continuously fed, completely mixed digestion

Strengths

In a continuously fed, completely mixed digestion system, a good contact between fresh input and bacteria is guaranteed. This leads to better process conditions. The continuous feeding also makes it possible to correct whenever undesirable conditions would arise (optimal process control). A continuous feeding also leads to a more stable biogas production and composition. It is also the most frequently used reactor design for treating wet waste products, so a firm knowledge and experience with this technology is available.

Weaknesses

A tank that is continuously fed with fresh input material also needs a continuous outflow of digested material in order to keep the reactor volume constant. Because complete mixing is guaranteed, this also increases the risk of a part of the input leaving the system before it has had the time to degrade completely (short circuiting of biomass). This way, a part of the biogas potential is lost. Continuous mixing calls for a miscible digestate, which makes this technology less suitable for treating drier waste products.

Opportunities

A continuously fed, completely mixed digester is the most frequently used design worldwide on a wide scale of waste products, although usually for waste with high moisture content. The good process control and available knowledge on this system makes it a safe choice.

Threats

In many countries, a lot of AD plants have been installed over the last 10 years, treating most of the available liquid waste products in that region. Alternative waste products are usually characterized by a much higher total solids content (waste from households, agriculture, landscaping), and are therefore less suitable for this reactor design. It is therefore likely that this reactor design will not be eligible for treating these newer waste products.

SWOT analysis of continuously fed, plug flow digestion

Strengths

Continuous feeding assures that the biogas production and composition will be much more constant compared to a batch fermentation process. In a plug flow design, short circuiting of biomass is reduced as far as possible, resulting in a hydraulic retention time equaling the sludge retention time. This assures that no fresh biomass leaves the system before complete degradation. It is the most frequently used system design for treating dry waste products.

Weaknesses

As the biomass moves through the reactor, it is not impossible that zones occur with suboptimal process conditions because of the lack of mixing. A high recycle rate can improve the process, as this simulates intensive mixing.

Opportunities

As plug flow digestion is the most frequently used with dry digestion, this makes it a very suitable technology for treating organic waste, and also for agricultural byproducts and waste from landscaping.

Threats

Most available waste streams worldwide are liquid of nature. As a consequence, only a limited number of dry plug flow reactors have been built when compared to wet, completely mixed reactors. Nevertheless, in the area of organic waste treatment, plug flow reactors have built up a lot of experience, which should allow it to be a competitive alternative for these waste products.

4.4.10 Anaerobic Digestion Within the TRANSBIO Project

During the European FP7 project TRANSBIO, AD of the biomass that remained after extraction of valuable bioproducts from fruit and vegetable waste was investigated. The results of this research are presented below.

Short-term evaluation of biogas potential

To evaluate the suitability of a substrate for AD, the following analyses are performed:

- Dry matter content (TS) (method according to ISO 11465:1993)
- Organic matter content (VS) (method according to DIN 38414-3)
- Kjeldahl nitrogen content (Kj–N) (method according to ISO 5663:1994)

The maximum amount of biogas that can be produced during the AD of organics is measured through a high-rate dry anaerobic fermentation in a batch lab-scale test and is a measurement of the maximum biogas production potential (BPP). The test also gives a first indication with regard to the rate of biogas production.

The test substance is mixed with a large amount of highly active anaerobic inoculum and fermented during 14 days (or unless the daily net biogas production is <1% of the total) at 52°C. The BPP is determined by the net amount of biogas produced by the test item, after subtraction of the biogas produced by the blank (inoculum only). At the end of the fermentation the biogas composition is determined (CH_4 and CO_2) using GC-TCD (Gas Chromatography – Thermal Conductivity Detector).

Taking into consideration the net amount of biogas produced and the composition of the biogas, one can calculate the mass of biogas produced per unit of organic matter (= BVS). The BVS is calculated as the weight of dry biogas produced per 100 g of volatile solids (= organic matter) in the test item fed to the digester.

The determination of the BPP of a test substance gives a first, preliminary indication of the feasibility of fermentation of the test substance, and will also give an idea of the volume of biogas produced and the rate of biogas production.

The BPP test is based on ISO 15985 (2004), ASTM D.5511-12, and DIN 38414.

The first BPP tests were performed on byproducts from the vegetable industry before bioproduct extraction to obtain an indication of their suitability for AD as such. Table 4.7 and Fig. 4.9 show the results of this test. The sweet corn has the highest biogas production per kg TS, followed by potato waste and green beans. Lettuce has the lowest biogas production potential. Methane content of most byproducts varies around 50%, which is indicative of carbohydrates. Only sweet corn and green bean have a slightly higher methane content in the biogas. The BVS is calculated as the amount of biogas (weight) that is produced from the organic matter, and is expressed as percent on organic matter. It is a way to express the anaerobic biodegradability of the substrate. This value is good for all tested substrates, with the highest degradability for the sweet corn and the lowest value for the lettuce.

Because the potato and sweet corn samples showed the most promising potential for AD, these samples were selected to evaluate the effect of bioproduct extraction on the biogas potential. Table 4.8 and Fig. 4.10 show the results of the biogas potential test on an untreated sweet corn

Table 4.7: Analysis Results of Biogas Potential Tests on Byproducts Before Bioproduct Extraction.

Parameter	Unit	Broccoli	Cardoon	Green bean	Lettuce	Potato	Sweet Corn
TS	%	82.3	87.4	89.6	86.2	90.1	91.7
VS	% on TS	88.7	84.6	93.9	86.0	93.4	95.8
Kj-N	g/kg TS	30.5	13.4	19.0	31.2	17.5	15.4
C/N	-	14.5	31.7	24.8	13.8	26.6	31.1
Biogas	NL/kg TS	498.1	465.1	579.8	446.7	568.4	628.4
CH_4	%	49.4	48.0	52.2	49.4	49.0	54.2
	NL CH_4/kg VS	277.6	264.1	322.6	256.3	298.3	355.4
BVS	% on VS	75.6	75.0	81.0	69.9	82.2	84.4

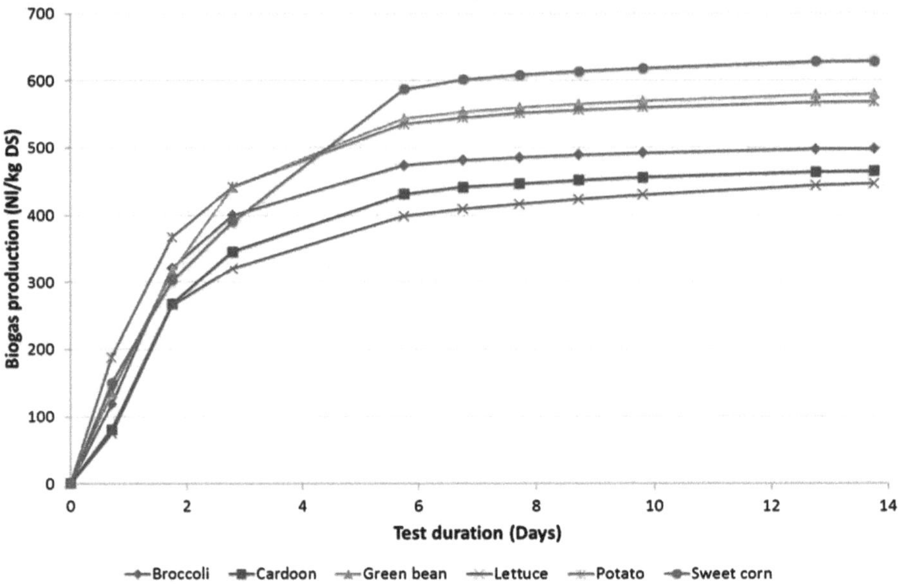

Figure 4.9
Evolution of biogas production on byproducts before bioproduct extraction.

Table 4.8: Analysis Results of Sweet Corn Products.

Parameter	Unit	Sweet Corn Untreated	Sweet Corn Hydrolyzed 1	Sweet Corn Hydrolyzed 2	Sweet Corn Hydrolyzed 3
TS	%	91.7	93.0	91.8	38.0
VS	% on TS	95.8	94.6	94.9	97.4
Kj-N	g/kg TS	15.4	14.5	12.0	17.6
C/N	-	31.1	32.6	39.4	27.6
Biogas	NL/kg	576.2	543.0	547.2	266.5
	NL/kg TS	628.4	584.0	596.2	700.5
CH_4	%	54.2	55.8	55.5	54.3
	NL CH_4/kg VS	355.4	344.3	348.9	390.2
BVS	% on VS	84.4	78.2	79.8	92.4

sample and three hydrolyzed sweet corn samples. As can be seen from both the table and figure, the characteristics of the sweet corn remain more or less unaltered after bioproduct extraction. A same conclusion can be drawn for the potato waste before and after bioproduct extraction (Table 4.9 and Fig. 4.11). It can therefore be concluded that AD remains a valuable treatment technology for vegetable waste biomass after high-value bioproducts were extracted.

Long-term evaluation of biogas production

From the short-term tests, it was clear that extraction of bioproducts from vegetable byproducts hardly influenced the biogas potential of the considered substrates (sweet corn and potato).

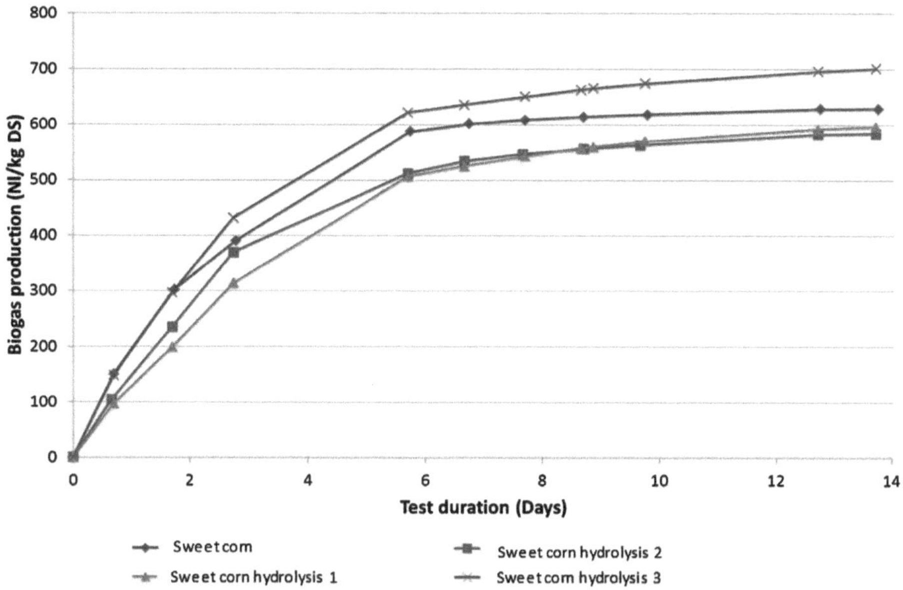

Figure 4.10
Evolution of the biogas production of sweet corn before and after bioproduct extraction.

Table 4.9: Analysis Results of Potato Products.

Parameter	Unit	Potato Waste Untreated	Potato Waste Hydrolyzed 1	Potato Waste Hydrolyzed 2	Potato Waste Hydrolyzed 3
TS	%	90.1	24.6	23.5	24.6
VS	% on TS	93.4	94.9	94.6	90.0
Kj-N	g/kg TS	17.5	20.2	19.4	23.9
C/N	-	26.6	23.5	24.4	18.8
Biogas	NL/kg	511.9	132.9	127.0	137.3
	NL/kg TS	568.4	539.8	541.0	557.7
CH_4	%	49.0	47.9	47.3	48.8
	NL CH_4/kg VS	298.3	272.5	270.3	302.7
BVS	% on VS	82.2	77.6	78.4	83.9

Therefore, solid fraction of sweet corn and potato pulp byproducts after hydrolysis and extraction of bioproducts were selected to evaluate the feasibility of AD using long-term, semicontinuous, lab-scale tests.

The goal of a semicontinuous fermentation test is to investigate what the long-term effects are of digestion of the substrates of interest. Important characteristics regarding the fermentation cannot always be deduced from the chemical characteristics of the substrates. This test considers characteristics such as nitrogen deficiency or surplus, deficiency or surplus of macro- or micronutrients, toxicity, etc. A lot of these effects only become visible after a

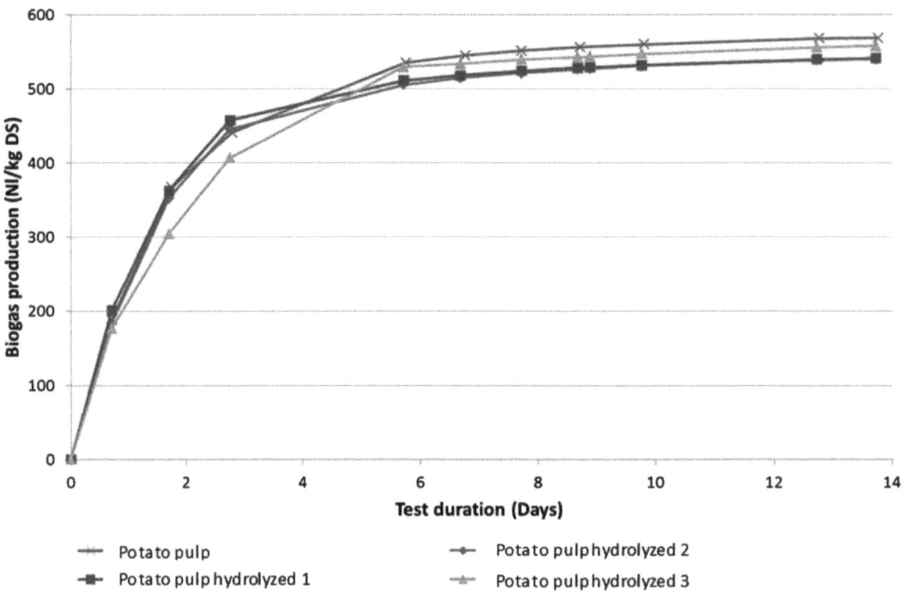

Figure 4.11
Evolution of the biogas production of potato waste before and after bioproduct extraction.

longer period of testing. The lab-scale test setup has been validated with full-scale digestion plants. It is therefore safe to say that results obtained during these long-term tests are representative for a full-scale plant.

The lab reactors are operated under a semicontinuous regime. Several process parameters (loading rate, residence time, average biogas production per kg of feeding, etc.) are tested and adjusted to optimal values.

The lab-scale reactors are fed semicontinuously (three times per week) with the test substrates. The biogas production is measured on a daily basis. The composition of the biogas is measured on a weekly basis:

- CH_4, CO_2, H_2S: GC-TCD
- NH_3: using Dräger tube (reaction principle: NH_3+bromophenol blue→blue reaction product)

A digestate sample is analyzed weekly for:

- pH: EN 13037:1999
- ammonium (NH_4^+–N): ISO 5664: 1984
- volatile fatty acids: GC-FID after extraction in ether solution

A digestate sample is analyzed biweekly for:

- dry matter content (TS): ISO 11465:1993
- organic matter content (VS): DIN 38414-3

Five test reactors (2 kg active reactor content) were set up and operated for >6 months:

- R1: 100% solid fraction of hydrolyzed sweet corn; thermophilic (47°C)
- R2: 100% solid fraction of hydrolyzed sweet corn; mesophilic (37°C)
- R3: 100% solid fraction of hydrolyzed potato pulp; thermophilic (47°C)
- R4: 100% solid fraction of hydrolyzed potato pulp; mesophilic (37°C)
- R5: 50% solid fraction of hydrolyzed sweet corn + 50% hydrolyzed potato pulp; thermophilic (47°C)

The sweet corn samples after bioproduct extraction showed the most promising results. At thermophilic conditions, a stable biogas production rate of 9 NL/kg_r/d could be maintained for 3 months. Also, mesophilic digestion resulted in a stable process, but the maximum biogas production rate that could be reached was limited to 6 NL/kg_r/d (due to the lower reaction speeds at mesophilic temperature).

Thermophilic digestion of potato samples after bioproduct extraction showed to be very unreliable and led to complete reactor failure after only 2 months. The relatively high potassium (K) content in this product was the cause for this instability. Mesophilic digestion, on the other hand, showed better resistance to high K levels, and a stable process could be maintained for >3 months at a biogas production rate of 6.5 NL/kg_r/d.

The thermophilic digestion of a combination of sweet corn and potato samples (both after bioproduct extraction) also allowed for a stable process which could be maintained for >3 months, but again due to the high K levels in the potato waste (despite "dilution" with sweet corn), the maximum biogas production rate at which the process remained stable was limited to 6 NL/kg_r/d. For codigestion of both products, it seems therefore that mesophilic digestion has the advantage, as it has a higher resistance to K toxicity and can reach the same biogas production rate as the thermophilic alternative (which is inhibited by K).

Fertilizer value of the digestate will be released at the conclusion of the project in the final TRANSBIO report to the EU commission, in 2016.

4.5 Conclusions

In the TRANSBIO project, the feasibility to apply AD as a technology to treat the byproducts after bioextraction from vegetable waste was studied. Both short-term and long-term lab-scale tests were performed to assess the potential and boundary conditions of the process. These test showed very promising results. First of all, the extraction of bioproducts hardly influenced the biogas potential of the studied materials. During continuous tests, a stable process could be proven for both the sweet corn and potato waste, although thermophilic digestion seems less suitable for treating the potato waste, due to its high K content. It may therefore be concluded that AD is a suitable process to generate energy for the extraction processes based on the waste generated during these processes. The digestate that was obtained after AD shows good fertilizer value and could therefore be recycled to the arable land for vegetable production, thus closing the nutrient cycle.

References

A Comparative Study of Waste Collection Systems in Mexico and Sweden, (Bachelor thesis of Nancy Contreras Valenzuela). University West, Sweden, 2005.

Activity report, 2013. Bundesgütegemeinschaft Kompost e.V. http://www.compostnetwork.info/wordpress/wp-content/uploads/2010/08/Activity-report-BGK_2013.pdf (consulted: January 22, 2014).

AgSTAR Database, www.epa.gov/agstar/projects.

Ahjar, M., Melin, T., 2006. Siloxane removal with gas permeation membranes. Desalination 200, 234–235.

Algemene Verwerkingstechnieken: Biologische Verwerking: Composting. EMIS. http://www.emis.vito.be/techniekfiche/algemene-verwerkingstechnieken-biologische-verwerking-compostering (consulted: January 08, 2014).

American Bio Tech, Alternative Composting Methods.

An Introduction to Anaerobic Digestion of Organic Waste, Remade Scotland, 2003.

AP Business Technology Consultancy. Integrated Facility Outline (consulted: July 09, 2015).

Bayr, et al., 2012. Mesophilic and thermophilic anaerobic codigestion of rendering plant and slaughterhouse wastes. Bioresource Technology 104, 28–36.

Bekon, Bassum Dry Fermentation Plant and Composting Plant.

Biogashandbuch Bayern, Bayerisches Staatsministerium für Umwelt, Gesundheit und Verbracherschutz, 2004.

Biologische Abfallbehandlung (Biological Waste Management), EF-Verlag für Energie- und Umwelttechnik, Berlin, 1995.

Center for Clean Policy, 2014. Promoting Net-zero emissions from the waste sector in Latin America through NAMAs. Policy report.

Code of Federal Regulations, Title 40 – Protection of Environment, National Archives and Records Administration, US Government.

Compost stability determination. In: Presentation of the 1996 Conference on Composting in the Carolinas, Myrtle Beach, South Carolina, October 23–25, 1996.

Composting in Business. http://www.offaly.ie/eng/Services/Environment/Publications/Composting_in_Business.pdf (consulted: January 13, 2014).

Converti, et al., 1999. Anaerobic digestion of the vegetable fraction of municipal refuses: mesophilic versus thermophilic conditions. Bioprocess Engineering 21 (4), 371–376.

Council Directive 1999/31/EC of 26 April 1999 on the landfill of waste. Official Journal of the European Communities 1999.

Deyerling, L.A., Fuchs, B.E. The GORE™ Cover System. A Leading Composting Technology for Organic Waste Treatment. http://archive.cityofpaloalto.org/civica/filebank/blobdload.asp?BlobID=15184 (consulted: January 13, 2014).

Directive 2008/98/EC of the European Parliament and the Council of 19 November 2008 on waste and repealing certain directives. Official Journal of the European Union 2008.

Directive 2009/28/EC of the European Parliament and of the Council of 23 April 2009 on the promotion of the use of energy from renewable sources and amending and subsequently repealing Directives 2001/77/EC and 2003/30/EC.

DTEnvironmental. Enviro-drum. http://www.dtenvironmental.com/Categories/Products/Enviro-Drum/ (consulted: January 08, 2014).

European Commission, 2005. EU Waste Policy – The Story behind the Strategy.

Eurostat: http://ec.europa.eu/eurostat/statistics-explained/index.php/Municipal_waste_statistics.

Evaluation of the performance of different anaerobic digestion technologies for solid waste treatment. In: CSCE/EWRI of ASCE Environmental Engineering Conference, Niagara, 2002.

Garcia-Peña, et al., 2011. Anaerobic digestion and codigestion processes of vegetable and fruit residues: process and microbial ecology. Bioresource Technology 102 (20).

Gomes, V.G., Hassan, M.M., 2001. Coalseam methane recovery by vacuum swing adsorption. Separation and Purification Technology 24, 189–196.

Guidelines for the Composting of Organic Waste, Ministry of the Environment, Rhineland-Palatinate, Germany, 1992.

Hagman, M., Hesse, E., Hentschel, P., Bauer, T., 2001. Purification of biogas – removal of volatile silicones. In: Proceedings of Scandinavia, 8th International Waste Management and Landfill Symposium, pp. 641–644.

Handboek Composteren en Vergisten van GFT-afval (Handbook on Composting and Anaerobic Digestion of VGF Waste). Ministerie van Volkshuisvesting, Ruimtelijke Ordening en Milieubeheer, Ministry of Housing, Spatial Planning and the Environment, Nederland, The Netherlands, 1991.

Handreichung Biogasgewinnung und -nutzung, Bundesministerium für Verbraucherschutz. Ernährung und Landwirtschaft, 2006.

Herstellerforum Bioabfall – Verfahren der Kompostierung und anaeroben Abfallbehandlung im Vergleich (Biowaste Producers Forum – Comparison of Composting and Anaerobic Digestion). Instituts für Abfallwirtschaft, Waste Management Institute, Witzenhausen, 1995.

Hoornweg, D., Bhada-Tata, P., 2012. WHAT A WASTE A Global Review of Solid Waste Management.

How American Homes Vary by the Year They Were Built, Working Paper No. 2011–18, US Census Bureau, 2011.

Information on Composting and Anaerobic Digestion, Organic Reclamation & Composting Association (ORCA), 1992.

JRC, 2008. End of Waste Criteria. http://susproc.jrc.ec.europa.eu/documents/Endofwastecriteriafinal.pdf (consulted: January 27, 2014).

Krich, K., Augenstein, D., Batmale, J.P., Benemann, J., Rutledge, B., Salour, D., 2005. Upgrading Dairy Biogas to Biomethane and Other Fuels, Biomethane from Dairy Waste – a Sourcebook for the Production and Use of Renewable Natural Gas in California, pp. 47–69.

Large-Scale Organic Materials Composting. NC State University. http://www.bae.ncsu.edu/topic/composting/pubs/ag593-large-scalecompost.pdf (consulted: January 13, 2014).

Ley General de CambioClimático 6 junio 2012 (General Law on Climate Change 6 June 2012), Mexican Government.

Ley General del EquilibrioEcológico y la Protección al Ambiente 28 enero 1988 (General Law for Ecological Equilibrium and Environment Protection 28 January 1988), Mexican Government.

Ley General para la Prevención y Gestión Integral de losResiduos 8 octubre 2003 (General Law for Waste Prevention and Waste Integral Management 8 October 2003), Mexican Government.

Ley para el Aprovechamiento de EnergiasRenovables y el Financiamiento de la TransiciónEnergética 28 noviembre 2008 (Law for the Use of Renewable Energy and Finance of Energy Transition 28 November 2008), Mexican Government.

Mexico's National Waste Management Program, EnvironLaw, 2009.

PAHO/WHO/IDB, 2010. Regional evaluation on urban solid waste management in Latin America and the Caribbean, Report of AIDIS.

Periurban Vegetable Project. Composting.

Productie van compost. WIPS nv. http://www.compost-wips.be/wie-zijn-wij/productie-compost/ (consulted: January 13, 2014).

Programa Nacional para la Prevención y Gestión Integral de losResiduos (National Program for Waste Prevention and Waste Integral Management). SEMARNAT, 2008.

Rutledge, B., 2005. Biomethane Production. California Biogas Industry Assessment – White Paper, pp. 8–12.

Schomaker, A.H.H.M., Boerboom, A.A.M., Visser, A., Pfeifer, A.E., 2000. Anaerobic Digestion of Agro-industrial Wastes: Information Networks – Technical Summary on Gas Treatment. AD-NETT.

SGC, Adding Gas from Biomass to the Gas Grid. Report SGC 118. Gastec nv, Danish Gas Technology Center a/s, Swedish Gas Center, 2001.

Temesi. Temesi Compost.

Turning Food Waste into Energy at the East Bay Municipal Utility District, Investigating the Anaerobic Digestion Process to Recycle Post-consumer Food Waste, US EPA.

UNEP International Environmental Technology Centre, 2006. Solid Waste Management (Volume II: Regional Overviews and Information Sources).
Use of microbial inoculants and organic fertilizers in agricultural production. In: Proceedings of the International Seminar on the Use of Microbial and Organic Fertilizers in Agricultural Production, 1994. Food and Fertilizer Technology Center, Taipei, Taiwan.
Velghe, F., Wierinck, I., 2013. Evaluatie van de vergisters in Nederland Fase II, Agentschap NL.
Weiland, P., 2010. Biogas production: current state and perspectives. Applied Microbiology and Biotechnology 85, 849–860.
Wellinger, A., Lindberg, A., 2000. IEA BioEnergy – Task 24: energy from biological conversion of organic waste. Biogas Upgrading and Utilization.
Zhang, et al., 2005. The influence of pH on hydrolysis and acidogenesis of kitchen wastes in two-phase anaerobic digestion. Environmental Technology 26, 329–339.
Ziemba, et al., 2010. Net energy production associated with pathogen inactivation during mesophilic and thermophilic anaerobic digestion of sewage sludge. Water Research 45 (16), 4758–4768.

Biomass Pretreatment Strategies (Technologies, Environmental Performance, Economic Considerations, Industrial Implementation)

C. Peral
Tecnalia, Miñano, Spain

5.1 Introduction

Agricultural wastes are the most abundant and renewable resources in the world. Management of this biomass is of major concern as not only accumulation could lead to environmental problems, but because of the loss of potentially valuable material which can be used to produce valuable products such as food, feed, fiber, and fuel.

These wastes, commonly called lignocellulosic biomass, are composed mainly of cellulose, hemicellulose, and lignin. Both the cellulose and hemicellulose fractions are polymers of sugars and thereby a potential source for fermentation processes. Lignin can be used for many purposes, including production of chemicals, heat, and energy.

Lignocellulosic biomass in its native form is recalcitrant to hydrolysis with cellulolytic enzymes. Therefore, the production of fermentable sugars from biomass is usually approached in two steps:

The first is a pretreatment stage in which the cellulose polymers are made accessible for further processing. In this step, hydrolysis of hemicellulose may occur, as well as separation of the lignin fraction (for future applications), depending on the class and severity of the applied treatment, cellulose hydrolysis to fermentable sugars using cellulase enzyme cocktails either produced in situ or commercially available.

The pretreatment process aims to break down the lignin structure and disrupt the crystalline structure of cellulose so that the acids or enzymes can easily access and hydrolyze the cellulose. An ideal pretreatment step by this definition should render lignocellulose completely susceptible to the action of cellulases.

Biotransformation of Agricultural Waste and By-Products. http://dx.doi.org/10.1016/B978-0-12-803622-8.00005-7

Main drawbacks of existing treatment processes are:

Insufficient separation of cellulose and lignin. Various types of pretreatments have been defined, but they improve the digestibility of lignocellulosic biomass, and each pretreatment has its own effect on the cellulose, hemicellulose, and lignin fractions.
Formation of fermentation inhibitors. Special attention must be paid to the formation of phenolic compounds from lignin degradation, as well as the formation of furfural and 5-hydroxymethylfurfural (HMF) from sugar degradation. This can be minimized by maintaining the process temperature and residence time as low and as short as possible.
High use of chemicals and/or considerable waste production. Safety handling and environmental impact are major concerns.
High use of energy and/or very long or costly downstreaming process. Pretreatment can be the most expensive and most complicated process in biomass processing.

Different pretreatment technologies published in literature are described in terms of the mechanisms involved, advantages and disadvantages, and economic assessment. Pretreatment methods for lignocellulosic biomass include physical, chemical, and biological methods and various combinations thereof. The selection of pretreatment depends on the objective of the biomass pretreatment.

Research is focused on converting biomass into its constituents improving efficiency and lowering costs in an environmentally sustainable way. Only a small number of pretreatment methods have been reported as being potentially cost effective. These include steam explosion, liquid hot water (LHW), and concentrate and dilute acid pretreatments.

This chapter reviews developments in biomass pretreatment technologies with emphasis on concepts and strategies for industrial applications.

5.2 Physical and Chemical Characteristic of Lignocellulosic Biomass

Agricultural wastes are also named as lignocellulosic materials or biomass. The term "lignocellulosic biomass" is used when referring to higher plants, softwood or hardwood. The main components of the lignocellulosic materials are cellulose, hemicellulose, and lignin, along with smaller amounts of pectin, protein, extractives (soluble nonstructural materials such as nonstructural sugars, nitrogenous material, chlorophyll, and waxes), ash, and water. Inside the lignocellulose complex, cellulose is responsible for the crystalline fibrous structure, hemicellulose is positioned between the cellulose chains (microfibrils), and lignin provides a structural role of the matrix.

Cellulose is the major structural component of cell walls, and it provides mechanical strength and chemical stability to plants. Solar energy is absorbed through the process of photosynthesis and stored in the form of cellulose. Hemicellulose is a copolymer of different C5 and C6

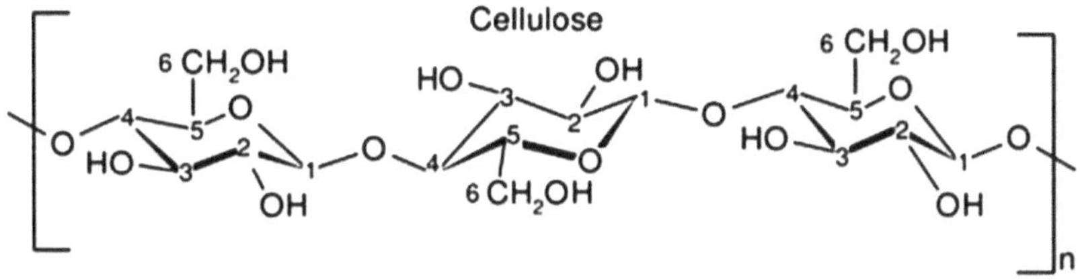

Figure 5.1
Structure of cellulose molecule.

sugars; meanwhile, lignin is polymer of aromatic compounds and forms a protective layer for the plant walls.

The composition of lignocellulose highly depends on its source. There is a significant variation from one plant to another, depending on whether it is derived from hardwood, softwood, or grasses. In addition, the ratios between various constituents within a single plant vary with age, stage of growth, and other conditions. The relative abundance of cellulose, hemicellulose, and lignin are key factors in establishment of the optimal pretreatment method for biomass processing (McKendry, 2002).

5.2.1 Cellulose

Cellulose is the main structural constituent of plant cell wall, and it is also present in bacteria, fungi, and algae. It is made up of a linear polymer chain, which appears in an organized fibrous structure. This linear polymer consists of cellobiose (4-O-β-D-glucopyranosyl-D-glucose) units linked to each other by β-(1,4)-glycosidic bonds. As cellobiose consists of two molecules of glucose, cellulose is more commonly considered as a polymer of glucose. This structure is shown in Fig. 5.1.

Cellulose in biomass is present in both crystalline and amorphous forms. The polymer is arranged in long, straight chains due to the nature of the bond between the glucose molecules. This allows the formation of hydrogen and Van der Waals bonds between the molecules of cellulose and results in cellulose be packed into microfibrils, several parallel chains attached to each other. An illustration of the arrangement of the cellulose molecules in parallel chains and the accompanying hydrogen bonding is given in Fig. 5.2.

Cellulose microfibrils are mostly independent, but the presence of covalent bonds, hydrogen bonding, and Van der Waals forces are responsible of cellulose structure. Hydrogen bonding within a cellulose microfibril determines "straightness" of the chain, but interchain hydrogen bonds might introduce order (crystalline) or disorder (amorphous) into the structure of the cellulose. Hemicelluloses and lignin cover the microfibrils, which are bundled together to form cellulose fibers.

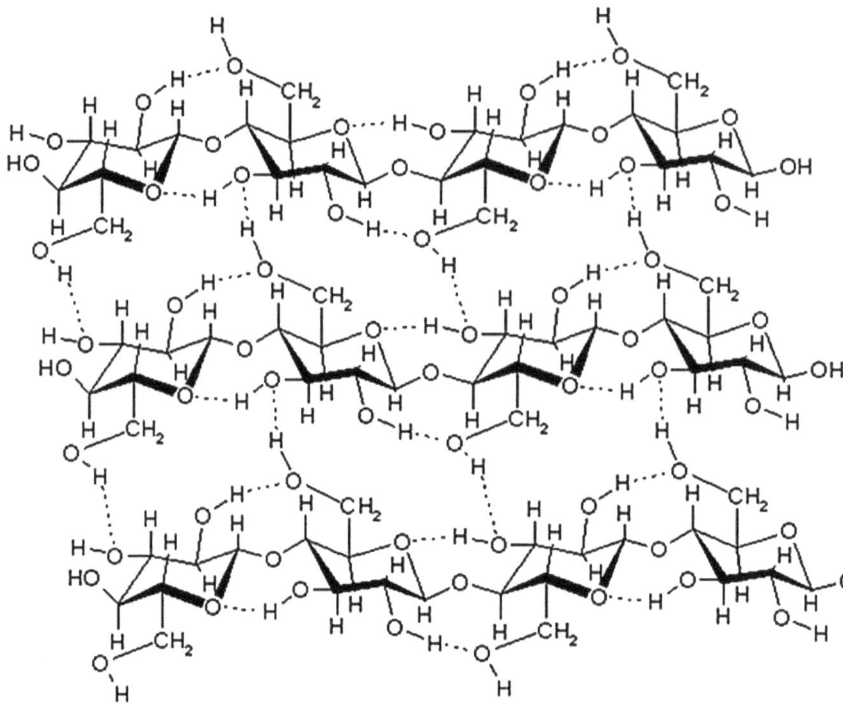

Figure 5.2
Hydrogen bonds between cellulose polymer chains.

Many properties of cellulose depend on its degree of polymerization (DP), i.e. the number of glucose units that make up one polymer molecule. The DP of cellulose chains in nature can range from 10,000 units in wood to 15,000 in native cotton.

Cellulose is insoluble in water, where it swells. It is also insoluble in dilute acid solutions at low temperature. The solubility of the polymer is strongly related to the degree of hydrolysis achieved, although solubilized molecule is different in structure that the native one. At higher temperatures, it becomes soluble due to the breakage of the hydrogen bonds that hold the crystalline structure.

Cellulose is also soluble in concentrated acids, but severe degradation of the polymer by hydrolysis is caused. In alkaline solutions, extensive swelling of cellulose takes place, as well as dissolution of the low-molecular weight fractions of the polymer.

Fermentable sugars can be produced from cellulose through the action of aid and enzymes breaking of the β-(1,4)-glycosidic bonds. Cellulose is more susceptible to enzymatic degradation in its amorphous form.

5.2.2 Hemicellulose

Hemicellulose is the second most abundant polymer of lignocellulosic biomass and differs from cellulose in that it is not chemically homogeneous. The term hemicellulose is a

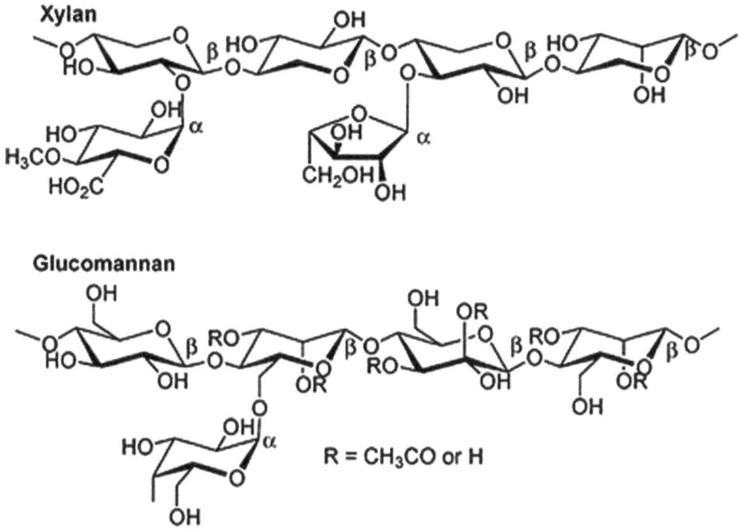

Figure 5.3
Structure of hemicellulose.

collective term; it is used to represent a family of polysaccharides linked in a branched polymer chain. The function of hemicellulose has been proposed as a bonding agent between lignin and cellulose.

The backbone of hemicellulose is either a homopolymer or a heteropolymer with short branches linked by β-(1,4)-glycosidic bonds and occasionally β-(1,3)-glycosidic bonds. These short branches consist of different monosaccharides that include pentoses (xylose, rhamnose, and arabinose), hexoses (glucose, mannose, and galactose), and uronic acids (4-O-methyl-glucuronic, D-glucuronic, and D-galacturonic acids). Methyl or acetyl groups are attached to the carbon chain to various degrees (Fig. 5.3).

Hemicelluloses have lower molecular weight compared to cellulose; they are more thermally sensitive (Hendriks and Zeeman, 2009) and are easily hydrolyzed. Hemicellulose is insoluble in water at a low temperature, but its hydrolysis is possible at a lower temperature than that of cellulose. The presence of acid or alkali highly improves the solubility of hemicellulose in water because xylan can be easily extracted in an acid or alkaline environment.

Key aspects of the structure and composition of hemicellulose are the lack of crystalline structure, mainly due to the highly branched structure, and the presence of acetyl groups connected to the polymer chain. They possess a high degree of polydispersity, polydiversity, and polymolecularity. However, the DP does not exceed the 200 units. These polymers do not aggregate, even when they cocrystallize with cellulose chains.

Hemicelluloses are believed to coat cellulose fibrils in the plant cells, and it has been proposed that at least 50% of hemicellulose should be removed to significantly increase cellulose digestibility. Parameters must be carefully optimized to avoid the formation of inhibitors products such as furfurals and hydroxymethylfurfurals. For this reason, pretreatment conditions must be a compromise to maximize sugar recovery, and for this reason, hemicellulose could be obtained either as a solid fraction or a combination of both solid and liquid fractions, depending of the type of pretreatment used.

5.2.3 Lignin

Lignin is the most complex natural polymer and the third most abundant in nature. It is present in plant cell walls conferring structural support, impermeability, resistance to impact, compression, and bending, and against microbial attack and oxidative stress. It plays an important role in the transport of water, nutrients, and metabolites in the plant cell.

It is an amorphous heteropolymer with a three-dimensional network of phenylpropanoid units (*p*-coumaryl, coniferyl, and sinapyl alcohol) held together by different kind of linkages. Alkyl–aryl, alkyl–alkyl, and aryl–aryl ether bonds link these monomers together (Hendriks and Zeeman, 2009) (Fig. 5.4).

Different kinds of feedstock contain different amounts of lignin and with different composition. In general, herbaceous plants (grasses) have the lowest contents of lignin, and softwoods have the highest. It has been identified that lignin from softwood is made up of more than 90% of coniferyl alcohol, whereas lignin in hardwood is made up of varying ratios of coniferyl and sinapyl alcohol.

Lignin is commonly accepted as the "glue" that binds the different components of lignocellulosic biomass together, thus making it insoluble in water. Due to its linkage to cellulose, lignin has been identified as a major problem for enzymatic and microbial hydrolysis of lignocellulosic materials. Biomass digestibility in enhanced with increasing lignin removal. Its main effects include (1) it is a physical barrier to enzymes' access to sugars; (2) nonspecific adsorption of enzymes to sticky lignin; (3) nonproductive binding of enzymes to lignin; and (4) lignin derivatives are toxic for microorganisms and strong inhibitors for fermentation.

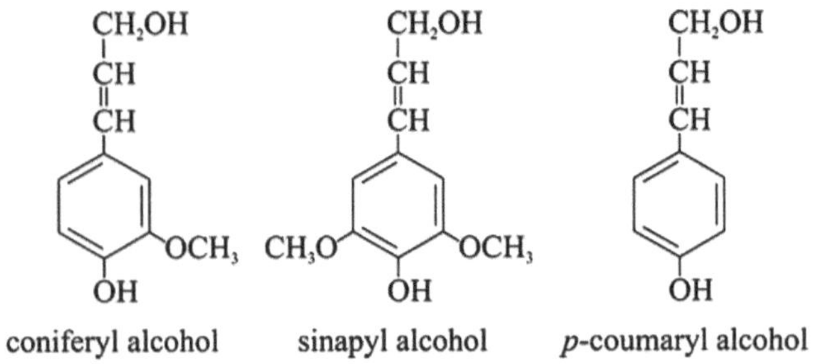

Figure 5.4
p-coumaryl-, coniferyl-, and sinapyl alcohol, building blocks of lignin.

Delignification causes disruption of lignin structure, resulting in biomass swelling, increase in internal surface area, and accessibility of enzymes to cellulose fibers.

5.3 Pretreatments

Pretreatment is a necessary tool to break down the structure of lignocellulosic biomass formed of cellulose, hemicellulose, and lignin to make more accessible carbohydrates for further transformation processes, as said earlier in this chapter. Pretreatments aim to remove lignin and hemicellulose, reduce the crystallinity of cellulose, and increase the porosity of biomass. A good pretreatment must avoid the degradation or loss of carbohydrates and the formation of inhibitors to the subsequent hydrolysis and fermentation processes while improving the formation of sugars or the ability of pretreated material to form sugars by hydrolysis, altogether in a cost-effective operational process. The choice of the best pretreatment depends, among other factors, on the foreseen utilization of the main biomass components (cellulose, hemicellulose, and lignin).

A general classification of the pretreatments groups them into the following categories: physical, chemical, and biological, and a combination of them with special mention to the physicochemical pretreatments (Harmsen et al., 2010). These combinatorial strategies are generally more efficient. Key features of different pretreatments are outlined in Table 5.1.

5.4 Physical Pretreatments

5.4.1 Mechanical Comminution

Mechanical pretreatment of biomass aims to enhance the digestibility of biomass. Coarse size reduction, cutting, shredding, chipping, grinding, or milling are among the different mechanical methods that can be used to decrease particle size, increase accessible specific surface area, increase pore size of particles and the number of contact points, and reduce the DP or crystallinity of the cellulose, although different fractions (cellulose, hemicellulose, and lignin) will not be separate.

Commonly, starting materials are presized during harvesting or preconditioning, using methods such as shredding, forage cutting, or chipping to sizes of about 10–50 mm. This is the minimum pretreatment needed prior to biomass processing. The size of material can be further reduced to 0.2–2 mm by milling or grinding through different machines: vibratory ball mills, hammers, knifes, balls, discs, colloids, and extruders.

Chipping is used to reduce heat and mass transfer limitations. Grinding and milling are more effective at reducing the particle size and cellulose crystallinity (Dumas et al., 2015). Vibratory ball milling is more effective than ordinary ball milling in reducing cellulose crystallinity. Disk milling, which produces fibers, is more efficient in enhancing cellulose hydrolysis than

Table 5.1: Effect of Various Pretreatments Methods on Composition of Lignocellulosic Biomass.

Pretreatment	Examples	Increase Accessible Surface Area	Decrystallizes Cellulose	Removes Hemicellulose	Removes Lignin	Alters Lignin Structure
Physical	Mechanical comminution		■			
	Extrusion					
	Liquid hot water	■		■		■
	Pyrolysis and torrefaction					
	Freeze					
	Irradiation		■		■	
Chemical	Acid hydrolysis	■	■	■		■
	Alkaline hydrolysis	■		■	■	■
	Organosolv	■		■	■	■
	Peroxide	■		■	■	■
	Ozonolysis				■	
Physicochemical	Steam explosion	■		■		■
	Ammonia fiber explosion (AFEX)	■	■	■	■	■
	Carbon dioxide (CO_2) explosion		■	■		
	Wet oxidation		■	■	■	
Biological	Fungal species					■
	Enzymatic			■		■
Combined	Alkaline + Mechanical			■	■	

hammer milling. The type and duration of milling, as well as the kind of biomass, determine the digestibility of the biomass (Hideno et al., 2013; Zakaria et al., 2015).

This method has some disadvantages in terms of the energy demand. The final particle sized desired and the biomass characteristics determine the energy requirements for mechanical comminution, which is usually very high. There is a critical size below which further reduction will not affect the biomass treatment significantly (Dumas et al., 2015; Liu, 2015). Initial moisture content and biomass composition have shown to be important parameters impacting the specific energy requirement (Barakat et al., 2015). In this sense, mechanical comminution is not considered as an attractive option in biomass pretreatment.

The combination of biological or chemical treatment prior to mechanical diminution has confirmed the feasibility of reduction of the energy consumption of mechanical processes (Fougere et al., 2015; Motte et al., 2015). Studies show that milling after chemical pretreatment can significantly reduce milling energy consumption, cost of solid liquid separation, and liquid to solid ratio, and does not result in the production of fermentation inhibitors (Zhu et al., 2009).

5.4.2 Extrusion Pretreatment

Extrusion pretreatment is a physical pretreatment in which materials are exposed to mixing, heating, and shearing, suffering physical and chemical modification. The shear forces applied in the extrusion process serve to remove the softened surface regions, exposing the interior to chemical and/or thermal action and therefore improving the cellulose conversion (Mood et al., 2013).

Factors such as extruder temperature, screw speed, feedstock particle size, and moisture content have been investigated to determine their influence in energy requirements of the pretreatment (Karunanithy and Muthukumarappan, 2011a).

Main advantages of this method include short residence time, moderate temperature, no formation of inhibitors such as furfural or 5-hydroxymethylfurfural (HMF), no need of washing step, no solid loss, rapid mixing, feasibility of scale-up, and possibilities of continuous operation (Karunanithy and Muthukumarappan, 2011b,c).

Recent studies about alkali-combined extrusion pretreatment indicated that combined pretreatment increased the number of pores in the biomass structure, giving improved sugar yields (Zhang et al., 2012).

5.4.3 Liquid Hot Water Pretreatment

LHW processes are biomass pretreatments based on the use of pressure to keep water at high temperatures (160–240°C). It is also referred in the literature as autohydrolysis, hydrothermolysis, hydrothermal pretreatment, aqueous fractionation liquefaction or extraction, solvolysis, aquasolv, steam pretreatment, or water prehydrolysis (Mosier et al., 2005).

The reaction is initiated by the hydro ions [H_3O^+] generated from the dissociation of water molecules. This process changes the biomass structure, resulting in the hydrolysis of hemicellulose and removal of a small portion of the lignin, which makes the cellulose more accessible for further hydrolysis while avoiding the formation of fermentation inhibitors that occurs at higher temperatures (Vallejos et al., 2015).

It is important to maintain the pH between 4 and 7 during the pretreatment because at this pH, the dissolved hemicellulose exists mainly in oligomeric form, and the formation of monosaccharides and the subsequent degradation products that further catalyze hydrolysis of cellulosic material are minimized.

Pretreatment of some biomass feedstock can be carried out under mild conditions (140–180°C), but for most raw biomass, it needs to be performed at higher temperatures (up to 190–230°C). The water and the biomass (1–8%) are brought in contact up to 5 min at most severe conditions. At this high temperature, sugar degradation may increase significantly. Between 40% and 60% of the total biomass can be dissolved in the process with the removal of 4–22% of the cellulose, 35–60% of the lignin and the majority of the hemicellulose (Rogalinski et al., 2008).

Water pretreatments provide an interesting alternative in the biomass treatment owing to the following reasons:

- There is no need to add a chemical such as sulfuric acid, lime, ammonia, or other catalyst. The whole process results are environmentally friendly.
- In comparison with acid pretreatments, the reactors do not require high cost materials or maintenance due to their low corrosion potential.
- The process itself is simplified. The use of lower temperatures with minimization of degradation products eliminates the need for a final washing step or neutralization because the pretreatment solvent here is water. Neither sludge handling nor acid recycling result.
- The biomass glucan content is not modified. The physicochemical modification caused by treatment on lignin and cellulose facilitates the further separation of different fractions.
- Hemicelluloses can be converted into hemicellulosic sugars at good yields with low byproduct generation.
- Economically, the low cost of the solvent itself is also an advantage for large-scale applications.

The main disadvantage of LWH pretreatment is related to the downstream processing. The amount of solubilized product is higher, while the concentration of these products is lower compared to other pretreatments. High energy is demanded due to the large volumes of water involved.

The recovery of hemicellulose from solution is impeded by high-lignin solubilization. A catalyst such as an acid can be added making the process similar to dilute acid pretreatment. However, degradation of sugars can result in undesirable inhibitory products. During pretreatment, the pH and pKa of water is affected by temperature, so KOH can be used to maintain the pH above 5 and below 7 to minimize the formation of monosaccharides that are degraded to fermentation inhibitors (Mosier et al., 2005).

5.4.4 Pyrolysis and Torrefaction

Pyrolysis is a thermal pretreatment of lignocellulosic biomass where raw material is heated in an inert atmosphere at temperatures between 350 and 650°C. It is usually employed to enhance the energy density of fuels produced from biomass. Nitrogen is the commonly used carrier gas to provide a nonoxidizing atmosphere. Torrefaction takes place at similar conditions to those of pyrolysis but at lower temperatures of 200–300°C, due to which, it is also called mild pyrolysis.

In addition to temperature treatment, time is another important factor in determining the performance of thermal pretreatment. This can be carried out between several minutes to hours.

Cellulose decomposes rapidly to gaseous products and residual char when biomass is heated above 300°C. At lower temperatures, decomposition is much slower, and resulting products are less volatile. Under severe conditions, hemicellulose is almost depleted completely, and cellulose is oxidized to a great extent. Lignin is the most difficult component to be degraded, and thus, its removal is very low under torrefaction conditions.

After thermal pretreatment, the properties of biomass are improved to a great extent. Main benefits from torrefaction are more uniform properties in the biomass, which include improved grindability and reactivity, higher energy density, lower atomic O/C and H/C ratios and moisture content, and higher hydrophobicity (Chen et al., 2015b).

Recent studies have demonstrated that torrefaction pretreatment causes mechanical disruption of biomass fibers, resulting in their size reduction as well as high solid product yield (Das and Sarmah, 2015).

Torrefaction and pyrolysis have been studied as pretreatment processes for biomass to fuel conversion. Pretreatment by torrefaction was found to be far more attractive than pyrolysis (Kumar et al., 2009).

5.4.5 Freeze Pretreatment

Freeze/thaw pretreatment is a novel approach for physical pretreatment of biomass. In this process, biomass is frozen in a conventional freezer at temperature below −20°C for a certain period of time (between 2 and 24 h) and then immediately thawed in hot water or at room temperature (Chang et al., 2011).

Treatment with freezing/thawing could be an efficient alternative for pretreatment of lignocellulosic biomass due to significantly increasing the enzyme digestibility of substrates (Smichi et al., 2015).

Only a few studies have been carried out, but despite the high cost involved, its attractive characteristics, i.e. lower negative environmental impact, application of less dangerous chemicals, and high effectiveness, make freeze/thaw process a promising pretreatment in biomass processing.

5.4.6 Irradiation

Irradiation pretreatment produces delignification, depolymerization, and destruction of the crystalline structure of cellulose.

Some authors have suggested treatments involving the use of gamma rays that give a larger surface area and lower crystallinity by cleaving the β-1,4 glycosidic bonds. Gamma irradiation after sulfuric acid pretreatment on wheat straw showed a great influence on enzymatic hydrolysis, owing to disruption of cellulose crystallinity, removal of hemicelluloses, and structural modification of lignin polymers (Hong et al., 2014). This method would be very expensive on a large scale with huge environmental and safety concerns.

Microwave irradiation could be an alternative to the conventional heating in order to modify the structure of cellulose, degrade and partially remove lignin and hemicelluloses, and enhance the enzymatic susceptibility of reducing sugars. The advantages of this method include short process time, high uniformity and selectivity, and less energy input than conventional heating. Microwave-assisted pretreatment has demonstrated the improvement in enzymatic hydrolysis of corn straw and rice husk (Diaz et al., 2015).

5.4.7 Pulsed Electric Field Pretreatment

Pulsed electric field (PEF) pretreatment is a physical pretreatment of lignocellulosic biomass that involves the application of a short burst of high voltage to a sample (biomass) situated between two electrodes. The sample can be either placed or transported between the electrodes, and the electric discharge is applied in the form of pulses. High intensity electric field produced structural changes in the cell membrane, resulting in an increase in mass permeability and mechanical rupture (Kumar et al., 2009). The creation of permanent pores in the cell walls facilitate the entry of acids (in the case of chemical treatments) or enzymes (biological processes) used to break down the cellulose into its constituent sugars, and thus PEF increases the hydrolysis rate.

Most important factors in PEF pretreatment include electric field strength, which is usually above 1 kV/cm (5–20 kV/cm), number of pulses, and treatment time, in the microsecond range. Major benefits of PEF pretreatment are that it can be carried out at ambient conditions, energy requirement is low, and the process itself is not very complex as it does not involve moving parts.

5.4.8 Ultrasonic Pretreatment

Ultrasonic pretreatment (USP) of lignocellulosic biomass has been studied at laboratory scale, although it is a well-established technique for industrial wastewater treatment. It promotes the pretreatment and conversion process through cavitation phenomenon. Ultrasonic energy allows destruction of the lignocellulosic structure and fractionation of biomass components, with increased yields of sugars, bioethanol, and gas products. Sonication promotes hydrolysis and leads to reduced reaction time, lower reaction temperature, and less amounts of solvents (Luo et al., 2014).

Experiments carried out on a model compound (carboxymethyl cellulose) showed that reaction time was dramatically increased (Imai et al., 2004). Other studies on lignocellulosic biomass in combination with hydrogen peroxide for bioethanol production gave higher yields of cellulose recovery and delignification (Ramadoss and Muthukumar, 2016).

5.5 Chemical Pretreatments

5.5.1 Acid Hydrolysis

Concentrated strong acids, such as H_2SO_4 and HCl, have been also used to treat lignocellulosic materials because they are powerful agents for cellulose hydrolysis, and there is no need for subsequent use of enzymes to reach acid hydrolysis and obtain fermentable sugars. Although they have advantages as flexibility in feedstock choice, high monomeric sugar yield, and mild temperature operation conditions, they are toxic, hazardous, and corrosive, which makes the pretreatment process very expensive.

Dilute acid pretreatment is a method that has received extensive research, both as a biomass pretreatment step to enhance biomass digestibility and as an alternative to enzymatic hydrolysis of cellulose. It is one of the most widely used processes due to its high efficiency. Dilute acid removes and recovers most of the hemicellulose as dissolved sugars (almost 100%), rendering the cellulose fraction more accessible to further hydrolysis. The method is especially suitable for low-lignin biomass, as the removal of lignin is insignificant, and it is retained in the solid matter (Lenihan et al., 2010).

Hydrochloric acid, nitric acid, phosphoric acid, and sulfuric acid have been tested for use in biomass pretreatment with sulfuric acid the most commonly used as the acid of choice, usually below 4 wt% as it is inexpensive and effective. As an alternative to inorganic acids, some organic acids (maleic acid, fumaric acid) can be used (Kootstra et al., 2009).

In general, there are two types of dilute acid pretreatment:

High temperature and continuous flow process for low-solids loading ($T > 160°C$, 5–10 wt% substrate concentration) (Cara et al., 2008).
Low temperature and batch process for high-solids loading ($T \leq 160°C$, 10–40 wt% substrate concentration) (Esteghlalian et al., 1997).

In some processes, an aqueous solution of biomass is heated to the desired temperature and pretreated using preheated sulfuric acid (concentrations below 4 wt%) in a stainless steel reactor. Reaction time ranges from a few seconds to minutes, depending on the temperature of the pretreatment. In another method, preconditioned and physically pretreated biomass is submerged in a circulating bath of acid heated to desired temperatures. Two-stage pretreatments can also be carried out (Wijaya et al., 2014).

Hydrolysis of hemicellulose then occurs, releasing monomeric sugars and soluble oligomers. Hemicellulose removal increases porosity of the cell walls and improves enzymatic digestibility. Dilute acid pretreatment can significantly improve hemicellulose and cellulose hydrolysis by varying the severity of the pretreatment.

However, dilute acid pretreatment presents important drawbacks:

During the process, furfural from xylose degradation and lignin degradation products are formed, which act as inhibitors in the fermentation step. A neutralization step with lime is used to both neutralize and detoxify the pretreatment stream, which adds cost to the process. In addition, the generation of calcium sulfate gypsum creates a disposal problem. Alternative detoxification options, such as ion exchange membranes, are expensive.

Biomass subjected to dilute acid pretreatment may be difficult to be fermented due if inhibitors have not been completely removed (Palmqvist and Hahn-Hagerdal, 2000a,b).

Due to the high temperatures used, along with the corrosive action of acids, expensive alloys resistant to hot acid corrosion must be used for the reactor construction, which results in the reactor cost being a significant element of the total capital cost. Elevated maintenance costs must also be considered.

Acid recovery is also very expensive, as it must be recovered to make the pretreatment economically feasible.

Environmental problems caused by its waste streams and their expensive treatment have led to the need of other options for biomass pretreatment.

The process' costs are higher than in other physicochemical pretreatments, such as steam explosion or ammonia fiber explosion.

5.5.2 Alkaline Hydrolysis

Some bases can be used for the pretreatment of lignocellulosic materials. Sodium, potassium, or calcium hydroxides are suitable alkaline pretreatment agents. Usually lime (calcium hydroxide) or sodium hydroxide is used.

In the alkaline pretreatment, the biomass is soaked in the alkaline solution and mixed at a mild temperature for a period of time. Alkali pretreatment utilizes lower temperatures and pressures than other pretreatment technologies. It can be carried out at ambient conditions, but this requires a very long processing time, hours or days rather than minutes.

Alkaline pretreatment disrupts the lignin structure and breaks the linkage between lignin and the other carbohydrates in lignocellulosic biomass. As the lignin is removed, the reactivity of remaining polysaccharides is increased. Some studies have shown that alkaline pretreatment mechanism is based on saponification of intermolecular ester bonds crosslinking xylan hemicelluloses and other components, such as lignin (Sun and Cheng, 2002). Alkaline pretreatment retains the hemicellulose in the biomass solids, but acetyl and the various uronic acid substitutions of hemicellulose are also removed (Galbe and Zacchi, 2012; Mosier et al., 2005). The addition of air/oxygen to the reaction mixture greatly improves delignification.

The low severity of the process prevents condensation of lignin, resulting in high-lignin solubility, especially for biomass with low lignin content. Due to the mild conditions, little degradation of sugars to furfural, HMF, and organic acids occurs. It causes less sugar degradation than the acidic pretreatments.

Prior to enzymatic hydrolysis, a neutralizing step is necessary, which leads to salts generation. They may be incorporated in the biomass and need to be removed or recycled. Besides removing lignin, the washing step also removes inhibitors, salts, furfural, and phenolic acids (Teixeira et al., 2014).

The process of lime pretreatment involves slurrying the lime (either quick lime (CaO) or slaked lime ($Ca(OH)_2$)) with water, spraying it onto the biomass material, and storing or heating the material for a period of time (Kim and Holtzapple, 2005, 2006). It can last a very long time. Alternatively, the pretreatment can be performed in a closed reactor with temperature and flow controls. Lime is recyclable using established technology (kiln process).

Alkali pretreatment is most effective with low lignin content biomass. To achieve good digestibility with high-lignin biomass, an oxidant needs to be added, and the process essentially turns into an alkaline wet oxidation process.

5.5.3 Organosolv Pretreatment

The pretreatment of lignocellulosic biomass with organosolvents involves the use of organic or aqueous organic solvent mixture with inorganic acid catalysts (HCl or H_2SO_4). Organic acids such as oxalic, acetylsalicylic, and salicylic acids can also be used as catalysts. Solvents commonly used are methanol, ethanol, acetone, and ethylene glycol. This type of pretreatment is usually carried out at high temperature (up to 200°C) and pressure (Chen et al., 2015a; Kabir et al., 2015).

The catalytic mechanism with alcohol is similar to that of the autohydrolysis pretreatment. The solvent hydrolyses lignin bonds as well as lignin–carbohydrate bonds. Lignin is extensively removed, and hemicellulose is almost completely solubilized, while cellulose remains in solid form. Pretreatment with organic acids occurs via the dissociation of hydrogen ions to accelerate delignification and hydrolysis of cellulose and dissolution of the lignin. In this case, the addition of a catalyst is not necessary at high temperatures. A significant part of soluble carbohydrates is further broken to byproducts such as furfural or HMF, which are inhibitory to fermentation microorganisms.

Main benefits of organosolv pretreatment include:

It is able to separate three different fractions: dry lignin, an aqueous hemicellulose stream, and a relatively very pure cellulose fraction. Due to this, it is often viewed as a fractionation method more than a pretreatment technology (Duff and Murray, 1996).
Production of high-quality lignin can be achieved. Dissolved lignin can be precipitated by evaporating the solvent or adding cold water. Organosolv lignin is sulfur free with high purity and can be used for the production of specialty chemicals (Hambardzumyan et al., 2015).

It is the only effective physicochemical pretreatment method for the pretreatment of high-lignin lignocellulose materials.

Organosolv pretreatment does not require significant size reduction of feedstock to achieve satisfactory cellulose conversion, making the process less energy intensive.

Disadvantages of organosolv pretreatment can be summarized as follows:

The use of flammable organic solvents at high temperature and the process operations involve the use of containment vessels. No leaks can be tolerated due to fire and explosion hazards and environmental, health, and safety concerns.

Solvents need to be recovered from the reactor, evaporated, condensed, and recycled to reduce costs (Viel et al., 2013). This is very necessary because the solvent may be inhibitory to growth of microorganisms, subsequent enzymatic hydrolysis, and fermentation or anaerobic digestion.

The cost of chemicals and catalyst makes organosolv pretreatment more expensive than other pretreatment processes.

5.5.4 Peroxide Pretreatment

Hydrogen peroxide can be used as an oxidative pretreatment for agro-industrial residues at operational temperatures of ≥100°C in alkaline solution (Alvarez-Vasco and Zhang, 2013). Decomposition of H_2O_2 is rapid in these conditions, and reactive radicals are produced, which cause lignin degradation. It becomes particularly unstable in the presence of certain metals such as manganese (Mn), iron (Fe), and copper (Cu), which makes necessary the addition of chelants to reduce the rate of decomposition (Lucas et al., 2012).

This pretreatment results in partial delignification of the cell wall of biomass and reduction of cellulose microfibril crystallinity. Delignification by this process on a large scale can therefore be costly. Current research has focused on reducing energy requirements and working temperatures. Dissolution of about 11% of lignin and most of the hemicellulose (70%) can be achieved in a solution of 1% H_2O_2 at 150°C (Ayenio et al., 2013).

5.5.5 Ozonolysis

Ozone pretreatment is used to reduce the lignin content of lignocellulosic biomass and wastes. Lignin consumes most of ozone during the degradation of the carbohydrate. The degradation is mainly limited to lignin, although hemicellulose is slightly affected, but cellulose is not (Li et al., 2015).

Ozone can be used to disrupt the structure of many lignocellulosic materials such as wheat straw, bagasse, peanut, pine, cotton straw, and poplar sawdust (Kumar et al., 2009).

Main advantages of ozonolysis pretreatment are that the reactions take place at room temperature and normal pressure; it does not generate toxic residues, unlike other chemical pretreatments, and it can be designed to minimize environmental concerns. Mayor drawback of ozonolysis is that the process is very expensive (Kumar et al., 2009).

5.5.6 Peroxyformic Acid

Peroxyformic acid pretreatment is a method for oxidative delignification through chemical pretreatment. Peroxyformic acid is generated in situ by mixing formic acid with hydrogen peroxide. Lignocellulosic biomass is added to this mixture and allowed to stand for several hours (Villaverde et al., 2012).

Formic acid acts as a solvent for lignin and breaks down hemicellulose chains. Peroxyformic acid produces oxidative delignification through increased solubility of lignin. The reaction temperature is increased and most delignification occurs. At the final stage, any remaining lignin is degrades (Singh et al., 2009).

This pretreatment is known as the Milox process, derived from "milieu pure oxidative pulping" (Ligero et al., 2010).

5.6 Physicochemical Pretreatments

5.6.1 Steam Explosion

Steam explosion is the most commonly applied physicochemical method of biomass pretreatment. In this method, biomass is treated with high-pressure saturated steam followed by a rapid decompression. Subsequently, an explosive reaction takes place, resulting in hemicellulose degradation and lignin matrix disruption. Cellulose crystallinity may decrease and substrate surface may increase, improving cellulose digestibility (Li et al., 2007).

Physically pretreated biomass (chipped, ground, or simply raw preconditioned) is treated with high-pressure saturated steam at temperatures of about 160–260°C and pressures between 0.7 and 4.8 MPa. The pressure is held for several seconds to a few minutes to promote hemicellulose hydrolysis before material is exposed to atmospheric pressure (Zhao et al., 2015).

As a result of the treatment, hemicellulose is hydrolyzed by acetic acid generated in the process and solubilized in the liquid phase, lignin is depolymerized, and the cellulose in the solid fraction becomes more accessible, hence increasing the digestibility of the lignocellulosic feedstock. Results depend on residence time, temperature, particle size, and moisture content, and they can be improved by addition of chemicals such as acid (H_2SO_4, CO_2, or SO_2) or alkali (Huang et al., 2015).

The main benefits of steam explosion pretreatment are to reduce the energy input for pumping and mixing of substances after pretreatment and the possibility to use coarse particles, thus avoiding the necessity of a biomass size conditioning step, which can represent up to one-third of total energy required in the process. Other attractive features are the no use of chemicals, which makes it environmental friendly, with less costs due to recycling, high recovery of sugars with low dilution, and the feasibility for industrial implementation. The soluble stream rich in carbohydrates can be easily removed and used for the production of high added value products.

Limitations related to steam explosion pretreatment are the enzyme and fermentation inhibitors generated during the pretreatment: furfural and hydroxymethylfurfural, weak acids, and phenolic compounds produced due to lignin breakdown (Wang et al., 2015). Several detoxification methods have been developed in order to reduce the inhibitory effect, which represents additional costs in the overall process. Another drawback is the incomplete destruction of lignin–carbohydrate matrix, resulting in the risk of precipitation and condensation of components. Required washing of hydrolysate may decrease overall yields. A major constraint is also that an efficient isolation of the lignin, and cellulose components must be carried out, as lignin needs to be fully removed depending on the subsequent product. Fractionation can be carried out by solvent extraction.

This method requires high amount of energy and involves elevated running costs, which must be kept in mind for industrial implementation on a large scale (Sharma et al., 2014).

5.6.2 Ammonia Fiber Explosion and Ammonia Pretreatment

Ammonia fiber explosion (AFEX) is a physicochemical pretreatment in which lignocellulosic biomass is treated with liquid ammonia, mild temperature, and high pressure for a period of time, and then pressure is swiftly reduced. The AFEX process is very similar to steam explosion pretreatment. The chemical effect of ammonia under pressure causes the biomass swelling, and therefore, an increase in accessible surface area, and at the same time, a decrease in cellulose crystallinity (Mosier et al., 2005).

Typically, biomass is brought in contact with liquid ammonia in a ratio of 1–2 kg of ammonia/kg of dry biomass at 60–90°C with a residence time of 10–60 min (usually 30 min) and pressure between 40 and 50 atm. The rapid release of temperature causes evaporation of the ammonia and a drop in the temperature of the system. Only a small amount of the hemicellulose is degraded to soluble oligomeric form, and the lignin distribution remains the same after AFEX pretreatment, although its structure is altered. Consequently, there is an increase in water holding capacity and digestibility of biomass.

Another type of processes utilizing ammonia are soaking aqueous ammonia (SAA) and ammonia recycle percolation (ARP). SAA is a modified version of AFEX in which biomass is treated in a batch reactor at moderate temperatures (25–60°C). Mild process conditions may minimize the formation of sugar degradation products and fermentation inhibitors, but duration could be up to 10–60 days. ARP uses ammonia percolating in a flow-through mode through a packed bed reactor. In this process, aqueous ammonia (10–15%) passes through biomass in a flow-through column reactor at elevated temperatures (150–190°C) and a flow rate of 1–5 mL/min with residence time of 10–120 min, after which the ammonia is recycled or recovered. To prevent evaporation, the reactor system must be slightly pressurized (Kim and Lee, 2005).

Delignification in the SAA pretreatment is highly selective, which means that most of carbohydrate remains in the solids after the process. The lignin is efficiently removed by minimizing the interaction with hemicellulose. An increase of surface area and pore size is achieved, and sugar loss is minimal after water washing (Kim et al., 2008). In the ARP process, at higher temperatures, some hemicellulose sugars are lost from the solids, which arises a concern about the recovery (Kim et al., 2003).

There are several advantages to AFEX/ARP pretreatment processes, which include the high selectivity of ammonia for reaction with lignin. In addition, ammonia is a widely used low-cost commodity chemical, and it is easily recovered and recycled, which makes a continuous industrial process more feasible. Ammonia is volatile and can be efficiently removed from the system, eliminating the need for the water washing step and increasing total sugar yields.

Although most of the ammonia is recyclable after the pretreatment, some is not, and some is consumed in the pretreatment as a result of reaction with the acidic groups and lignin in biomass. However, this portion of lost ammonia could be a nitrogen source for the fermentation microorganisms. Together with the fact that the use of mild temperatures and short residence time reduces the formation of fermentation inhibitors leads to a hydrolysate compatible with fermentation organisms without the need for conditioning.

Disadvantages of AFEX/ARP pretreatments include the fact that recycling ammonia consumes a large amount of energy and contributes significantly to the process costs, and there are also safety and environmental concerns with the use and onsite storage of ammonia in pilot as well as in industrial-scale facilities. Some studies have been done to reduce this process costs (Sendich et al., 2008). In addition, AFEX/ARP processes have not been very effective in the pretreatment of high-lignin-containing biomass.

5.6.3 Carbon Dioxide Explosion Pretreatment

Carbon dioxide (CO_2) explosion (CDE) pretreatment method is similar to steam and AFEX. It was developed as an attempt to improved lignocellulosic biomass pretreatment by a green method with lower temperature and cost.

The method involves the use of supercritical CO_2 under pressure to enhance the digestibility of lignocellulosic biomass. High-pressure CO_2 (15–30 MPa) is injected in a batch reactor, heated to the desired temperature (up to 200°C), held for a certain time (several minutes), and then liberated by an explosive decompression (Narayanaswamy et al., 2011). It is believed that CO_2 forms carbonic acid when dissolved in water, which helps in the hydrolysis of hemicellulose as well as cellulose. Moreover, the low temperature used prevents any decomposition of monosaccharides by the acid. Rapid release of pressure results in disruption of biomass structure and increases accessible surface area, enhancing the digestibility of lignocellulosic biomass (Alinia et al., 2010).

This pretreatment offers the benefit of acid catalysts without the use of an acid like sulfuric acid. Other attractive features are the no generation of toxins or chemicals, the high solids capacity, and the use of low temperatures. CDE improves the yield of enzymatic hydrolysis, although it is lower than steam or AFEX explosion (Sun and Cheng, 2002). As CDE takes place at low temperatures, sugar degradation does not occur, and it does not cause the formation of inhibitory compounds.

However, equipment capable to be used with high-pressure conditions elevates the capital costs in large-scale applications. In addition, CDE pretreatment is not effective on biomass with low moisture content (Kim and Hong, 2001).

5.6.4 Wet Oxidation Pretreatment

Wet oxidation pretreatment (WOP) was industrially used for wastewater treatment and soil remediation by oxidation of soluble suspended materials, using oxygen at high temperature (150–350°C) and pressure (Zerva et al., 2003). It was later presented for pretreatment of lignocellulosic biomass as an alternative to steam explosion.

The process involves the use of water and oxygen or air at elevated temperature and pressure. This pretreatment converts hemicellulose and lignin to oxidized compounds such as low molecular weight carboxylic acids, alcohols, or even CO_2 and H_2O. Most of hemicellulose is dissolved, and delignification is reached to a certain level.

Wet oxidation benefits the lower production of furfural and HMF, strong inhibitors in the fermentation step. However, for the same reason, the hemicellulose sugars are largely lost, and thus the overall process yield is reduced (Banerjee et al., 2009). In addition, the exothermal behavior of the process is difficult to control.

In comparison with other pretreatment processes, crystalline structure of cellulose is opened during WOP, and therefore, it has proved to be efficient for treating lignocellulosic materials (Martin et al., 2008).

WOP is often developed in combination with other pretreatments: alkaline, with the addition of an alkali such as Na_2CO_3, or by use of hydrogen peroxide or even oxygen. But none of these treatments has improved significantly the results to be used for industrial application in biomass processing.

5.6.5 Ionic Liquid Pretreatment

A new approach on physicochemical pretreatment of biomass is the use on ionic liquids (ILs). They are salts that are in the liquid phase at temperature as low as room temperature. There are a high variety of different ILs. Most ILs used in biomass fractionation are imidazolium salts, although 1-allyl-3-methylimidazonium chloride (AMIMCl) and 1-butyl-3-methylimidazonium chloride (BMIMCl) have also presented good behavior.

ILs are considered "green" cellulose solvents. They can disrupt the biomass structure by solubilizing large amounts of cellulose at temperatures below 100°C with the possibility of recovering nearly 100% of the ILs. The dissolution mechanism suggests that these ILs compete with lignocellulosic components for hydrogen bonding, therefore disrupting its three-dimensional structure (Yao et al., 2015; Zhang et al., 2015). This class of salts modifies macro- and microstructure of cellulose, resulting in decreased crystallinity. This renders a pretreated material that can be easily hydrolyzed into monomeric sugars with high yields and rates for enzymatic hydrolysis (Silva et al., 2011).

This new class of solvents with low melting points, high polarities, high thermal stabilities, negligible vapor pressure, consisting entirely of ions (Zavrel et al., 2009), and environmentally friendly, can be recovered and reused by various methods such as pervaporation, reverse osmosis, salting out, and ionic exchange. They are very versatile and can function as selective solvents of lignin or cellulose. That would result in separation of lignin and increase of cellulose accessibility under ambient conditions. The formation of inhibitor compounds could also be avoided.

Application of ILs has opened up a new method of biomass pretreatment and fractionation. Despite their potential, there are still some uncertainties due to the very limited experience. Nevertheless ILs are still too expensive to be used for biomass pretreatment at industrial scale.

5.7 Biological Pretreatment

Biological pretreatment is a safe and environmentally friendly pretreatment associated with some fungi and bacteria capable of producing enzymes that can convert lignocellulosic biomass into more accessible compounds for hydrolysis. Microorganisms such as brown-, white-, and soft-rot fungi are used to degrade lignin, hemicelluloses, and polyphenols. Brown rots mainly attack cellulose, while white and soft rots attack both cellulose and lignin. Lignin degradation mechanism occurs through lignin degradation enzymes such as lignin peroxidases, polyphenol oxidases, manganese-dependent peroxidases, and laccases (Eggert et al., 1997; Lundell and Hatakka, 1994).

Biological pretreatment benefits include low energy demand, mild environmental conditions, and no chemical requirement.

On the other hand, biological pretreatment is too slow for industrial purposes. The residence time of 2 to 5 weeks required for sufficient delignification, the need of maintain careful growth conditions, and the large amount of space to perform biological pretreatments are the main drawbacks to making this method attractive on an industrial scale, especially in comparison to other physical and chemical methods. Direct application of enzymes has also been considered, but it is very expensive. Some of the carbohydrates are also consumed by the microorganism, which contributes to give a very low rate of hydrolysis.

5.7.1 Pretreatment With Fungal Species

White-rot fungi are the most effective for biological pretreatment of lignocellulosic biomass, with *Phanerochaete chrysosporium* achieving the highest efficiency among the known species of white-rot fungi due to its high growth and lignin degradation properties. Many studies have been carried out to analyze the impact of white-rot fungi on biomass degradation (Castoldi et al., 2014; Dinis et al., 2009; Ntougias et al., 2015; Saheed et al., 2015).

Various factors must be taken into account on biological pretreatment performance, such as particle size, moisture content, pretreatment time, and temperature, as they could affect degradation and yield (Shi et al., 2009).

Some white-rot fungi can utilize lignin at a rate similar to cellulose and hemicellulose, whereas others preferentially degrade lignin and hemicellulose. White-rot fungi produce multienzyme complexes to degrade cellulose and hemicelluloses, and when biomass fractions are simultaneously degraded, enzymes and radical mechanisms are developed (Singh and Singh, 2014).

Studies about enzyme production in brown-rot fungi have been done on a chemically defined liquid medium under controlled conditions to produce a particular enzyme. But in solid-state fermentation, as in the natural environment, rot fungi grow under different conditions, and hence the growth of fungi in solid-state fermentation (fungus immobilized) and in liquid submerged fermentation is different, and therefore the enzymes produced are different. Solid-state fermentation allows reduction in cost because agricultural or industrial waste biomass can be used as substrate, and higher enzyme productivity can be achieved through immobilization on agricultural residue. Therefore, selection of rot fungi and growth conditions are important to optimize yield.

Recent works have tried to make biological pretreatments more selective through the use of genetically manipulated fungi. Cellulose promoting enzymes have been deactivated, and therefore, higher sugar yields are reached, although hemicellulose degradation is increased with these mutant fungi.

White-rot and brown-rot fungi and their enzymes are being increasingly used in a huge number of biotechnological applications, which include industrial processing of pulp and textiles, bioenergy and biofuel production, waste treatment, and detoxification technologies (Singh and Singh, 2014), although there is room for improvement.

5.7.2 Enzymatic Pretreatment

Enzymatic pretreatment procedures employ pure enzymes derived from the same organisms as in pretreatment with fungi. Laccases and manganese peroxidases, as well as other peroxidases, are extensively used enzymes to treat the lignocellulosic substrate (Ravindran and Jaiswal, 2015). Laccase enzymes can be used as detoxifying agents for phenol removal in bioethanol production, and although laccases enhance microbial growth and improve fermentation capability, they also reduce the glucose recovery yield (Oliva-Taravilla et al., 2014). The effect of enzymatic pretreatment with laccase and versatile peroxidase on a variety of biomass was studied (Schroyen et al., 2015). It was observed that enzymatic pretreatment can help to degrade the biomass matrix and lignin barrier as for the high amount of phenolic compounds released. Substrate composition and nature proved to be of great interest to optimize yields.

5.8 Combined Methods

Combinatorial pretreatment strategies are generally more effective in enhancing the biomass digestibility and are often employed in designing leading pretreatment technologies. These treatments are a combination of physical, chemical, biological, and enzymatic treatments used in order to get the best of all treatments, hence minimizing the disadvantages of individual treatments.

5.8.1 Alkaline Pretreatment and Gamma Irradiation

Irradiation pretreatment can cause delignification, depolymerization, and destruction of the crystalline structure of cellulose by cleaving the β-1,4 glycosidic bonds. Alkaline process alone decreases lignin and hemicellulose percentage with swelling in the cellulose content. Combined treatment of irradiation and sodium hydroxide has better effects in increasing the values of organic matter digestibility. The irradiation may have broken the structure of the lignocellulose so that the NaOH solution could enter easily into the lignocellulose complex; thus the rate of reaction was increased.

Studies about irradiation of biomass for bioethanol production have shown that the concentration of reducing sugar of hydrolysates significantly increased as a result of gamma irradiation. Results indicate that the combined method of gamma irradiation with acid hydrolysis can significantly improve the saccharification (Hong et al., 2014; Yoon et al., 2012).

5.8.2 Alkaline and Mechanical Pretreatment

Combinatorial pretreatment of mechanical and alkaline methods consists of a continuous mechanical pretreatment such as milling or extrusion of lignocellulosic biomass with the aid of an alkali. The main effect of the alkaline pretreatment methods is the removal of lignin, whereas cellulose and part of the hemicelluloses remain in the solid fraction. After combined pretreatment, the accessibility of cellulose is improved, resulting in higher delignification and better enzymatic hydrolysis (Harmens et al., 2010).

5.8.3 Alkaline Pretreatment and Wet Oxidative Pretreatment

Lime pretreatment is a low-cost alkaline physicochemical pretreatment that has been reported to enhance the digestibility of lignocellulosic biomass. The pretreatment utilizes aqueous $Ca(OH)_2$ at low temperatures and pressures as a pretreatment agent to solubilize hemicellulose and lignin. Oxygen is introduced at high pressures to enhance the pretreatment. Studies resulted in better yield than without oxygen (Klinke et al., 2002; Martin et al., 2007).

This method benefits of low reagent cost, safety, and ease of recovery. The process is simple, and energy requirements are low. Great amounts of biomass could simply be pretreated without the need for a special vessel or using a simple design for pilot plants.

Main disadvantages to combined pretreatment are that it is not very effective for removing lignin in high-lignin biomass such as softwood. The oxidizing agent is not very selective; as a result, losses in hemicellulose and cellulose may occur, as well as the formation of inhibitors from oxidation of lignin. In addition, downstream processing uses huge volumes of washing water, and it is very expensive.

5.8.4 Combinatorial Biological Pretreatment

Biological pretreatment with other pretreatment technologies has also been studied, such as organosolvents and AFEX. The advantages of biological pretreatment include low energy requirements and mild environmental conditions. However, the rate of hydrolysis is very low.

5.9 Industrial Implementation of Lignocellulosic Biomass Pretreatments

There are a number of key features in the selection of the best pretreatment of lignocellulosic biomass. The choice of pretreatment should take into account the foreseen utilization of the main biomass components (cellulose, hemicellulose, and lignin). The ideal pretreatment process should have a low capital and operational cost. It should be effective on a wide range and loading of lignocellulosic biomass, and should result in the efficient recovery of most the lignocellulosic components in separate fractions. There should be no significant sugar degradation into toxic products and yeast and bacterial fermentation compatibility of the sugars.
It should have minimum heat and power requirements or be performed in a manner that energy invested could be used for other purposes (Agbor et al., 2011).

Other features that represent the basis of comparison of different pretreatment options include generation of higher-value lignin coproducts, possibility of obtaining solubilized hemicellulose sugars (Galbe and Zacchi, 2012; Mosier et al., 2005).

All these factors make it difficult to evaluate and compare pretreatment technologies. However, as an integrated part of an industrial system or biorefinery, it is necessary to do a thorough economic analysis to determine the best pretreatment method suitable for a particular local feedstock for an industrial process.

Many pretreatment methods have been studied and are still in development. Only a small number of pretreatment methods have been reported as being potentially cost effective. These include steam explosion, LHW, and concentrated and dilute acid pretreatments. Other pretreatments, such as ILs and organosolv pretreatment, also appear as promising options:

> Steam explosion: It is an effective pretreatment for hardwoods and agricultural residues but less effective for softwoods, in which case the use of an acid catalyst is necessary. Steam explosion is being used in some plants, such as Abengoa commercialization plant in Hugoton (US), NREL pilot plant in Golden (US), SEKAB pilot plant in Sweden, and Iogen demo plant in Ottawa (CA). Steam explosion is also applied commercially in the masonite process for the production of fiber board and other products (Galbe and Zacchi, 2012; Mosier et al., 2005).
> Dilute acid pretreatment is being used to commercially produce furfural. The technique is being developed by NREL for subsequent commercial application, and both TNO pilot plant in the Netherlands and SEKAB demo plant in Poland use this technology.
> Organosolv pretreatment: In Canada, Fibria Cellulose (former Lignol Innovations) pretreats hardwood and softwood residues with solvents to extract high-quality lignin for the production of value-added products.

In the USA, AFEX pretreated spent grains and lignocellulosics are being considered as feed for livestock at DOE bioenergy research centers (Agbor et al., 2011).

At the moment, the production of ethanol and bioproducts from lignocellulose is growing rapidly. The International Energy Agency (IEA) has a Bioenergy Task 39 entitled "Commercializing 1st and 2nd generation Liquid Biofuels from Biomass." A report by Bacovsky shows demonstration and commercialization plants worldwide for the production of advanced biofuels and bioproducts and applied pretreatment methods (Bacovsky, 2009). These advanced biofuels technologies include conversion technologies for biomass, which are still in the research and development, pilot, or demonstration phase, although many of them are already being operated in commercialization plants. Four main pathways are defined: biochemical conversion, conversion in biorefineries, thermochemical conversion, and chemical technologies. Biochemical pathway includes biomass pretreatment, hydrolysis, fermentation, and distillation to obtain ethanol. An updated summary was explained by Bacovsky (2014) in the sixth Stakeholder Plenary Meeting of EBTP.

Main data from these reports have been extracted (biochemical pathway) and summarized in the Table 5.2, along with data about biorefinery industry status in the United States as well as in other countries published in Biofuels Digest (Lane, 2015) with special attention to the applied pretreatment method. However, many of the technology are under license or patented, and little information is released.

Table 5.2: List of Main Biorefinery Facilities, Status and Pretreatment Strategies.

Company	Location	Country	Feedstock	Product	Type	Status	Start-up	Pretreatment
Abengoa Bioenergy Corporation	Babilafuent, Salamanca	Spain	Cereal straw (mostly barley and wheat)	Ethanol	Commercial	Operational	2008	Steam explosion
	Hugoton	United States	Corn stover, wheat straw, switchgrass	Ethanol + 18 MW power	Commercial	Under construction	2013	
	York	United States	Corn stover	Ethanol	Pilot	Operational	2007	
	Arance	France	Agricultural and forest residues	Ethanol	Demo	Planned	2013	
Amyris, Inc.	Campinas	Brazil	Sugarcane	Diesel-type hydrocarbons	Demo	Operational	2009	
	Emeryville	United States	Sugarcane	Diesel-type hydrocarbons	Pilot	Operational	2008	
	Piracicaba	Brazil	Sugarcane	Diesel-type hydrocarbons	Commercial	Operational	2010	
	Brotas	Brazil	Sugarcane	Diesel-type hydrocarbons	Commercial	Operational	2012	
	Pradópolis	Brazil	Sugarcane	Diesel-type hydrocarbons, farnesene	Commercial	Planned	2013	
	Decatur	United States	Corn dextrose	Diesel-type hydrocarbons	Commercial	Operational	2011	
	Leon	Spain	Sugar beet, dextrose	Diesel-type hydrocarbons	Commercial	Operational	2011	
Beta Renewables	Rivalta Scrivia	Italy	Corn stover, straw, husk, energy crops (Giant Reed), woody biomass	Ethanol, various chemicals	Pilot	Operational	2009	Enzymatic pretreatment
	Crescentino	Italy	Lignocellulosics: Straw, energy crops (giant reed)	Ethanol	Commercial	Operational	2012	
BioAmber		France	Corn	Succinic acid	Commercial	Operational	2010	
		Canada	Glucose from agriculture	Succinic acid	Commercial	Operational	2015	

Borregaard Industries AS	Sarpsborg	Norway	Sugarcane bagasse, straw, wood, energy crops, other lignocellulosics	Ethanol, lignin, various chemicals	Demo	Operational	2012	Acidic calcium bisulfite cooking liquor
	Sarpsborg	Norway	Sulfite spent liquor from spruce wood pulping	Ethanol	Commercial	Operational	1938	
Cane Technology Center (CTC)	Sao Paulo	Brazil	Sugarcane bagasse	Ethanol, sugars	Commercial	Operational	2012	
Chempolis Ltd	Oulu	Finland	Nonwood and nonfood lignocellulosic biomass such as straw, reed, empty fruit bunch, bagasse, corn stalks, as well as wood residues	Ethanol, various chemicals	Demo	Operational	2008	Selective fractionation by novel biosolvent
Clariant	Straubing (München)	Germany	Wheat straw and other agricultural residues	Ethanol	Demo	Operational	2012	
DuPont	Vonore	United States	Lignocellulosics: Corn stover, cobs and fiber, switchgrass	Ethanol	Demo	Operational	2010	
Fiberight LLC	Lawrenceville	United States	Municipal solid waste	Ethanol, power	Demo	Operational	2012	
	Blairstown	United States	Municipal solid waste	Ethanol, power	Commercial	Operational	2013	

Continued

Table 5.2: List of Main Biorefinery Facilities, Status and Pretreatment Strategies.—cont'd

Company	Location	Country	Feedstock	Product	Type	Status	Start-up	Pretreatment
GranBio, commercial-izing Beta renewables technology		Brazil	Sugarcane bagasse and straw	Ethanol	Commercial	Operational	2013	
Inbicon (DONG energy)	Kalundborg	Denmark	Wheat straw	Ethanol, c5 molasses	Demo	Operational	2009	Hydrother-mal
	Fredericia	Denmark	Straw	Ethanol, c5 molasses	Pilot	Operational	2003	
	Fredericia	Denmark	Ethanol, c5 molasses	Ethanol, c5 molasses	Pilot	Operational	2005	
INEOS Bio	Vero Beach	United States	Vegetative waste, waste wood, garden waste, citrus waste	Ethanol + 6 MW power	Commercial	Operational	2013	Gasification
Iogen Corporation	Ottawa	Canada	Wheat/oat/barley straw, corn stover, sugarcane bagasse, and other agricultural residues	Ethanol	Demo	Operational	2004	Steam explosion 2015 large-scale commercial implementa-tion of technology in Brazil
LanzaTech	Shanghai	China	Industrial flue gasses	Ethanol	Demo	Operational	2012	Gas fermentation
	Parnell	New Zealand	Industrial flue gasses	Ethanol	Pilot	Operational	2008	
	Beijing	China	Industrial off gas	Ethanol	Demo	Under construction	2013	
	Georgia	United States	Woody biomass, biomass syngas	Ethanol	Commercial	Planned	2013	
	Aurangabad	India	Any gas contain-ing carbon monoxide from municipal waste	Ethanol, electricity	Demo	Planned	2013	

Fibria Cellulose SA (former Lignol Innovations Ltd)	Burnaby	Canada	Hardwood and softwood residues	Ethanol, lignin	Pilot	Operational	2009	Organosolv
Green Biologics	Ohio	United States	Corn stover, bagasse, woody biomass, sugarcane	Butanol, acetone	Pilot	Operational	2012	
	Iowa	United States			Demo	Operational	2015	
	Minnesota	United States			Commercial	Under construction	2016	
Mascoma Corporation	Rome	United States	Wood chips, switchgrass, and other raw materials	Ethanol, lignin	Demo	Operational		Enzyme diges-tion + fermen-tation
Myriant	Lake Providence	United States	Multiple feedstock	Lactic acid, succinic acid	Commercial	Operational	2013	
NREL (National Renewable Energy Laboratory)	Golden, Colorado	United States	Dry biomass	Ethanol	Pilot	Operational	1994/ 2011	Chemical impregna-tion + steam explosion
POET-DSM Advanced	Scotland	United States	Agricultural residues	Ethanol	Pilot	Operational	2008	
Biofuels	Emmetsburg	United States	Agricultural residues	Ethanol, biogas	Commercial	Under construction	2013	
Raizen	Sao Paulo	Brazil	Sugarcane bagasse	Ethanol, sugars, energy	Commercial	Under construction	2015	
Reverdia	Cassano Spinola	Italy		Succinic acid	Commercial	Operational	2013	
Segetis	Minnesota	United States	Cellulosic biomass	Levulonic acid	Pilot	Operational	2009	
	Minnesota	United States			Demo	Operational	2012	
	Minnesota	United States			Commercial	Under construction	2016	

Continued

Table 5.2: List of Main Biorefinery Facilities, Status and Pretreatment Strategies.—cont'd

Company	Location	Country	Feedstock	Product	Type	Status	Start-up	Pretreatment
SEKAB	Goswinowice	Poland	Wheat straw and corn stover	Ethanol	Demo	Operational	2014	Dilute acid pretreatment in one step
	Örnsköldsvik	Sweden	Primary wood chips, sugarcane bagasse, wheat, corn stover, energy grass, recycled waste etc., have been tested	Ethanol	Pilot	Operational	2004	
TNO	Zeist	Netherlands	Wheat straw, grass, corn stover, bagasse, wood chips	Pretreated biomass	Pilot	Operational	2002	Dilute acid + superheated steam Concentrated acid
Weyland AS	Bergen	Norway	Lignocellulose—various feedstock, mostly spruce and pine	Ethanol, lignin	Pilot	Operational	2010	
ZeaChem	Boardman	United States	Poplar trees, wheat straw	Ethanol, various chemicals	Commercial	Planned	2014	Biological pretreatment
	Boardman, Oregon	United States	Poplar trees, wheat straw	Ethanol, diesel-type hydrocarbons, various chemicals, gasoline-type fuel, jet fuel	Demo	Operating	2011	

5.10 Conclusions

Lignocellulosic biomass, especially agricultural wastes, is the most renewable and abundant source in the world. It could be an ideally inexpensive source of potentially valuable material, which can be used to produce valuable products such as food, feed, fiber, and fuel. These wastes are composed mainly of cellulose, hemicellulose, and lignin. Both the cellulose and hemicellulose fractions are polymers of sugars and thereby a potential source for fermentation processes. Lignin can be used for production of chemicals, heat, and energy.

The intrinsic structure of the biomass is primarily responsible for its enzymatic hydrolysis. Therefore, a pretreatment step is necessary. Physical pretreatments (often referred to as mechanical) reduce crystallinity of cellulose and reduce particle size, making material handling easier and increasing surface/volume ratio. On the other hand, these processes have high energy requirements and capital costs. Chemical methods involving the use of chemicals as alkalis, peroxide, or organic solvents are mainly focused on lignin removal, which results in enhanced enzymatic degradability of cellulose. Chemical pretreatments with acids lead hydrolysis of (hemi)cellulose to sugars. Biological pretreatments employ microorganisms, which degrade lignin and hemicellulose. Nevertheless, the pretreatment is usually a combination of several pretreatments, mechanical and chemical processes being the most used.

The choice of pretreatment must take into account several factors, the most critical being the foreseen utilization of the main biomass components. The conditions employed in the selected pretreatment will affect the subsequent enzymatic hydrolysis and the biomass processing itself. Pretreatment can be the most expensive unit cost in biomass processing. Only a small number of pretreatment methods have been proved to be cost effective at industrial production level.

Although industrial processing of biomass for biofuel and several biochemical has preceded a detail understanding of pretreatment, necessary optimization of biorefinery processes requires intense research, and coordinated effort between involved parties have to be carried out. There is still much room for improvement.

References

Agbor, V.B., Cizek, N., Sparling, R., Berlin, A., Levin, D.B., 2011. Biomass pretreatment: fundamentals toward application. Biotechnology Advances 29, 675–685.

Alinia, R., Zabihi, S., Esmaelzadeh, F., Kalajahi, J.F., 2010. Pretreatment of wheat straw by supercritical CO_2 and its enzymatic hydrolysis for sugar production. Biosystems Engineering 107 (1), 61–66.

Alvarez-Vasco, C., Zhang, X., 2013. Alkaline hydrogen peroxide pretreatment of softwood: hemicellulose degradation pathways. Bioresource Technology 150, 321–327.

Ayeni, A.O., Hymore, F.K., Mudliar, S.N., Deshmukh, S.C., Satpute, D.B., Omoleye, J.A., 2013. Hydrogen peroxide and lime based oxidative pretreatment of wood waste to enhance enzymatic hydrolysis for a biorefinery: process parameters optimization using response surface methodology. Fuel 106, 187–194.

Bacovsky, D., 2014. Overview of advanced biofuels technologies. In: 6th Stakeholder Plenary Meeting of EBTP, Brussel.

Bacovsky, D., Ludwiczek, N., Ognissanto, M., Worgetter, M., 2009. Status of Advanced Biofuels Demonstration Facilities in 2012, a Report to IEA Bioenergy Task 39. From: http://demoplants.bioenergy2020.eu/.

Banerjee, S., Sen, R., Pandey, R.A., Chakrabarti, T., Satpute, D., Giri, B.S., Mudliar, S., 2009. Evaluation of wet air oxidation as a pretreatment strategy for bioethanol production from rice husk and process optimization. Biomass and Bioenergy 33 (12), 1680–1686.

Barakat, A., Monlau, F., Solhy, A., Carrere, H., 2015. Mechanical dissociation and fragmentation of lignocellulosic biomass: effect of initial moisture, biochemical and structural proprieties on energy requirement. Applied Energy 142, 240–246.

Cara, C., Ruiz, E., Oliva, J.M., Saez, F., Castro, E., 2008. Conversion of olive tree biomass into fermentable sugars by dilute acid pretreatment and enzymatic saccharification. Bioresource Technology 99 (6), 1869–1876.

Castoldi, R., Bracht, A., Rodriguez de Morais, G., Baesso, M.L., Carvalho Gomes Correa, R., Peralta, R.A., Peralta Muniz Moreira, R.F., Teixeira de MorasPolizeli, M.L., Giatti Marques de Souza, C., Peralta, R.M., 2014. Biological pretreatment of *Eucalyptus grandis* sawdust with white-rot fungi: study of degradation patterns and saccharification kinetics. Chemical Engineering Journal 258, 240–246.

Chang, K.L., Thitikorn-amorn, J., Hsieh, J.F., Ou, B.M., Chen, S.H., Ratanakhanokchai, K., Huang, P.J., Chen, S.T., 2011. Enhanced enzymatic conversion with freeze pretreatment of rice straw. Biomass and Bioenergy 35 (1), 90–95.

Chen, H., Zhao, J., Hu, T., Zhao, X., Liu, D., 2015a. A comparison of several organosolv pretreatments for improving the enzymatic hydrolysis of wheat straw: substrate digestibility, fermentability and structural features. Applied Energy 150, 224–232.

Chen, W.H., Peng, J., Bi, X.T., 2015b. A state-of-the-art review of biomass torrefaction, densification and applications. Renewable and Sustainable Energy Reviews 44, 847–866.

Das, O., Sarmah, A.K., 2015. Mechanism of waste biomass pyrolysis: effect of physical and chemical pre-treatments. Science of the Total Environment 537, 323–334.

Diaz, A.B., Moretti, M.M., Bezerra-Bussoli, C., da Costa CarreiraNunes, C., Blandino, A., da Silva, R., Gomes, E., 2015. Evaluation of microwave-assisted pretreatment of lignocellulosic biomass immersed in alkaline glycerol for fermentable sugars production. Bioresource Technology 185, 316–323.

Dinis, M.J., Bezerra, R.M., Nunes, F., Dias, A.A., Guedes, C.V., Ferreira, L.M.M., Cone, J.W., Marques, G.S.M., Barros, R.N., Rodrigues, M.A.M., 2009. Modification of wheat straw lignin by solid state fermentation with white-rot fungi. Bioresource Technology 100 (20), 4829–4835.

Duff, S.J.B., Murray, W.D., 1996. Bioconversion of forest products industry waste cellulosics to fuel ethanol: a review. Bioresource Technology 55, 1–33.

Dumas, C., Silva Ghizzi Damasceno, G., Barakat, A., Carrere, H., Steyer, J.P., Rouau, X., 2015. Effects of grinding processes on anaerobic digestion of wheat straw. Industrial Crops and Products 74, 450–456.

Eggert, C., Temp, U., Eriksson, K.E.L., 1997. Laccase is essential for lignin degradation by the white-rot fungus *Pycnoporus cinnabarinus*. FEBS Letters 407 (1), 89–92.

Esteghlalian, A., Hashimoto, A.G., Fenske, J.J., Penner, M.H., 1997. Modeling and optimization of the dilute-sulfuric-acid pretreatment of corn stover, poplar and switchgrass. Bioresource Technology 59, 129–136.

Fougere, D., Clarke, K., Zhao, Y., Li, K., 2015. Chemical–mechanical pretreatment of wood: reducing downsizing energy and increasing enzymatic digestibility. Biomass and Bioenergy 80, 17–29.

Galbe, M., Zacchi, G., 2012. Pretreatment: the key to efficient utilization of lignocellulosic materials. Biomass and Bioenergy 46, 70–78.

Hambardzumyan, A., Foulon, L., Bercu, N.B., Pernes, M., Maigret, J.E., Molinari, M., Chabbert, B., Aguie-Beghin, V., 2015. Organosolv lignin as natural grafting additive to improve the water resistance of films using cellulose nanocrystals. Chemical Engineering Journal 264, 780–788.

Harmsen, P.F.H., Huijgen, W.J.J., Bermudez Lopez, L.M., Bakker, R.R.C., 2010. Literature Review of Physical and Chemical Pretreatment Processes for Lignocellulosic Biomass. BioSynenergy Project, Energy Research Centre of the Netherlands (ECN).

Hendriks, A.T., Zeeman, G., 2009. Pretreatments to enhance the digestibility of lignocellulosic biomass. Bioresource Technology 100 (1), 10–18.

Hideno, A., Kawashima, A., Anzoua, K.G., Yamada, T., 2013. Comparison of the enzymatic digestibility of physically and chemically pretreated selected line of diploid-*Miscanthus sinensis Shiozuka* and triploid-*M.* × *giganteus*. Bioresource Technology 146, 393–399.

Hong, S.H., Lee, J.T., Lee, S., Wi, S.G., Cho, E.J., Sing, S., Lee, S.S., Chung, B.Y., 2014. Improved enzymatic hydrolysis of wheat straw by combined use of gamma ray and dilute acid for bioethanol production. Radiation Physics and Chemistry 94, 231–235.

Huang, Y., Wei, X., Zhou, S., Liu, M., Tu, Y., Li, A., Chen, P., Wang, Y., Zhang, X., Tai, H., Peng, L., Xia, T., 2015. Steam explosion distinctively enhances biomass enzymatic saccharification of cotton stalks by largely reducing cellulose polymerization degree in *G. barbadense* and *G. hirsutum*. Bioresource Technology 181, 224–230.

Imai, M., Ikari, K., Suzuki, I., 2004. High-performance hydrolysis of cellulose using mixed cellulase species and ultrasonication pretreatment. Biochemical Engineering Journal 17 (2), 79–83.

Kabir, M.M., Rajendran, K., Taherzadeh, M.J., Horvath, I.S., 2015. Experimental and economical evaluation of bioconversion of forest residues to biogas using organosolv pretreatment. Bioresource Technology 178, 201–208.

Karunanithy, C., Muthukumarappan, K., 2011a. Influence of extruder and feedstock variables on torque requirement during pretreatment of different types of biomass – a response surface analysis. Biosystems Engineering 109 (1), 37–51.

Karunanithy, C., Muthukumarappan, K., 2011b. Optimization of alkali soaking and extrusion pretreatment of prairie cord grass for maximum sugar recovery by enzymatic hydrolysis. Biochemical Engineering Journal 54 (2), 71–82.

Karunanithy, C., Muthukumarappan, K., 2011c. Optimization of switchgrass and extruder parameters for enzymatic hydrolysis using response surface methodology. Industrial Crops and Products 33 (1), 188–199.

Kim, K.H., Hong, J., 2001. Supercritical CO_2 pretreatment of lignocellulosic enhances enzymatic cellulose hydrolysis. Bioresource Technology 77, 139–144.

Kim, S., Holtzapple, M.T., 2005. Lime pretreatment and enzymatic hydrolysis of corn stover. Bioresource Technology 96 (18), 1994–2006.

Kim, S., Holtzapple, M.T., 2006. Effect of structural features on enzyme digestibility of corn stover. Bioresource Technology 97 (4), 583–591.

Kim, T.H., Kim, J.S., Sunwoo, C., Lee, Y.Y., 2003. Pretreatment of corn stover by aqueous ammonia. Bioresource Technology 90 (1), 39–47.

Kim, T.H., Lee, Y.Y., 2005. Pretreatment and fractionation of corn stover by ammonia recycle percolation process. Bioresource Technology 96 (18), 2007–2013.

Kim, T.H., Taylor, F., Hicks, K.B., 2008. Bioethanol production from barley hull using SAA (soaking in aqueous ammonia) pretreatment. Bioresource Technology 99 (13), 5694–5702.

Klinke, H.B., Ahring, B.K., Schmidt, A.S., Thomsen, A.B., 2002. Characterization of degradation products from alkaline wet oxidation of wheat straw. Bioresource Technology 82 (1), 15–26.

Kootstra, A.M.J., Beeftink, H.H., Scott, E.L., Sanders, J.P.M., 2009. Comparison of dilute mineral and organic acid pretreatment for enzymatic hydrolysis of wheat straw. Biochemical Engineering Journal 46 (2), 126–131.

Kumar, P., Barrett, D.M., Delwiche, M.J., Stroeve, P., 2009. Methods for pretreatment of lignocellulosic biomass for efficient hydrolysis and biofuel production. Industrial & Engineering Chemistry Research 48 (8), 3713–3729.

Lane, J., January 25, 2015. U.S. Biorefinery Industry Is Emerging in Biofuels Digest.com. http://www.biofuelsdigest.com/bdigest/2015/01/25/a-u-s-biorefinery-industry-is-emerging/.

Lenihan, P., Orozco, A., O'Neill, E., Ahmad, M.N.M., Rooney, D.W., Walker, G.M., 2010. Dilute acid hydrolysis of lignocellulosic biomass. Chemical Engineering Journal 156 (2), 395–403.

Li, C., Wang, L., Chen, Z., Li, Y., Wang, R., Luo, X., Cai, G., Li, Y., Yu, Q., Lu, J., 2015. Ozonolysis pretreatment of maize stover: the interactive effect of sample particle size and moisture on ozonolysis process. Bioresource Technology 183, 240–247.

Li, J., Henriksson, G., Gellerstedt, G., 2007. Lignin depolymerization/repolymerization and its critical role for delignification of aspen wood by steam explosion. Bioresource Technology 98 (16), 3061–3068.

Ligero, P., Vega, A., Villaverde, J.J., 2010. Delignification of *Miscanthus* × *Giganteus* by the Milox process. Bioresource Technology 101 (9), 3188–3193.

Liu, S., 2015. A synergetic pretreatment technology for woody biomass conversion. Applied Energy 144, 114–128.

Lucas, M., Hanson, S.K., Wagner, G.L., Kimball, D.B., Rector, K.D., 2012. Evidence for room temperature delignification of wood using hydrogen peroxide and manganese acetate as a catalyst. Bioresource Technology 119, 174–180.

Lundell, T., Hatakka, A., 1994. Participation of Mn(II) in the catalysis of laccase, manganese peroxidase and lignin peroxidase from *Phlebia radiata*. FEBS Letters 348, 291–296.

Luo, J., Fang, Z., Smith Jr., R.L., 2014. Ultrasound-enhanced conversion of biomass to biofuels. Progress in Energy and Combustion Science 41, 56–93.

Martin, C., Thomsen, M.H., Hauggaard-Nielsen, H., Thomsen, A.B., 2008. Wet oxidation pretreatment, enzymatic hydrolysis and simultaneous saccharification and fermentation of clover-ryegrass mixtures. Bioresource Technology 99 (18), 8777–8782.

Martin, C., Klinke, H.B., Thomsen, A.B., 2007. Wet oxidation as a pretreatment method for enhancing the enzymatic convertibility of sugarcane bagasse. Enzyme and Microbial Technology 40 (3), 426–432.

McKendry, P., 2002. Energy production from biomass (part 1): overview of biomass. Bioresource Technology 83, 37–46.

Mood, S.H., Golfeshan, A.H., Tabatabei, M., Jouzano, G.S., Najafi, G.H., Gholami, M., Ardjmand, M., 2013. Lignocellulosic biomass to bioethanol, a comprehensive review with a focus on pretreatment. Renewable and Sustainable Energy Reviews 27, 77–93.

Mosier, N., Wyman, C., Dale, B., Elander, R., Lee, Y.Y., Holtzapple, M., Ladisch, M., 2005. Features of promising technologies for pretreatment of lignocellulosic biomass. Bioresource Technology 96 (6), 673–686.

Motte, J.C., Sambusit, C., Dumas, C., Barakat, A., 2015. Combination of dry dark fermentation and mechanical pretreatment for lignocellulosic deconstruction: an innovative strategy for biofuels and volatile fatty acids recovery. Applied Energy 147, 67–73.

Narayanaswamy, N., Faik, A., Goetz, D.J., Gu, T., 2011. Supercritical carbon dioxide pretreatment of corn stover and switchgrass for lignocellulosic ethanol production. Bioresource Technology 102 (13), 6995–7000.

Ntougias, S., Baldrian, p, Ehaliotis, C., Nerud, F., Merhautova, V., Zervakis, G.I., 2015. Olive mill wastewater biodegradation potential of white-rot fungi–Mode of action of fungal culture extracts and effects of ligninolytic enzymes. Bioresource Technology 189, 121–130.

Oliva-Taravilla, A., Moreno, A.D., Demuez, M., Ibarra, D., Tomás-Pejo, E., Gonzalez-Fernandez, C., Ballesteros, M., 2014. Unraveling the effects of laccase treatment on enzymatic hydrolysis of steam-exploded wheat straw. Bioresource Technology 175C, 209–215.

Palmqvist, E., Hahn-Hagerdal, B., 2000a. Fermentation of lignocellulosic hydrolysates. I: inhibition and detoxification. Bioresource Technology 74, 17–24.

Palmqvist, E., Hahn-Hagerdal, B., 2000b. Fermentation of lignocellulosic hydrolysates. II: inhibitors and mechanisms of inhibition. Bioresource Technology 74, 25–33.

Ramadoss, G., Muthukumar, K., 2016. Mechanistic study on ultrasound assisted pretreatment of sugarcane bagasse using metal salt with hydrogen peroxide for bioethanol production. Ultrasonics Sonochemistry 28, 207–217.

Ravindran, R., Jaiswal, A.K., 2015. A comprehensive review on pretreatment strategy for lignocellulosic food industry waste: challenges and opportunities. Bioresource Technology. http://dx.doi.org/10.1016/j.biortech.2015.07.106.

Rogalinski, T., Ingram, T., Brunner, G., 2008. Hydrolysis of lignocellulosic biomass in water under elevated temperatures and pressures. The Journal of Supercritical Fluids 47 (1), 54–63.

Saheed, O.K., Jamal, P., Karim, M.I.A., Alamd, Z., Muyibi, S.A., 2015. Utilization of fruit peels as carbon source for white rot fungi biomass production under submerged state bioconversion. Journal of King Saud University – Science. http://dx.doi.org/10.1016/j.jksus.2015.08.002.

Schroyen, M., Vervaeren, H., Vandepitte, H., Van Hulle, S.W.H., Raes, K., 2015. Effect of enzymatic pretreatment of various lignocellulosic substrates on production of phenolic compounds and biomethane potential. Bioresource Technology 192, 696–702.

Sendich, E.N., Laser, M., Kim, S., Alizadeh, H., Laureano-Perez, L., Dale, B., Lynd, L., 2008. Recent process improvements for the ammonia fiber expansion (AFEX) process and resulting reductions in minimum ethanol selling price. Bioresource Technology 99 (17), 8429–8435.

Sharma, S., Kumar, R., Gaur, R., Agrawal, R., Gupta, R.P., Tuli, D.K., 2014. Pilot scale study on steam explosion and mass balance for higher sugar recovery from rice straw. Bioresource Technology 175C, 350–357.

Shi, J., Sharma-Shivappa, R.R., Chinn, M., Howell, N., 2009. Effect of microbial pretreatment on enzymatic hydrolysis and fermentation of cotton stalks for ethanol production. Biomass and Bioenergy 33 (1), 88–96.

Silva, A.S., Lee, S.H., Endo, T., Bon, E.P.S., 2011. Major improvement in the rate and yield of enzymatic saccharification of sugarcane bagasse via pretreatment with the ionic liquid 1-ethyl-3-methylimidazolium acetate ([Emim] [Ac]). Bioresource Technology 102 (22), 10505–10509.

Singh, A.P., Singh, T., 2014. Biotechnological applications of wood-rotting fungi: a review. Biomass and Bioenergy 62, 198–206.

Singh, P., Gupta, N., Anthwal, A., 2009. Pre-treatment of agro-industrial residues. In: Singh, P., Pandey, A. (Eds.), Biotechnology for Agro-industrial Residues Utilisation. Springer, pp. 13–33.

Smichi, N., Messaoudi, Y., Moujahed, N., Gargouri, M., 2015. Ethanol production from halophyte *Juncus maritimus* using freezing and thawing biomass pretreatment. Renewable Energy. http://dx.doi.org/10.1016/j.renene.2015.07.010.

Sun, Y., Cheng, J., 2002. Hydrolysis of lignocellulosic materials for ethanol production: a review. Bioresource Technology 83, 1–11.

Teixeira, R.S.S., Silva, A.S., Moutta, O., Ferreira-Leitao, V.S., Barros, R.R.O., Ferrara, M.A., Bon, E.P.S., 2014. Biomass pretreatment: a critical choice for biomass utilization via biotechnological routes. BMC Proceedings 8 (Suppl. 4), O34.

Vallejos, M.E., Felissia, F.E., Kruyeniski, J., 2015. Kinetic study of the extraction of hemicellulosic carbohydrates from sugarcane bagasse by hot water treatment. Industrial Crops and Products 67, 1–6.

Viell, J., Harwardt, A., Seiler, J., Marquardt, W., 2013. Is biomass fractionation by organosolv-like processes economically viable? A conceptual design study. Bioresource Technology 150, 89–97.

Villaverde, J.J., Li, J., Ligero, P., Ek, M., de Vega, A., 2012. Mild peroxyformic acid fractionation of *Miscanthus* × *giganteus* bark. Behaviour and structural characterization of lignin. Industrial Crops and Products 35 (1), 261–268.

Wang, W., Ling, H., Zhao, H., 2015. Steam explosion pretreatment of corn straw on xylose recovery and xylitol production using hydrolysate without detoxification. Process Biochemistry. http://dx.doi.org/10.1016/j.procbio.2015.06.001.

Wijaya, Y.P., Putra, R.D.D., Widyaya, V.T., Ha, J.M., Suh, D.J., Kim, C.S., 2014. Comparative study on two-step concentrated acid hydrolysis for the extraction of sugars from lignocellulosic biomass. Bioresource Technology 164, 221–231.

Yao, Y., Li, Y., Liu, X., Zhang, X., Wang, J., Yao, X., Zhang, S., 2015. Mechanistic study on the cellulose dissolution in ionic liquids by density functional theory. Chinese Journal of Chemical Engineering. http://dx.doi.org/10.1016/j.cjche.2015.07.018.

Yoon, M., Choi, J., Lee, J.W., Park, D.H., 2012. Improvement of saccharification process for bioethanol production from *Undaria* sp. by gamma irradiation. Radiation Physics and Chemistry 81 (8), 999–1002.

Zakaria, M.R., Hirata, S., Fujimoto, S., Hassan, M.A., 2015. Combined pretreatment with hot compressed water and wet disk milling opened up oil palm biomass structure resulting in enhanced enzymatic digestibility. Bioresource Technology 193, 128–134.

Zavrel, M., Bross, D., Funke, M., Buchs, J., Spiess, A.C., 2009. High-throughput screening for ionic liquids dissolving (lingo-)cellulose. Bioresource Technology 100, 2580–2587.

Zerva, C., Peschos, Z., Poulopoulos, S.G., Philippopoulos, C.J., 2003. Treatment of industrial oily wastewaters by wet oxidation. Journal of Hazardous Materials B97, 257–265.

Zhang, S., Keshwani, D.R., Xu, Y., Hanna, M.A., 2012. Alkali combined extrusion pretreatment of corn stover to enhance enzyme saccharification. Industrial Crops and Products 37 (1), 352–357.

Zhang, Y., Xu, A., Lu, B., Li, Z., Wang, J., 2015. Dissolution of cellulose in 1-allyl-3-methylimizodalium carboxylates at room temperature: a structure-property relationship study. Carbohydrate Polymers 117, 666–672.

Zhao, X., Moates, G.K., Wilson, D.R., Ghogare, R.J., Coleman, M.J., Waldron, K.W., 2015. Steam explosion pretreatment and enzymatic saccharification of duckweed (*Lemna minor*) biomass. Biomass and Bioenergy 72, 206–215.

Zhu, J.Y., Wang, G.S., Pan, X.J., Gleisner, R., 2009. Specific surface to evaluate the efficiencies of milling and pretreatment of wood for enzymatic saccharification. Chemical Engineering Science 64 (3), 474–485.

CHAPTER 6

Tobacco Seed Oil for Biofuels

P. Poltronieri
CNR-ISPA, Lecce, Italy

6.1 Introduction

Tobacco is an annual plant with the harvest in the same year of the sow, allowing farmers to plant even two or more crops in the same year where the climate conditions are favorable. It is the major nonalimentary plant in the world with a production extension higher than 4 million hectares in the whole world. Tobacco, among the agricultural plants, is valued mainly for its leaves through transformation of leaves into smoking products. Considering its nicotine content and the harm posed for human health, there have always been regulations aimed to control its production.

The evolution of the *Nicotiana* genus into different habitats, initially through natural selection and polyploidization and through human-driven selection, has brought to the development of new varieties selected on the basis of leaf properties. The tobacco plant presents a very large leaf area, a small inflorescence and a ratio of aerial part/roots that is the highest observed among agricultural plants.

Recently, alternative uses of tobacco have been proposed in addition to its use as a plant easily modified with transgenes: the production of alimentary proteins through purification from leaves and the extraction of pharmacologically useful active ingredients.

The economic lifeline of millions of people world over depends on tobacco; therefore, this crop needs to be sustained by taking advantage of its potential for alternative uses. To convert this threat into an opportunity and to sustain the crop, it is imperative to intensify the research efforts toward channeling tobacco into nonconventional and economically viable alternative uses.

Few publications suggest further use for tobacco as a source of the seed by-product for oil extraction. In particular, several authors stated that "the seed is a by-product of the leaf production" (Giannelos et al., 2002), proposing the possibility of using seeds for the production of fuels, describing methods for the extraction of oil form tobacco seeds that uses solvents, and indicating that the oil extracted from tobacco may not be used as such as biodiesel due the high iodine value in it.

Biotransformation of Agricultural Waste and By-Products. http://dx.doi.org/10.1016/B978-0-12-803622-8.00006-9

Another work (Usta, 2005) described transesterification of tobacco seed oil (TSO) to make biodiesel fuel, estimating the worldwide production of seed deriving from tobacco's cultivation for leaves and presenting a protocol for oil extraction from seed through the use of solvents.

The technological processes for oil extraction comprise mechanical (pressure) method and solvents extraction. In practice, the two systems are often combined. In general, the mechanical extraction is carried out on seeds containing more than 20% of fat material (e.g. rape and sunflower) wherein the seeds dimensions are favorable for the pressing technique. Generally, the possibility of extracting oil mechanically facilitates the direct extraction in the seed production sites, hence also at the farm's level, with small plants. For lower quantities of fat material, chemical extraction is used and can be applied also to the oil cake, leftover of the mechanical extraction, in order to recover the remaining 6–12% of oil left after the mechanical treatment. The mechanical extraction further produces the protein oil cake, whereas the chemical one produces flour. The latter, used in animal feeding, weighs in a critical way upon the production and processing of oily seeds economy. In certain cases, the production is bound to the protein flour request (e.g. soybean). The crude oil may subsequently be rectified with a series of physicochemical treatments (e.g. pH adjustment, filtration, degumming, discoloration, etc.), depending on the intended use.

The mass balance of the entire process varies from species to species. Considering sunflower seeds with an oil content of 42%, from a ton of seeds 420 kg of crude oil, 580 kg of oil cake are obtained, with a yield of 390 kg of refined oil and 30 kg of process residuals. Taking into account that the average yield of sunflower seeds is about 2.6 t/ha (+/−15%), it can be calculated that the yield/hectare of oil is equal to one ton. This relation is valid also for other species, in particular for rapeseed. Vegetable oils may be used directly as fuel oils for heat production (ovens or boilers) or mechanic energy production (engines), utilizing their gross calorific value that is about 8500 kcal/kg or, after transesterification, to transform them into biodiesel with an iodine value that has to be equal or lower than 120.

The publication by Patel et al. (1998) estimated the production of tobacco seed as a by-product of leaves in India equal to 1171 kg/ha with a content of oil of 38% by weight. In literature, other authors proposed to select varieties for TSO content and quality (Velikovic et al., 2006; Usta et al., 2011; Mohammad and Tahir, 2014). Several tobacco varieties were assessed for their TSO content (Chiririwa et al., 2014). Isolation and separation of the chemical constituents of the TSO by thin layer chromatography is performed and the TSO characterized by Fourier transform infrared spectroscopy (FTIR); chemical composition of TSO is analyzed using gas chromatography. The tobacco seeds varieties studied by Chiririwa yielded glyceride oil content around 34.17–36.55%. Fatty acid content showed a main presence of palmitic (8.83–9.26%), oleic (9.97–11.69%), and linoleic acid (64.38–68.49%), determining physical properties such as density (925.63 kg/m^3) and viscosity of 94.30. The oil had iodine value of 133.71, too high to be used as fuel.

6.1.1 Feedstock Biomasses

Sustainable biomass feedstock is the key to sustainable biofuels. The impact of bioenergy on social and environmental issues may be positive or negative, depending on local conditions and the design and implementation of specific projects. That means that a specific feedstock may be sustainable, or unsustainable, depending on how and where the feedstock is produced.

Plant production must be optimized with respect to energy inputs and highly efficient conversion of biomass. Conversion processes such as fermentation must be optimized for optimal conversion of feedstock to useful products, which must then be optimized for different end uses as mixtures for motor fuel, additives, and chemical feedstocks. In turn, internal combustion engines need to be redesigned to run on different formulations with unprecedented fuel efficiency. All of these activities must be held together within an overriding framework of sustainability and economic competitiveness. Of particular interest are the systems-based analyses of sustainability at global and at local levels. The present studies address centrally important issues such as the availability of land, access to production inputs such as water and sunlight, and global trade. There is sufficient land available for cultivating bioenergy crops, and the potential of lignocellulose production and conversion can meet a substantial proportion of transport fuels. The present efforts focus on two potential bioenergy feedstocks: the so-called "first-generation" feedstocks oils, starch, and sugars, and the "second-generation" feedstock lignocellulose. It is clear that first generation feedstocks need to be developed to meet current objectives in Europe. AB Agri, in Leeds, has focused on efforts on coproduction as an essential component of ethanol production from wheat grains. Starch is fermented to ethanol, fibrous material is processed to ruminant feed, and the protein rich fraction is used for nonruminant feed.

Sugar beet, sorghum, corn, and sugarcane are first-generation sugar or starch bioethanol feedstocks. Second-generation cellulosic ethanol is primarily produced from lignocellulosic biomass such as perennial grasses like switchgrass, miscanthus, giant cane, eucalyptus, willow, and poplar. Progress in developing poplar, willow, and miscanthus as sources of lignocellulose has been described. In the closely related willow, a crop more specifically suitable for growth in Northern Europe, breeding populations and selected lines have been established that exhibit stunning yield increases.

Oil producing plants such as *Jatropha curcas*, sunflower, and palms offer similar opportunities: they are grown sustainably, and seeds and fruits can provide both animal fodder and an oil that may readily be made into diesel fuel. The various feedstock sources, from lignocellulosic feedstock biomasses to multipurpose starch and sugar crops to algae, have been detailed in Fig. 6.1; the figure shows the processes available to produce sugars, ethanol, and final products to be obtained, such as renewable transport fuels as gasoline components and fuels for diesel engines.

Noteworthy, feedstock biomass can potentially be increased by changing the environmental growth conditions such as carbon allocation in the form of carbon dioxide, oxygen, water

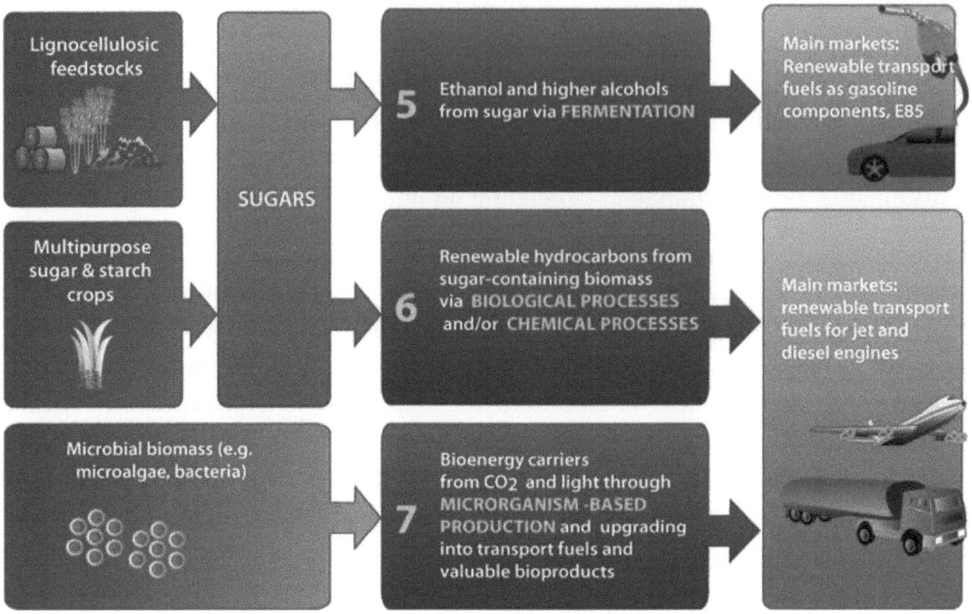

Figure 6.1
Feedstock biomass options for production of ethanol, fuels, and valuable chemicals.

supply, and soil nutrient content. Selection criteria for agronomic traits have been based on high biomass yield and type of lignocellulosic composition. Other traits include (1) photosynthetic efficiency; (2) long canopy duration with a short perennial life span; and (3) translocation of photosynthesized carbohydrate into structural lignocellulose. Such traits are common in perennial species such as miscanthus and switchgrass. The recalcitrance of perennial crops to *Agrobacterium*-mediated transformation is the main bottleneck so that scientists need to improve the transformation potential through further research.

6.1.2 Tobacco Mutagenesis and Genetic Studies

An approach to increase the feedstock biomass is applying functional genomics and mutation studies on model plant systems such as tobacco. In these studies, genes involved in phytohormone metabolism, cell-cycle machinery, or cellulose and hemicellulose biosynthesis pathways have been identified, targeted, and altered for enhanced plant growth and by-products. There have been numerous attempts in recent years to manipulate the cellulose and hemicellulose biosynthetic pathways. For instance, a recent study analyzed transgenic tobacco lines downregulated, using antisense sequences, for the lignin biosynthetic enzymes cinnamate-4-hydroxylase (se4h), cinnamoyl-CoA-reductase (ccr), and lignin-specific peroxidase (prx), and compared with UDP-glucuronic acid decarboxylase (uxs) involved in nucleotide sugar pathway of xylan biosynthesis. It was shown that lines downregulated in prx showed threefold improvement in enzymatic saccharification, and lines down regulated in uxs improved saccharification by 50% when compared to wild type (Cook et al., 2012). Secondary walls presented great differences in xylem of transgenic lines of tobacco altered for lignin or xylan content, with potential industrial applicability to improve biomass utilization (Cook et al., 2012).

In one mutational study, lignin biosynthetic enzyme, O-methyltransferase, has been suppressed in order to increase overall biomass accumulation in transgenic tobacco, while no changes in lignin deposition were reported (Blaschke et al., 2004). In other studies, ccr was downregulated in transgenic tobacco, resulting in an overall drop in lignin composition and simultaneous increase in xylan and cellulose composition in tobacco secondary cell wall (Chabannes et al., 2001), while enhanced enzyme saccharification was obtained by downregulation of a lignification-specific peroxidase (Kavousi et al., 2010). In another study, lignin accumulation was delayed in transgenic tobacco with altered expression of phenylpropanoid pathway associated enzyme class II se4h (Blee et al., 2001).

Plant height (PH) is one of the most important agronomic traits in tobacco. The genetic basis of PH has been linked to chromosomes 6 and 12 (Cheng et al., 2015). Inheritance and transgeneration transmission studies were done using the F2 and F3 populations derived from NC82 (P1) and Kang88 (P2) crosses to identify quantitative trait loci (QTL) affecting PH in tobacco. Two main quantitative trait loci (M-QTL), designated qPH-6 and qPH-12, were mapped using linkage mapping (LM). The QTL qPH-12 was identified as stably expressed in different tobacco generations as well as in various environments. The results showed the advantages of using QTL analysis in tobacco through a combination of LM and association mapping, for the understanding of the inheritance of PH in tobacco.

Plant hormones are responsible for regulating growth and development throughout their life span. Phytohormones interact with each other and other signaling pathways affecting plant growth. Gibberellins (GAs) and brassinosteroids have been reported to play a vital role in growth associated with stem elongation and thickness. In one study, transgenic tobacco was engineered to express *GA 20-oxidase* gene from *Arabidopsis*, showed enhanced biomass accumulation owing to increased PH and higher lignification of vessels, with GA mediating deposition of lignin (Biemelt et al., 2004). Silencing *GA 2-oxidase*, a GA-deactivating enzyme, abolished the catabolism of GA, allowing greater GA accumulation (Dayan et al., 2010). Silenced tobacco plants showed an improvement in growth characteristics, compared with the wild type and *GA 20-oxidase* overexpressing plants. Moreover, the number of xylem fiber cells in the silenced lines exceeded that of *GA 20-oxidase* overexpressing plants, potentially making *GA 2-oxidase* silencing more profitable for the wood and fiber industries. In another study, elevated growth rate in tobacco has been reported by overexpression of *D-type cyclin* (*CycD2*) gene from *Arabidopsis*, responsible for cell division and proliferation (Cockcroft et al., 2000). The transgenic tobacco was described having taller stem and elevated overall growth rate with early flowering timing.

During a transformation experiment by mutagenesis using T-DNA insertion, Kuhar obtained a transgenic tobacco plant that grew to the size of a small tree (Kuhar, 2014). The high biomass producing transgenic tobacco (*Nicotiana tabacum* cv. Xanthi) line, termed giant recombinant (GR), has the potential for a new class of energy crops by converting normal plants to high biomass producing plants. To characterize the GR line, Kuhar analyzed lignocellulosic composition (cellulose, hemicellulose, and lignin) relative to the nontransgenic control line and the growth rate of the plants. The GR line accounted for 240% more biomass than the untransformed line within 135 days of its germination. There were significant differences in chemical composition within GR line and relative to control line. The GR characteristics are likely due to a disruption or activation of an unknown gene, which could lead to develop feedstock plants for biobased energy.

6.1.3 Selection of Tobacco Varieties for Biotic Stress, Pathogen, and Nematode Resistance

Tobacco varieties show a high variability in resistance to plant pests. The objective of breeding programs has been the development of tobacco varieties that have significantly higher levels of pest resistance, combined with improved yield potential. Agronomic traits, including yield, plant type, leaf quality, and disease resistances are determined for each entry. To be acceptable for release as a new variety, a test entry must compare favorably to two standard or "check" varieties to evaluate and approve new breeding lines before made available to burley producers.

Black shank, *Phytophthora nicotianae* (Pn), is prevalent throughout the burley-producing region of the United States and is responsible for millions of dollars in lost profits for tobacco farmers. Development of varieties with increased resistance to black shank has been shown feasible (Miller et al., 2000). KT 200, a new black shank-resistant burley hybrid, was released by the Kentucky–Tennessee Tobacco Improvement Initiative in April 2000. Although current management is effective, unsolved problems are pathogen persistence and also costs for treatments (Holdcroft, 2013).

Hybridization has been used to develop flue-cured tobacco cultivars heterozygous for a single, dominant gene (*Php*) originally transferred from *Nicotiana plumbaginifolia* (Johnson et al., 2009). The *Php* gene provides complete resistance to race 0 of the black shank.

The successful development of blue mold resistant tobacco varieties required offshore breeding nurseries established in Guatemala and Mexico. Since blue mold generally occurs throughout Central America each year, establishment of nurseries in these countries increased the probability of having significant blue mold pressure to screen for resistance in breeding materials.

The incidence of *Fusarium* wilt in tobacco crops is also increasing each year. The majority of varieties planted today have not been thoroughly evaluated for resistance to this disease; many appear to have little or no resistance to *Fusarium* wilt. Germplasm and breeding materials need to been screened for resistance to the disease. Resistant germplasm has been used to incorporate *Fusarium* wilt resistance into existing varieties.

Concerning tobacco cyst nematode (TCN), *Globodera tabacum solanacearum*, four *Nicotiana* species, *Nicotiana glutinosa* L., *Nicotiana paniculata* L., *Nicotiana plumbaginifolia* Viv., and *Nicotiana longiflora* Cav., are known to possess resistance to TCN. Hybrids developed from *N. plumbaginifolia* are highly resistant to TCN. In the past, few TCN-resistant cultivars were available, but the yield and leaf quality from these cultivars were lower than in TCN-susceptible cultivars. The *Php* gene also significantly reduces TCN population densities (Johnson et al., 2009). These hybrid cultivars combine resistance to black shank and TCN with high yield and leaf quality characteristics (Crowder et al., 2003).

The application of the systemic acquired resistance (SAR)-inducing compound acibenzolar-S-methyl (ASM), as well as the addition of a mixture of plant growth promoting (PGPR) *Bacillus* strains, are able to suppress TCN reproduction by an average of 60% in oriental (cv. Xanthi NN) and flue-cured tobacco (cv. K326) (Parkunan et al., 2009). Tobacco plants treated with SAR-inducing agents and plant growth–promoting bacteria were challenged with TCN under greenhouse conditions. Cultivars possessing the *Php* gene (*Php+*) were compared with *Php-* cultivars to assess the effects of resistance mediated via *Php* gene versus induced resistance to TCN. Administration of bacterial strains consistently reduced nematode reproductive ratio on both *Php+* and *Php*-cultivars, but the effect of ASM across *Php*-cultivars was limited. The results showed that PGPR bacteria consistently reduced TCN reproduction in all flue-cured tobacco cultivars tested, while SAR affects TCN only in *Php+* cultivars (Parkunan et al., 2009).

The selection of varieties highly resistant to pathogens is based on a protocol with specific agronomical in field procedures. After nursery management techniques, plantlets are put in squared areas with replicates, numbered and challenged with pests and nematodes (see Fig. 6.6). For each treatment, a protocol for fungicide or parasite treatment is applied to cultivated areas to avoid inhibitory effects from other pests. A treatment for weed control facilitates the proper growth of the plantlets. At the adult stage, the varieties that show better growth in the presence of pests are selected. Then, hybrids are crossed to allow introgression of resistances to viruses and fungi.

6.1.4 Bioeconomy of Biofuels

The ever-increasing population of both the developing and developed nations of the world and the consequent increase in their diesel consumption and the nonrenewability of diesel source (petroleum), as well as the adverse environmental effects of diesel burning, are factors that compel authorities to find alternatives to petroleum diesel.

Global agricultural is expected to slow over the next 10 years, and cereal production is projected to be 15% higher by 2023 compared to 2013 period, outpaced by growth in livestock and biofuels. Cereals remain still at the core of human nutrition, but there is a shift to diets higher in fats, sugar, and protein. Biofuel production and consumption is expected to grow by more than 50%, led by sugar-based ethanol and biodiesel.

Renewable energy sources include wind power, solar power (thermal, photovoltaic, and concentrated), hydroelectric power, tidal power, geothermal energy, biomass, and the renewable part of waste.

The use of renewable energy has many potential benefits, including a reduction in greenhouse gas (GHG) emissions, the diversification of energy supplies, and a reduced dependency on fossil fuel markets (in particular, oil and gas). Biodiesel life cycle analysis (LCA) shows it affects a 78% reduction in CO_2 (GHG) emissions relative to petro diesel. The unpredictable price fluctuations of crude oil in the international market have also been a major source of concern in total dependence on diesel fuel. Reports by Rudolf Diesel in the early1900s showed that vegetable oils could be used as diesel fuel, and other papers reported on the use of vegetable oils in diesel engines in the 1940s and after the end of world war II (Tatti and Sirtori, 1937; Amrute, 1947).

The growth of renewable energy sources may also have the potential to stimulate employment through the creation of jobs in "green" technologies. The primary production of renewable energy within the EU-28 in 2013 was 192 million tons of oil equivalent, a 24.3% share of total primary energy production from all sources. Among renewable energies, the most important source in the EU-28 was biomass and renewable waste, accounting for two-thirds (64.2%) of primary renewables production in 2013. The share of renewable energy in gross final energy consumption is identified as a key indicator for measuring progress under the Europe 2020 strategy for smart, sustainable, and inclusive growth.

At the European level, there is an urge to increase the production of alternative biofuels. The current European target is for renewable fuels to make up 10% of the energy used in transport by 2020. The biggest cause of apprehension, considering the principle of greenness, is that biofuels are made from food crops or from plants grown on land that might otherwise produce such crops, hurting food supplies. According to this aspect, a committee of the European Parliament posed a limit to the use of "first-generation" biofuels. The new proposed targets in EU Commission define that only seven-tenths of renewable energy will originate from first-generation fuels. The difference of three-tenths will be made up by second-generation fuels, advanced fuels based on waste products, and other feedstocks that do not affect food production. That translates in European demand for advanced biofuels to reach 14 billion liters by 2020. Only two types of advanced fuels are capable of large-scale production today in Europe. The first one is based on turning waste cooking oil and other fats into diesel. Europe already has 2 billion liters of capacity to process these by-products. The second type of plants is producing ethanol from cellulose by enzymatic hydrolysis.

A number of EU projects addressed the feedstock issue, e.g. the ITAKA project, improving the readiness of existing technology and infrastructure, focusing on camelina and cooking oil.

Oil from plants is extracted, hydrocracked, and products are isomerized to obtain paraffins and isoparaffins.

Several EU projects have been funded to provide solutions to the challenge of feedstock processing. In one such project, the EMPYRO consortium was led by BTG Bio-oil for combined heat and power (CHP) and acetic acid at AkzoNobel Global, the Netherlands, a multinational manufacturing corporation, active in healthcare products, coatings, and chemicals. The EMPYRO pyrolysis oil plant of Biomass Technology Group in the Netherlands was the first EU plant to sign a long-term supply contract of bio-oil to replace fuel oil. In EMPYRO, full conversion configurations have been estimated for a range of economies of scale, 1675 and 1350 MW LHV of bio-oil. The economic competitiveness was found to increase with increasing scale. A cost of production of FT liquids of 78.7 euro/MWh was obtained based on 80.12 euro/MWh of electricity, 75 euro/t of bio-oil, and 116.3 million euro/y of annualized capital cost (Ng and Sadhukhan, 2011).

Another project, the LED consortium, was led by Abengoa industries to produce from straw and maize bioethanol and renewable hydrocarbons. The CHEMREC Bio-DME project has been the first project to demonstrate the conversion of black liquor to bio-dimethyl ether through the production of synthesis gas, which is converted to second-generation biofuels. Black liquor is a waste product resulting from the conversion of pulpwood into paper pulp. Dimethyl ether is an advanced biofuel produced by catalytic dehydration of methanol or from syngas. Above −25°C or below 5 bar, DME is a gas. Hence, its use as a transport fuel is similar to that of liquid propane gas.

InteSusAl is a still ongoing EU project including three European algae biofuel projects with a common LCA Approach. Three large-scale algae production facilities are under development. These will be the largest facilities built in Europe with a productivity of 90 ton/ha of dry matter algal biomass per year for each facility (30 ha in total).

Infinite Fuels GmbH works on the development and market introduction of a unique technology for transformation of renewable electricity, biomass, and waste into sustainable hydrocarbons serving as basic chemicals and fuels. The incubator for startup KIC InnoEnergy Germany has signed five new German startups under its Business Creation Accelerator program (KIC InnoEnergy's Highway). The companies were selected by a committee of experts for their ability to innovate as well as potential to drive Europe's move toward sustainable energy with high-performance and efficient products. See http://www.kic-innoenergy.com/22638/.

Swedish Biofuels AB is devoted to the synthesis of fully synthetic kerosene (FSK). Swedish Biofuels produces alternative fuels for aviation; these must be drop-in fuels, meaning that they do not require any modification to the aircraft, the engine, the fuel system, the distribution network, or logistics. Lignocellulosic residues are hydrolyzed into sugars and charcoal, and processed into syngas, that together with alcohols, are hydrocracked and isomerized to obtain paraffins and isoparaffins, with aromatics and cycloparaffins.

There are several types of biofuels:

SKA is synthetic paraffinic kerosene with aromatics, used as a blendstock with conventional jet fuel.
FSK falls within the conventional jet fuel specification to be used as neat jet fuel.
SIP, a synthesized isoparaffin, C15, is obtained from farnesan, a sugar molecule (component for blending with conventional jet fuel)
Hydroprocessed oils and fats (hydrotreated renewable jet, HRJ/ hydrotreated esters and fatty acids, HEFA) are converted into kerosene-like fuel (SPK), a synthetic paraffinic kerosene used as a blendstock with conventional jet fuel.
ATJ is a fuel produced from C2–C5 alcohols, as single alcohol or a multicomponent mixture, into kerosene-like fuel and kerosene (SPK, SKA, FSK).

Overview of the process to produce SPK: Coal and biogas are gasified into syngas; then *n*-paraffins are synthesized through the Fischer–Tropsch (FT) process and hydrocracked and isomerized to obtain paraffins and isoparaffins as SPK. In the FT process, the purified syngas is processed through a fixed-bed tubular reactor where it reacts with a proprietary catalyst to form three intermediate FT products, a Heavy Fraction FT Liquids (HFTL) product, a Medium Fraction FT Liquids (MFTL) product, and a Light Fraction FT Liquids (LFTL) product, commonly called Naphtha. The Naphtha is recycled to the partial oxidation unit with remaining tail gas to be reformed to hydrogen and carbon monoxide.

"Production of fully synthetic paraffinic jet fuel from wood and other biomass," BFSJ 612763, is a project in the EU 7th Framework Program (2007–2013), involving Swedish biofuels.

Full-scale commercial plant size was estimated to be 200,000 ton/y of motor fuel, of which jet fuel would make up 100,000 ton/y. The business plan is to deploy three commercial units in the 10 years following the project, subject to market acceptance, safety, and financial risks. With a good political and economic environment, up to 600,000 tons/y of advanced biofuels can be produced by 2030 using Swedish Biofuels ATJ technology. Production is economic at various production volumes, e.g. processing 2,500,000 m^3/y of humid, low-grade wood residues. A wide range of biomass suitable for process is available. Biological fuel capacity is 30 tons/y (3300 US gallons/y), jet is 14.4 tons/y, gasoline is 10.5 tons/y, and diesel is 5.1 tons/y.

A third type of biofuel is under development using municipal solid waste (MSW) as source of lignocellulose.

Key market drivers for waste as feedstock are of various nature, enumerated here: increased scarcity of urban landfill space and societal desire for waste diversion; turning carbon waste into a useful building block for the chemical and petrochemical industry; low-cost, nonland using, unconventional feedstocks for biofuels and renewable chemicals; renewable fuels mandates around the world; consumer pull for renewable and biobased products; and focus on carbon footprint and GHG emissions reduction.

The potential for transforming garbage (estimated values are positioned around 254 million metric tons/year in Europe) into chemicals and fuels (375 L of cellulosic ethanol per metric ton) is delineated in Fig. 6.2.

There are ongoing strategic alliances with EU and partners around the world by Enerkem biorefineries (full-scale commercial biorefinery in Edmonton, and two facilities in Quebec, Canada), a producer of biofuels and renewable chemicals from MSW, agriculture biomass, plastics, petcoke, biosolids from pulp and paper industry, forest biomass, and wood pellet.

Enerkem has a proprietary clean technology developed in-house. The thermochemical process converts MSW feedstock into low-carbon renewable transportation fuels including jet fuel and diesel.

Fulcrum, US, http://fulcrum-bioenergy.com/, is a pioneer in the development of a reliable and efficient process for transforming everyday household garbage into low-carbon transportation fuels including jet fuel and diesel. The low-cost process reduces the dependence on imported oil, creates new clean energy jobs, and significantly reduces GHG emissions compared to traditional petroleum production. Fulcrum has established industrial partnerships with US Renewables Group and Rustic Canyon Partners, two leading venture capital firms in the clean energy space. In addition, Waste Management, Inc., and Cathay Pacific Airways have become equity partners in the Company.

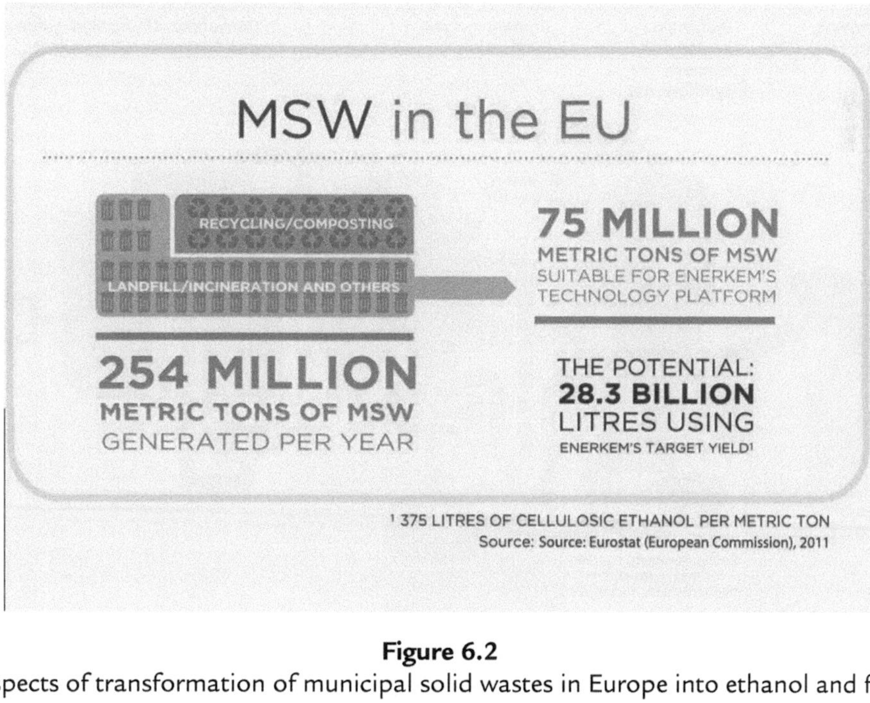

Figure 6.2
Prospects of transformation of municipal solid wastes in Europe into ethanol and fuels.

ThermoChem Recovery International has licensed to Fulcrum their highly efficient and economic gasification system for the conversion of the carbon-rich residues into synthetic gas (syngas) (Fig. 6.3). During the gasification process, the prepared MSW feedstock rapidly heats up upon entry into the steam-reforming gasifier and almost immediately converts to syngas. The syngas is further cooled in a packed gas cooler scrubber. The cleaned syngas is then processed through an amine system to capture and remove sulfur and carbon dioxide. The syngas then enters the secondary gas cleanup section that contains compression to increase syngas to the pressure required by the FT process. The syngas is catalytically converted, thus synthesizing the renewable fuel constituents.

Utilizing this transformation process, municipalities will be able to convert the garbage into 30 million gallons per year of clean renewable fuel. A number of facilities are under construction across North America with the annual capacity to produce hundreds of millions of gallons of renewable transportation fuel while eliminating trash landfilled annually across North America.

In 2015, Fulcrum announced that it had awarded an engineering, procurement, and construction contract to Abengoa for the construction of the Sierra BioFuels waste-to-transportation fuel plant. Abengoa will construct Sierra under the fixed-price contract that also includes cost, schedule, and plant performance guarantees.

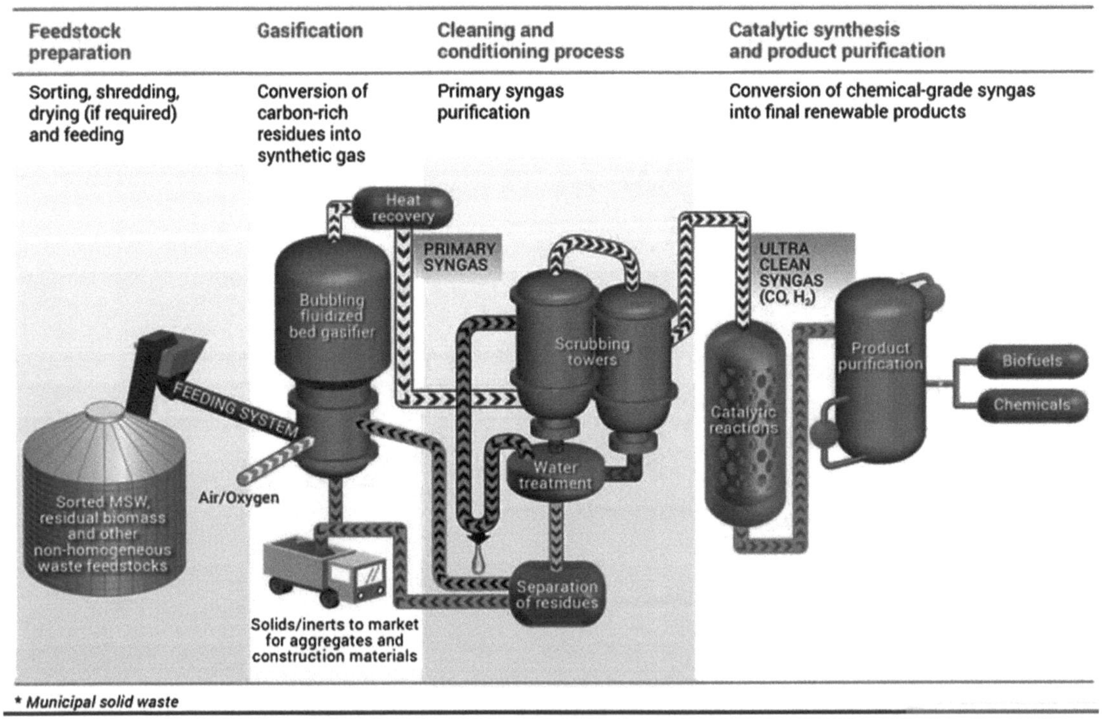

Figure 6.3
Scheme of production of biofuels and chemicals starting from municipal solid wastes.

In the US, United Airlines (UA) has announced the first stable use in the tract from Los Angeles–San Francisco by a new jet fueled with biokerosene. The required amount of fuel at this stage is 180 million liters each year. In the agreement with UA, Fulcrum will transform MSW, cooked oils, and fats derived from animal wastes to produce a biofuel that will be blended with traditional fuels.

6.1.5 Biokerosene Deployment and Sustainability Issues

Noteworthy, biofuel policies increase demand for agricultural land use. The macroeconomy and population growth increase the demand for food and consequently imply an agricultural land use expansion mitigated by yields growth and conversion of managed forests or other natural, nonagricultural ecosystems into agriculture.

Aviation Climate Change Commitments are exemplified in Fig. 6.4, showing the diagram of CO_2 emissions between 2005 and 2050. The graphic shows the improvement on evolution of GHG emission applying a carbon neutral growth action. There are several targets to be accomplished, such as an improvement of 1.5% fuel efficiency per year from 2009 to 2020, a carbon-neutral growth from 2020, and a reduction of net emissions by 50% by 2050 as compared to 2005 levels.

Air transport moves over 2.4 billion passengers annually, dumping 677 million tons of carbon dioxide into the atmosphere. While these emissions are small compared with other industry sectors, these industries have viable alternative energy sources. The power generation industry can look to wind, hydro, nuclear, and solar technologies to make electricity without producing much CO_2. Cars and buses can run on hybrid, flexible fuel engines or electricity. The primary objective of using biofuel is to reduce emissions. Carbon dioxide absorbed by plants during its growth is roughly equivalent to the amount of carbon produced when the fuel is burned. This would allow biofuel to be carbon neutral over its life cycle. See http://ec.europa.eu/energy/en/topics/renewable-energy.

Since 2001, a rapid growth of biofuel production has been observed, driven by the Renewable Energy Directive (RED) and high crude oil prices as well as by growing interest in reducing GHG emissions. Considering the CO_2 emission at 2005 and the projections estimated up to 2050, there are very few options to reduce the level of GHG emissions, from "no action," determining the doubling of GHG emissions, to a "carbon neutral" growth, to a net emission trajectory with a reduction of 50% by 2050, adopting economic measures such as substitution of aviation fuel GHG emission by using increasing amounts of renewable fuels (Fig. 6.4).

However, deforestation to provide more land for agriculture and to provide lignocellulose poses a great treat to climate change, since it contributes to 25% up to 50% of GHG increase. Blocking deforestation by all means is important, since deforestation has an impact on CO_2 emission greater than all the GHG produced by fuel use around the world.

The EU position on agricultural land use for biofuels. Since the 2009 version of RED, the EU posed some limits on use of agricultural lands for biofuels, in agreement with the alarm

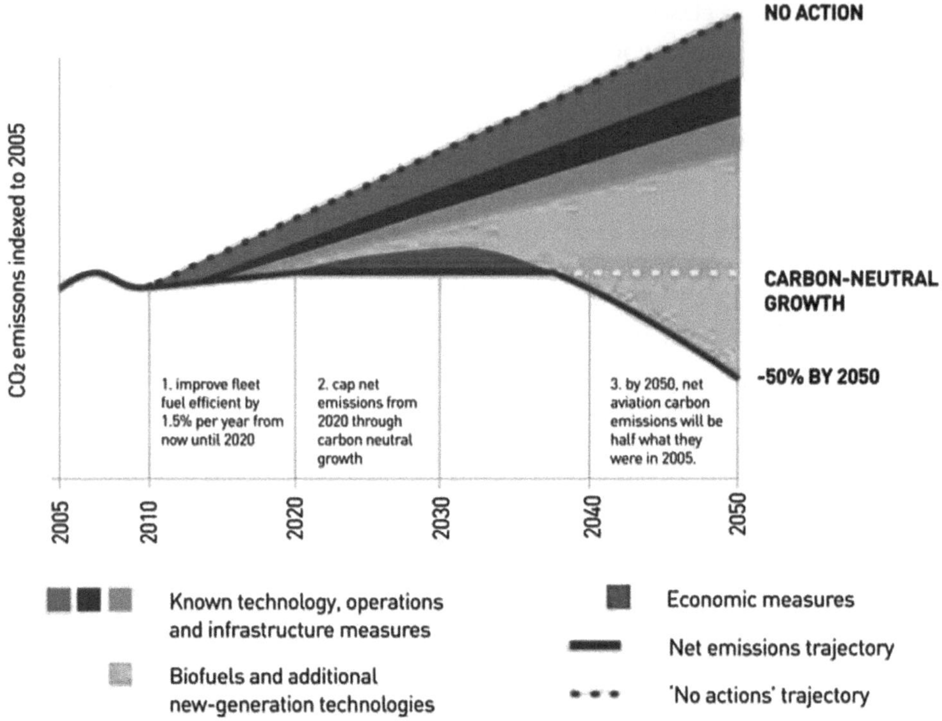

Figure 6.4
Diagram of CO_2 emissions between 2005 and 2050. The graphic shows the evolution of GHG emission under two contrasting economic measures and policies: no action, and carbon neutral growth: by 2050 a net aviation carbon emission trajectory will reach a −50% reduction.

launched by reducing emissions from deforestation and forest degradation (REDD, Bali, 2007), a set of recommendations designed to reduce the emissions of GHGs from deforestation and forest degradation (Myers, 2007; van Meijl et al., 2006).

Focus was on controlling effects of direct land use change, like preventing conversion of rain forests, peat land, or biodiverse areas; growth of feedstock on agricultural land was not considered a problem. Since 2007, major currents of thought opposing the RED initiative are focusing on the debate on indirect land use change (ILUC) and the maizification problem, the high demand of maize cropping for alternative energies and feed use instead of use of agriculture land for food production (Nowicki et al., 2007). Revision of RED has been carried on to set up new rules; contribution from cereals, sugars, and oil crops grown as main crops primarily for energy purposes on agricultural land has been limited to 7% of transport energy use. Cellulosic material (e.g. short rotation coppice, switchgrass) is now supported. Typically, the feedstock plants grow on marginal or degraded lands, so they are not expected to be used on agricultural lands, avoiding the ILUC issue. See http://ec.europa.eu/energy/renewables/studies/renewables_en.htm.

The expectation is that new RED will support use of lignocellulosic material as feedstock; it will sustain biofuel produced on marginal land, devoid of conflict with sustainability definitions. The US Energy Independence and Security Act requires that feedstock crops used for biofuel be harvested from agricultural land cleared or cultivated prior to December 2007. The intention was to protect nonagricultural areas from direct land use change. Currently there are no binding sustainability requirements in place in Canada and other countries not adhering to RED.

Resource Technology Strength. Bioenergy and advanced biofuel investments are constantly progressing. European production technology has shown to be a critical component of new plants outside Europe. EU technology providers are present in these investments. Therefore, the EU technology base continues to be very strong. However, European production capacity planning and investments remain weak, and regulatory uncertainties remain.

The European Biofuels Flight Path Initiative (EBFPI) and the European Biofuels Technology Platform (EBTP). The EU Commission has launched the EBFPI with the objective to reach the target of using 2 MTons of aviation biofuels in 2020, corresponding to about 4% of EU fuel consumption. By 2015, EBFPI will set up financial mechanisms, secure sustainable feedstock production to feed three refineries, construct three new refineries, and launch biofuel production. By 2018, EBFPI will start regular commercial flights using biojet fuel blends, construct four additional refineries, and construct two additional refineries producing algal and microbial oil-based aviation biofuels. By 2020, a full deployment of at least 2 million tons of biofuels per annum for EU aviation is envisaged.

In the US, Boeing has partnered up with other stakeholders to promote "Farm to Fly" biofuel programs that include the Midwest Aviation Sustainable Biofuels Initiative (MASBI), along with United Airlines, UOP (a Honeywell company), Airlines for America (A4A) Inc., the Chicago Department of Aviation, the Federal Aviation Administration (FAA), and the Clean Energy Trust. See http://www.triplepundit.com/2012/05/united-boeing-uop-join-big-push-biofuels/.

The US National Bioeconomy Blueprint is designed to create jobs and stimulate investment by using federal resources to speed the transition from fossil fuel dependency into a more sustainable, healthful, and diversified mix of fuels, chemicals, and other products.

Boeing has been looking for partners at various levels, including fuel from plants grown in the desert using saltwater, and it is optimistic that a range of biokerosene promising to be both cleaner than standard fuels and with a greater energy density—essentially offering more power for less weight, a crucial property for aviation—soon will be certified for aviation use.

Currently, these alternative fuels for transport are marketed by Neste Oil and by ENI. In 2014, Neste Oil produced approximately 1.3 million tons (1.6 billion liters) of renewable NEXBTL diesel from waste and residues. There is a big potential for aviation since three refineries in function, one in Italy, one in Rotterdam, and one near Helsinki, currently produce around 4 billion liters of biokerosene. This amount for aviation corresponds to 2% of fuel use globally.

Boeing Airlines have conducted more than 1500 passenger flights using biofuel since the fuel was approved in 2011. Alternative aviation biofuel reduces carbon emissions by 50–80% compared to petroleum jet fuel through its life cycle.

South African Airways (SAA) is partnering with Boeing aerospace company and Amsterdam-based SkyNRG to make sustainable aviation biofuel from a new type of tobacco devoid of nicotine, the Solaris variety "energy tobacco," in a pioneering project that could make aviation more environmentally friendly while advancing rural development in southern Africa. SAA says that the cost of the tobacco-based product matches that of jet fuel refined from fossil sources. SAA airlines expects to use 20 million liters of jet biofuel by the end of 2017, blended 50–50 with conventional fuel, and 500 million liters by the end of 2022. SkyNRG's mission is to create structural supply and demand for sustainable and affordable jet fuels. It's executing this mission in the short term via cofunded green routes (like AMS–JFK) and in the long term via developing regional supply chains that represent a real affordable alternative for fossil fuels called BioPorts. The company is working with fuel technology partners to create the best fit for a region in the world. SkyNRG is the world market leader for sustainable jet fuel, having supplied to more than 20 carriers worldwide. Since 2011, the company is expanding into the marine and heavy trucking segment as well. These segments, like aviation, have no other alternative than to use sustainable fuels to significantly reduce their carbon footprint.

6.1.5.1 The Civil Aviation Approach to Certification

Aircrafts can only use fuel they are certified to use. Approval of a new fuel with new properties would require individual certification of all and every aircraft on this fuel, making this prohibitively expensive. Rather than certify all aircrafts as compatible with a new fuel, the new fuel would be approved as being compatible with existing Jet A and Jet A-1 fuel. Appropriate blends with approved fuel can then be used by every civil aircraft approved to use Jet A and A-1 fuels. By agreement with European authorities, approval process is coordinated by US standards body ASTM, referring to the relevant specification, such as ASTM D7566. This defines required properties of blends of conventional (ASTM D1655) kerosene and synthetic material. The specification annexes define required properties of the neat synthetic fuels (currently three): FT, with maximum blend ratio 50%, HEFA, with maximum blend ratio 50%; and farnesane (SIP), with maximum blend ratio 10%. Fuel meeting ASTM D7566 by definition becomes ASTM D1655 kerosene and can use the same infrastructure as conventional fuel. In North America, no paper trail for synthetic component is required. Physical traceability will soon be lost in North American pipeline systems. In Europe, documentation of synthetic percentage is still required, but generic.

Implications for biokerosene. In case of aviation kerosene, a registered substance is defined as "being produced from crude oil sources." Kerosene not produced from crude oil still needs a separate registration. This makes the process expensive and time-consuming, costing several 100,000 euros. Full registration of biokerosene so far has only been performed by Neste Oil for HEFA.

Whereas ASTM D7566 conforming blends are within the experience base for conventional kerosene, the synthetic components are dissimilar. Main components of conventional kerosene are *n*-alkenes, isoalkanes, cycloalkanes, and aromatics. HEFA- and FT-kerosene consist almost solely of *n*-alkenes and isoalkanes. Other pathways currently up for approval produce fuel solely consisting of aromatics, fuel consisting of cycloalkanes and aromatics, or even fully synthetic fuel containing all main components of conventional kerosene (SIP kerosene is an extreme case, consisting solely of C15 isoalkanes).

In the aviation approach, blend is the key unit. Certification is released on the basis of blend properties that need to be within experience base for conventional kerosene. If D7566 conditions are met, blend by definition becomes D1566 fuel. Information of synthetic component is completely lost (US) or only generically preserved (Europe). Properties of individual blend components are defined only for quality control purposes. The rationale is that only the blend will ever end up in an aircraft.

6.1.5.2 The European Chemical Agency (ECHA) Point of View

All chemical substances produced in or imported into Europe need to be registered at ECHA. In case of blends of two or more substances, all individual substances of the blend need to be ECHA-registered. Registration is by producer and/or importer. In principle, every producer/importer must undertake individual application, proving their own data for substance properties and safety for human and environment. For chemically identical substances, coregistration with already registered substances is required. The regulation requires payment of a coregistration fee to previous registrants to defray their initial costs. However, identity of substances is dependent on definition in registration consortia and acceptance by ECHA. The ECHA approach defines individual blend components as the key units. All components must be registered. The properties of individual components are relevant for registration. The rationale is that all information must be available for the regulator. The workers or the environment might potentially be exposed to the neat synthetic component (e.g. when blending). This raises two issues. The first issue is the potential impact on trading kerosene. As synthetic fuel is increasingly produced worldwide, it will join fossil kerosene distribution networks. Synthetic fuel may quite possibly not be ECHA-registered if not produced for the European market. However, a nonregistered blend component may physically be present in kerosene imported into Europe if information on synthetic content is generic or lost. This is a violation of EU laws. The second issue is the potential multitude of registrations. The number of pathways and producers are currently increasing almost exponentially. Even for fundamentally the same pathways (e.g. ATJ), exact approaches of producers are different, resulting in different proportions of kerosene components, and hence in nonidentical synthetic fuels. This may potentially require each constituent to be registered individually. This translates in a duplication of work and costs.

6.1.5.3 Military Certification Constraints

ASTM approval of a synthetic fuel means that blend with synthetic fuel conforming to specification is considered Jet A/Jet A-1 fuel, can be used by any civil aircraft certified for use of Jet A/Jet A-1, and is accepted both within North America and Europe. This, however, does not imply certification for military use. The military is using its own fuel specifications. A military equivalent to Jet A-1 exists (JP-8), and NATO aircraft are qualified on it. However, changes to ASTM specification do not automatically cross-read to military specification. The military makes use of various additives not present in civil used fuels. They have different operational requirements (e.g. afterburner), while the principal requirement is a single fuel policy, i.e. fuel must also be suitable for tanks and trucks. Separate military certification is currently required, while no central coordination of synthetic fuel approval within NATO is present.

Approval is performed by military authorities of individual nations. The US Armed Forces have introduced advanced regulatory rules so that FT- and HEFA-kerosene have been approved for all military equipment. In other NATO members, a different certification is presently required. Some (e.g. Sweden) are also well engaged, though less experienced, while other states have barely started. The tendency is to approve equipment for the use of fuel, not to approve the fuel. The authorities request the manufacturer of equipment to make the approval. Often the national producer is not involved in international fuel certification efforts. These factors have adverse effects on interoperability. Military infrastructure is intended to be for all NATO partners. The US Armed Forces increasingly move to include blends with synthetic material in their fuel supply. This creates an issue for NATO partners using US infrastructure. The current situation is already affecting joint exercises in the US.

Another issue to be solved is the closure of European military infrastructures for synthetic fuels. European dual use (civil and military) infrastructures cannot be used for synthetic fuels. This closure also affects fuel farms of dual use airports, causing logistical issues for synthetic fuels.

6.1.6 Bioeconomics of Tobacco Seeds Oil for Aircraft Fuel: Energy Tobacco Variety Solaris

Solaris is a nicotine and GMO-free crop variety that yields significant amounts of sustainable oil (as feedstock for biojet fuel) and high quality animal feed (http://www.projectsolaris.co.za/).

In 2002, Prof. Corrado Fogher and his research team at Plantechno carried out a careful work of research and development aimed at creating a variety of tobacco suitable for energy applications characterized by high seed and biomass production and negligible amounts of nicotine. Prof. Fogher did several lab tests in order to proof the scientific concept on a new, stable variety of energy tobacco.

"Energy Tobacco" Solaris variety, GMO free, has been registered at the Ministry of Agriculture of Italy with the patent RM2007A000129 in 2007. In 2008, Prof. Corrado Fogher and a group of industrial partners started Sunchem and deposited on an international level the new industrial patent "Energy Tobacco" (international patent PCT/IB/2007/053412). Patent has been already granted in 38 countries, among which are USA, South Africa, West Africa countries, EuroAsia, among others.

The industrial patent covers the following claims: a group of claims covers the mutagenized tobacco plant characterized in that it has a much higher seed production than the average seed production of the currently existing tobacco plants. A second group claims the use of the plants for the production of seeds for the production of the products indicated in the patent: fuel, supplements, oil, oil cakes, etc. A third group claims the production method of the plant; therefore, they also protect the necessary procedure to create the desired plants, at least in the methods. A fourth group protects the seeds of the plant and their use for producing tobacco oil, fuel oils, biodiesel, animal food supplements, solid fuels, and dietary supplements. A fifth group protects the oil extraction method from tobacco seeds wherein the oil yield is equal to between 70% and 95% of the oil through the pressing step and any other steps. A sixth group covers the TSO that can be obtained with the extraction method according to previous claims characterized in that it has an iodine level of less than or equal to 120.

The tobacco plants for energy applications, contrary to the tobacco for the cigarette industry, maximize the production of flowers and seeds to the detriment of the leaves production, and biomass for biogas production. The variety is extremely robust, able to grow in various climates and soils, and can be cultivated on marginal lands that cannot be used for food production.

Plants have been followed throughout development and various parameters recorded. Inflorescences from the most productive plants have been collected and evaluated. The plants have been then bagged for self-pollination. The rest of the plants were cleared out.

Various lines of tobacco have been grown to either select highly productive and stable Solaris lines or to evaluate their characteristics to continue the breeding of new varieties. Tests on sites with different soil characteristics (clay and sandy) have been performed where plants from the following lines are being grown, and their development and yields followed carefully:

Selection of tobacco lines in 2013–2015:

- Lines N1, N2, and N3: Three Solaris lines, selected in 2012 for producing between 700 and 900 g of seed over two harvests, were chosen as the starting point for the selection of very productive plants in ideal conditions of minimal competition. Between 280 and 300 plants per line were transplanted in each field.

- Lines A1, A2, and A3: Three lines originated with an intraspecific cross aiming at increasing oil production.
- Lines Ca and Cb: Lines originated with an intraspecific cross between high-yield seed varieties.
- Line PN1S25 produced 1059 capsules, 229 g seed.
- Lines II1 and II17: Lines originated with an interspecific cross using wild species and colchicine treatment (polyploidization). Their phenotype has been evaluated under field conditions.
- Line G1: Tobacco line from Plantechno's germplasm, characterized by extremely large capsules.
- Line PLT103: A tobacco variety originally used in the crosses that were undertaken in the generation of the Solaris variety. Grown for comparison with the different lines grown this year.
- Line O1: Early variant of Solaris obtained in 2007. Grown for comparison with the different lines grown this year.
- Lines PLTA4, PLTA56, PLTA60, PLTA61, PLTA65, and PLTA88: Germplasm lines grown to evaluate their phenotype under field conditions. Data from past years cropping confirmed high seed production.

The content in fatty acids of some of the tobacco varieties selected is shown in Table 6.1. The table shows new varieties and their hybrids, including plants obtained with mutagenesis and the genetic intervention carried out by introducing some of the fatty acids biosynthesis genes in order to change the acidic composition; the oil of the three last columns has a iodine title suitable for the transformation of said oil into biodiesel.

6.1.6.1 Ongoing Research at Plantechno

Seed-specific overexpression and expression of carefully designed RNA interference (RNAi) constructs to increase and decrease the activity of genes involved in triglycerides (fatty acid

Table 6.1: Content in Fatty Acids of Some Tobacco Varieties Selected.

	PLT103 (%)	PLT256 (%)	PLT318 (%)	PLT335 (%)
Palmitic acid	6.31	8.26	7.15	17.20
Palmitoleic acid	0.11	0.18	0.18	1.25
Stearic acid	2.58	5.20	8.50	12.50
Oleic acid	12.62	22.58	25.56	53.27
Linoleic acid	77.48	58.78	52.00	6.45
Linolenic acid	0.65	4.15	5.25	7.80
Arachidic acid	0.13	0.85	0.80	0.85
Eicosanoic acid	0.13	0.58	0.56	0.68

After Mutagenesis, or Engineered and Selected In Order To Change the Fatty Acids Metabolic Pathway, and Selected for the Stability of the Character.

triesters) (TAGs) biosynthesis to boost oil production or modify its composition. Over 20 constructs were obtained and transformed into Solaris tobacco. Analysis of oil composition has started for T1 seeds obtained from some of these lines.

Some high oleic–lower linoleic varieties and some with lower palmitic acid and higher oleic acid were obtained. The effect on seed oil in stable homozygous lines is expected to be even higher.

Line 72 produced an oil with 47.46% oleic acid and 41.11% linoleic acid. In a homozygous line generated after crossing line 72 and line 92, Plantechno researchers obtained seed oil with a composition for oleic acid of about 64% and 29% for linoleic acid. The TSO shows a high gross calorific value, low sulfur content, and the low viscosity when compared to other vegetable oils (See Table 6.2).

The composition of this modified oil not only should grant it a higher resistance to oxidation (even without need to addition antioxidants), but also is well balanced to produce a more valuable winter biodiesel in accordance with the latest UNI standards for biodiesel.
UNI EN 14213: Heating fuels. Fatty acid methyl esters (FAME). Requirements and test methods;
UNI EN 14214: Automotive fuels. Fatty acid methyl esters (FAME) for diesel engines. Requirements and test methods;
UNI: Ente Nationale Italiano di Unificazione (Italian standards body).

The use of TSO is also envisaged in the production of liquid or solid fuels, biodiesel, industrial lubricants, plastic materials such as linoleum, dietary supplements for animal feeds, dietary supplements for human use.

6.1.7 Sustainability of Production and Exploitation of Solaris Tobacco; Protein Cake as Feed

Based on the market analysis, Sunchem Holding decided in 2009 a strategic partnership involving medium-large companies operating in the oil/fuel processing and distribution field (Seasif Holding; Alpha Trading Spa, Italy; Argos Oil Ltd., Holland; Diester group, France;

Table 6.2: Characteristics of the Tobacco Oil Obtained by Pressing the Seed and Filtering.

Determination	Results	Mis. Un.	Methods
Flash point	236.0	°C	UNI EN ISO 27 19 2005
Sulfur	<0.01	% m/m	ISO 8754 1992
Ashes	0.005	% m/m	EN ISO 6245 2002
Viscosity at 50°C	21.630	mm 2/s	UNI EN ISO 3104 200
Melting point	−18	°C	ISO 3016 1994
Calorific value	9670	Kcal/Kg	ASTM D240-97
Volume mass at 15°C	925.0	Kg/m^3	UNI EN ISO 3675 2002
Saponification value	193.6	mg KOH/g	ASTM D94-02

Terasol Energy, LLC and M&V Consultacoes, Brazil). Sunchem Holding holds the exclusive rights to the employment and development on an international scale of the industrial patent "Solaris." Since 2009, Sunchem Holding concentrated its efforts on research on tobacco plants for biofuels. In agro operations, the Solaris project deals with seeding production, planting, fertilizing, harvesting, processing, and transport. For these operations, Sunchem is involved in both commercial and community farming.

Sunchem has made an alliance with SkyNRG and South Africa Airlines (SAA) for the exploitation of Solaris. SkyNRG expanded production of the hybrid Solaris as an energy crop that farmers could grow instead of traditional tobacco. South African Airways (SAA), Boeing, and SkyNRG are developing fuel based on the oil from seeds of tobacco. The specially bred Solaris strain of tobacco is nicotine-free, with a high number of seeds produced and big inflorescences (Fig. 6.5).

SAA and SkyNRG companies have teamed up for a pilot project that has seen about 50 ha in Limpopo province planted with energy tobacco Solaris plants in a sustainable production of TSO and other valuable by-products (biomass, oil cake feedstock). Project Solaris is a pilot for this smallholder program. Solaris is a crop, and crops need some level of irrigation, fertilizer, pesticides, and other inputs. The fields are grown by hard-up farmers, who gain two cash crops a year from the tobacco, and then have money for seeds and fertilizer for a third food crop. The seeds' residue becomes animal feed. From 1 ha cultivation of energy of tobacco, it is possible to obtain an average of seed production from 4 to 10 tons, with multiple harvests per year (based on climate condition). Seed contains around 40% of oil that, by cold

Figure 6.5
Comparison of Solaris variety (left) inflorescence size with that of control tobacco (right). The differences in size of inflorescences and seed content establish Solaris as a variety producing a high quantity of tobacco seed oil.

press extraction, gives 33–34% of raw oil and 65% of protein cake. The cost of the tobacco-based product matches that of jet fuel refined from fossil sources (Fig. 6.6).

The seed cake is rich in protein. These properties are essential when considering the applications of the oil, which range from the fuel industry to pharmaceuticals. The by-products, such as seed cake, can be used as stock feeds in livestock nutrition. The research can also be diversified using cheaper and nontoxic methods such as oil press machines for extractions, designing of reactors for large-scale extraction, and purification of the oil and also the economics, where there is need to conduct cost benefit analysis and identify the markets. After threshing the inflorescence to obtain seed, dry biomass represents another valuable product. Therefore, the economy of Solaris plants is made of crude oil (energy production, biodiesel, niche market, and jet fuel for aviation), oil cake (animal feet because of a good profat mix), and fresh and dry biomass (electricity, biogas).

The value added of energy tobacco is: it is a nonfood crop (but edible due to the cake), it has high sustainability criteria, and it shows flexibility in different climatic conditions.

Regarding the high-yield productivity, it is possible to count on profit margin along the value chain, from agriculture to industrial phase and market.

Biofuels plantations have been blamed for deforestation and other land use change. Campaigns have been raised to warn about the problems in case airlines will demand larger quantities of

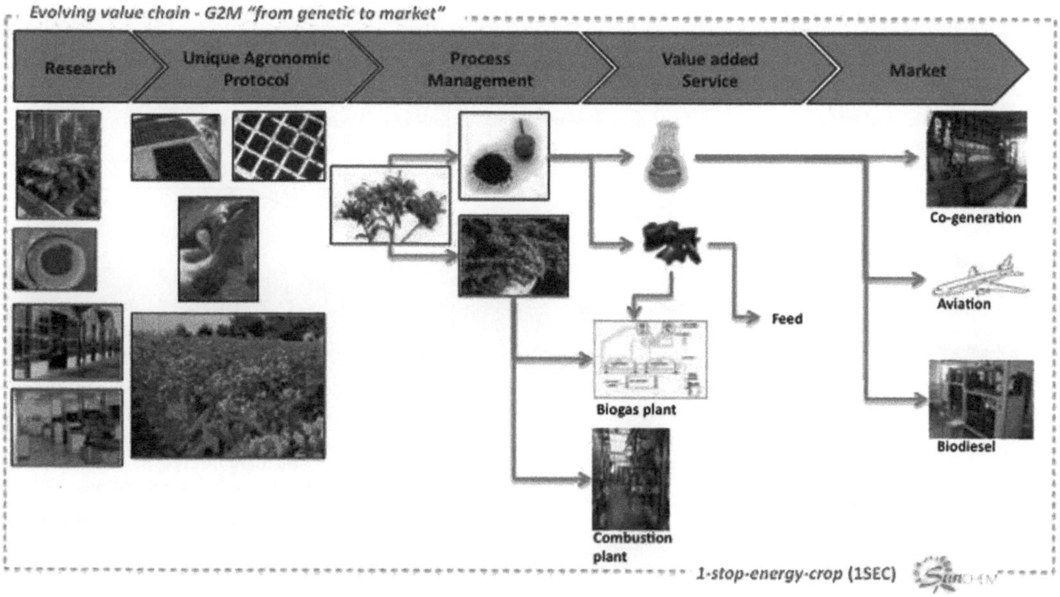

Figure 6.6
Solaris plants from genetics to market (G2M) flow. From research to agronomy studies, to process management, added value services, to market.

alternative fuels. In South Africa, this is not a problem since about 14% of the arable land is underutilized or unutilized. If just a small percentage of that 14% were used for Solaris or other similar feedstocks, this would provide enough fuel for all of SAA's needs (Fig. 6.7).

It will not displace essential food crops and will not affect total land footprint to produce the fuel quantity needed by SAA.

Solaris-based jet fuel will meet roundtable and sustainable materials (RSBs) minimum CO_2 life cycle reduction threshold of 50%, and when produced in an optimized supply chain setup, savings are expected to reach 75% compared to fossil jet fuel.

6.1.7.1 Roundtable of Sustainable Biomaterials (RSBs)

The RSB is a multistakeholder organization that has developed a global sustainability standard and certification scheme for biomaterials, including biofuels and biomass. Through third-party certification, RSB verifies the sustainability of biofuel operations according to environmental and social principles. RSB is widely recognized as the most robust scheme for ensuring sustainability and is especially strong in ensuring that operations promote rural development and enhance food security at the local level. The global aviation industry, including Boeing and South African Airways, is committed to using only RSB-certified biofuels.

Taking the product to the final customers, Solaris-based jet fuel has met RSB's minimum CO_2 life cycle reduction threshold of 50%, and when produced in an optimized supply chain setup,

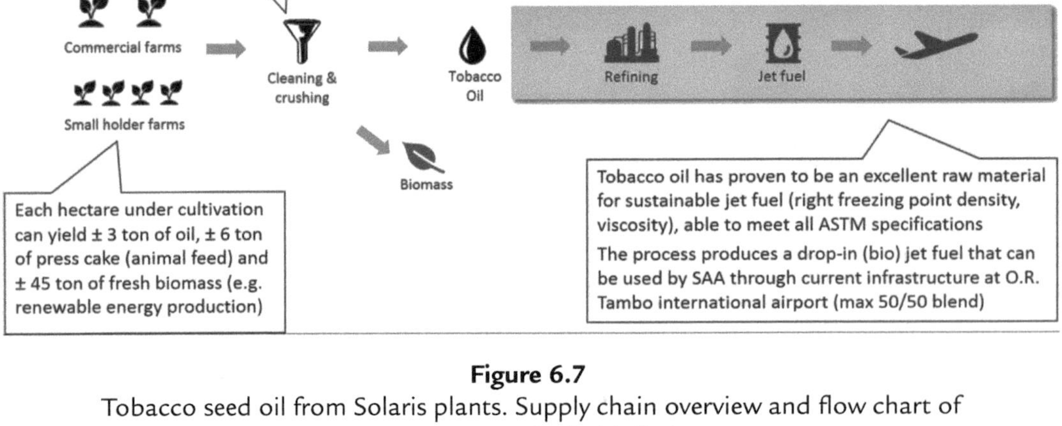

Figure 6.7
Tobacco seed oil from Solaris plants. Supply chain overview and flow chart of processes to obtain biofuel.

savings are expected to reach 75%, compared to fossil jet fuel. Project Solaris is currently undergoing RSB certification, starting with an initial 50 ha, which has been audited in May 2015. As Project Solaris grows, additional farms and processing plants will undergo RSB certification. Through certification, it is shown that the end products of Solaris not only reduce carbon emissions, but also improve the livelihoods of the communities in the Limpopo province. RSB certification is being supported by a smallholder program initiated by Boeing, South African Airways, and RSB. These partners launched a program in March 2014 to expand opportunities for smallholder farmers in Southern Africa to grow crops that produce sustainable fuels. This program is set up to help farmers with small plots of land to certify their product and gain access to markets for sustainable biofuels and biomaterials.

The development of Solaris production in South Africa can offer the following socioeconomic and environmental benefits: increased land productivity (additional biomass production and increased land use efficiency), development of rural economies and job creation, education and enablement of regional communities, agro innovation and knowledge transfer, shift from fossil fuel to renewable energy sources (e.g. diversification energy sources, energy security), displacement/substitution of less appealing industries (e.g. traditional tobacco industry), and GHG reductions.

6.2 Conclusion

Sustainable biofuel technologies have a key role to play in an energy-efficient, decarbonized transport sector around the world to provide reduced GHG emissions and energy dependency. Several workshops and conferences are planned for 2015 and 2016, covering the state-of-the-art knowledge in a diversity of topics from microbial hydrocarbon production and lignocellulosic ethanol to sustainability and policy. The 7th EBTP Stakeholder Plenary Meeting, SPM7, will take place in Brussels in the first quarter of 2016. Meanwhile, the EBTP continues to work closely with the European Commission (EC) and stakeholders in all member states to accelerate the deployment of innovative, sustainable biofuel technologies. In the tobacco model plant, advances have shown the feasibility to produce alternative jet fuel from tobacco seeds using the nongenetically modified Solaris variety and exploiting the by-products (biomass, oil cake) for a sustainable bioeconomy. Additionally, novel hybrids and more productive varieties will be introduced to large-scale field cultivation for implementing the production of biofuels.

References

Amrute, P.V., 1947. Ground-nut oil for diesel engines. Australasian Journal of Engineering 60-61.

Biemelt, S., Tschiersch, H., Sonnewald, U., 2004. Impact of altered gibberellin metabolism on biomass accumulation, lignin biosynthesis, and photosynthesis in transgenic tobacco plants. Plant Physiology 135 (1), 254–265.

Blaschke, L., Legrand, M., Mai, C., Polle, A., 2004. Lignification and structural biomass production in tobacco with suppressed caffeic/5-hydroxy ferulic acid-O-methyl transferase activity under ambient and elevated CO_2 concentrations. Physiologia Plantarum 121 (1), 75–83.

Blee, K., Choi, J.W., O'Connell, A.P., Jupe, S.C., Schuch, W., Lewis, N.G., Bolwell, G.P., 2001. Antisense and sense expression of cDNA coding for CYP73A15, a class II cinnamate 4-hydroxylase, leads to a delayed and reduced production of lignin in tobacco. Phytochemistry 57 (7), 1159–1166.

Chabannes, M., Barakate, A., Lapierre, C., Marita, J.M., Ralph, J., Pean, M., Danoun, S., Halpin, C., Grima-Pettenati, J., Boudet, A.M., 2001. Strong decrease in lignin content without significant alteration of plant development is induced by simultaneous down-regulation of cinnamoyl CoA reductase (CCR) and cinnamyl alcohol dehydrogenase (CAD) in tobacco plants. The Plant Journal 28 (3), 257–270.

Cheng, L., Yang, A., Jiang, C., Ren, M., Zhang, Y., Feng, Q., Wang, S., Guan, Y., Luo, C., 2015. Quantitative trait loci mapping for plant height in tobacco using linkage and association mapping methods. Crop Science 55 (2), 641–647.

Chiririwa, H., Hapanyengwi, A., Muzenda, E., 2014. Tobacco seed oil as an economically viable alternative for the tobacco industry. In: Int. Conf. Chemical Engineering & Advanced Computational Technologies (ICCE-ACT'2014) November 24–25, 2014 Pretoria (South Africa), pp. 36–39. http://dx.doi.org/10.15242/IIE.E1114017.

Cockcroft, C.E., den Boer, B.G., Healy, J.M., Murray, J.A., 2000. Cyclin D control of growth rate in plants. Nature 405 (6786), 575–579.

Cook, C.M., Daudi, A., Millar, D.J., Bindschedler, L.V., Khan, S., Bolwell, G.P., Devoto, A., 2012. Transcriptional changes related to secondary wall formation in xylem of transgenic lines of tobacco altered for lignin or xylan content which show improved saccharification. Phytochemistry 74, 79–89.

Crowder, B.J., Wilkinson, C.A., Johnson, C.S., Eisenback, J.D., 2003. Inheritance of resistance to tobacco cyst nematode in flue-cured tobacco. Crop Science 43, 1305–1312.

Dayan, J., Schwarzkopf, M., Avni, A., Aloni, R., 2010. Enhancing plant growth and fiber production by silencing GA 2-oxidase. Plant Biotechnology Journal 8 (4), 425–435.

Giannelos, P.N., Zannikos, F., Stournas, S., Lois, E., Anastopoulos, G., 2002. Tobacco seed oil as an alternative diesel fuel: physical and chemical properties. Industrial Crops Products 16, 1–9.

Holdcroft, A.M., 2013. Alternative Methods of Control for *Phytophthora Nicotianae* of Tobacco. Theses and Dissertations–Plant Pathology. Paper 5.

Johnson, C.S., Wernsman, E.A., LaMondia, J.A., 2009. Effect of a chromosome segment marked by the *Ph-p* gene for resistance to *Phytophthora nicotianae* on reproduction of tobacco cyst nematodes. Plant Disease 93 (3), 309–315.

Kavousi, B., Daudi, A., Cook, C.M., Joseleau, J., Ruel, K., Devoto, A., Bolwell, G.P., Blee, K.A., 2010. Consequences of antisense down-regulation of a lignification-specific peroxidase on leaf and vascular tissue in tobacco lines demonstrating enhanced enzyme saccharification. Phytochemistry 71 (5–6), 531–542.

Kuhar, P.S., 2014. Characterization of Genetically Modified High Biomass Producing Tobacco Plant. Biological Systems Engineering–Dissertations, Theses, and Student Research. Digital Commons University of Nebraska-Lincoln 1 (1), 6–58. http://digitalcommons.unl.edu/biosysengdiss/44/.

Miller, R.D., Kennedy, B.S., Ritchey, E.L., 2000. Tobacco breeding and genetics. Tobacco Research 31–34.

Mohammad, M.T., Tahir, N.A., 2014. Evaluation of chemical compositions of tobacco (*Nicotiana tabacum* L) genotypes seeds. Annual Research Review Biology 4 (9), 1480–1489.

Myers, E.C., December 2007. Policies to reduce emissions from deforestation and degradation (REDD) in tropical forests (PDF). Resources Magazine. 7. http://www.rff.org/Publications/Pages/PublicationDetails.aspx?PublicationID=17519.

Ng, K.S., Sadhukhan, J., 2011. Techno-economic performance analysis of bio-oil based Fischer-Tropsch and CHP synthesis platform. Biomass Bioenergy 35 (7), 3218–3234.

Nowicki, P., van Meijl, H., Knierim, A., Banse, M., Helming, J., Margraf, O., Matzdorf, B., Mnatsakanian, R., Reutter, M., Terluin, I., Overmars, K., Verhoog, C., Weeger, C., Westhoek, H., 2007. Scenar 2020-Scenario Study on Agriculture and the Rural World. European Commission, Directorate-General Agriculture and Rural Development, Brussels.

Patel, J.A., Patel, B.K., Chakraborty, M.K., 1998. Production potential and quality aspects of tobacco seed oil. Tobacco Research 24, 44–49.

Parkunan, V., Johnson, C.S., Eisenback, J.D., 2009. Effects of *Php* gene-associated versus induced resistance to tobacco cyst nematode in flue-cured tobacco. Journal of Nematology 41 (4), 261–266.

REDD, Bali Climate Change Conference, 2007. http://unfccc.int/meetings/cop_13/items/4049.php.
Tatti, E., Sirtori, A., 1937. Use of peanut oil in injection, high-compression, high-speed automobile motors. Energia Termica 5, 59–64; Chemical Abstracts 32, 2318 (1938).
Usta, N., 2005. Use of tobacco seed oil methyl ester in a turbocharged indirect injection diesel engine. Biomass Bioenergy 28, 77–86.
Usta, N., Aydogan, B., Con, A.H., Uguzdogan, E., Ozkal, S.G., 2011. Properties and quality verification of biodiesel produced from tobacco seed oil. Energy Conversion and Management 52 (5), 2031–2039.
van Meijl, H., van Rheenen, T., Tabeau, A., Eickhout, B., 2006. The impact of different policy environments on land use in Europe. Agriculture, Ecosystems Environment 114, 21–38.
Velikovic, V.B., Lakicevic, S.H., Stamenkovic, O.S., Todorovic, Z.B., Lazic, M.L., 2006. Biodiesel production from tobacco (*Nicotiana tabacum* L.) seed oil with a high content of free fatty acids. Fuel 85 (17–18), 2671–2675.

CHAPTER 7

Fermentative Production Methods

L.A. Prado Barragán, J.J.B. Figueroa, L.V. Rodríguez Durán
Iztapalapa Metropolitan Autonomous University, Mexico City, Mexico

C.N. Aguilar González
Autonomous University of Coahuila, Coahuila, Mexico

C. Hennigs
Naturstoff-Technik GmbH, Osterholz-Scharmbeck, Germany

7.1 Introduction

Over the years, fermentation techniques have gained immense importance due to their economic and environmental advantages. Ancient techniques have been updated and refined to maximize productivity at the same time that the fermenters' technology has also been developed with new processes control devices. Two broad fermentation methods have come into view: submerged fermentation (SmF) and solid-state fermentation (SSF). The first utilizes free-flowing liquid substrates or culture broths, while SSF uses solid substrates like bran, bagasse, soya pulp, and rice. Both fermentation methods have been upgrade on different parameters concerning to the nature of substrates used, process and instrumental controls, and microorganism selection, either with or without genetic modification. Each method may yield better results than the other, so it is highly desirable to select the best fermentation method for a particular product and facility's availability.

7.2 Submerged Fermentation

Microbial processes occurring with or without air are termed fermentation. Fermentation has been widely used for the production of a wide variety of substances that are highly beneficial to individuals and industry. Two different fermentation techniques have emerged as a result of this development: SmF and SSF. Microorganisms are cultivated on both solid (agar plates for small scale) as well as on liquid media (flasks or bioreactors for large-scale cultivation up to several hundred cubic meters).

Submerged fermentation is the cultivation of microorganisms in a liquid nutrient broth. Industrial enzymes and other metabolites can be produced using this process. This involves

Biotransformation of Agricultural Waste and By-Products. http://dx.doi.org/10.1016/B978-0-12-803622-8.00007-0

growing selected microorganisms (bacteria, yeasts, and fungi) in closed vessels containing a broth of nutrients (the fermentation medium). The required oxygen is introduced by a suitable gassing system. The microorganisms digest the nutrients and release the desired enzymes into the cultivation solution.

Due to the development of large-scale fermentation technologies, the production of microbial enzymes accounts for a significant proportion of the biotechnology industry's total output. The industrial fermentation process takes place in large vessels (fermenters) with volumes of up to 1000 m^3. The sterile fermentation media contain nutrients based on renewable raw materials like maize, sugars, or soy.

Most industrial enzymes are secreted by microorganisms into the fermentation medium in order to break down the carbon and nitrogen sources. Batch-fed and continuous fermentation processes are common. In the batch-fed process, sterilized nutrients are added to the fermenter during the growth of the biomass. Therefore, in the continuous process, sterilized liquid nutrients are fed into the fermenter at the same flow rate as the fermentation broth is leaving the system. This is called steady-state production. Parameters like temperature, pH, oxygen consumption, and carbon dioxide formation are measured and controlled to optimize the fermentation process. During enzyme harvesting from the fermentation medium, insoluble products, e.g. microbial cells or fragments, must be removed. This is normally done by centrifugation. As most industrial enzymes are extracellularly (secreted by cells into the culture medium), they remain in the fermented broth after the biomass has been removed.

7.2.1 Biotechnological Potential

Humans have been using fermentation to produce food and beverages for millenniums. For example, fermentation is used for preservation in a process that produces lactic acid, as well as for producing alcoholic beverages.

Nowadays, submerged fermentation processes are applied at large (several thousand cubic meters)-scale for industrial purposes. They include the practice of using cells such as microorganisms, or components of cells like enzymes, to generate industrially useful products in sectors such as chemicals, food and feed, detergents, paper and pulp, textiles, and biofuels. Bioconversion via enzymatic pathways or fermentation strategies with microorganisms represents a sustainable alternative for the implementation of cleaner processes and is currently favored by consumers as well as the industry. These processes are using renewable raw materials and may contribute to lowering greenhouse gas emissions and moving away from a petrochemical-based economy.

By using residues from the food industry as substrates, industrial fermentation contributes to reduction of waste as well as the recovery of valuable substances from residues by bioconversion.

One key issue for the worldwide industry is waste reduction and recycling of valuable substances, by-products, and residues. All industry branches are required to optimize production processes to reduce the exhaustion of raw materials. In this context, biotechnological processes do comprise an enormous potential wherever applicable.

7.2.2 Types of Bioreactors

Several bioreactor types in various scale sizes are available for submerged fermentation processes of which all show certain (dis)advantages for specific applications.

Batch stirred tank

A batch stirred tank reactor is the simplest type of reactor (Fig. 7.1). If applied as batch reactor, the system is composed of a reactor and a mixer such as a stirrer, a turbine wing, or a propeller. The sterile substrate is inoculated with the respective organisms or parts thereof (e.g. enzymes), which convert the nutrients into products that remain in the bioreactor until the end of the cultivation. Depending on the fermentation requirements, the tank

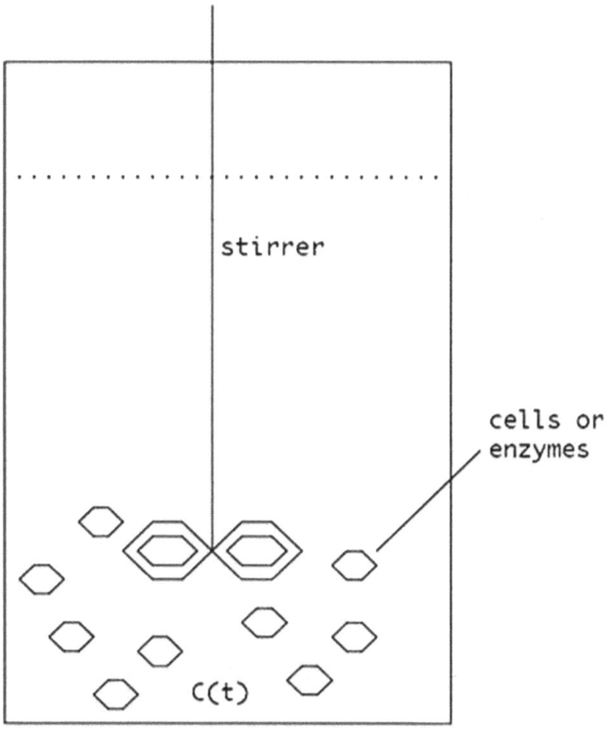

Figure 7.1
Principle of a batch stirred tank reactor.

can be equipped with gassing equipment at the bottom of the vessel. Attention must be paid to the sterile air inlet to be ensured by an air filter. By varying the flow rate and the speed of the mixer, the oxygen partial pressure can be controlled. Another option for controlling the process parameter is by adapting the temperature of water heated or cooled double jacket.

Application, advantages, and disadvantages:

This reactor type is useful for substrate solutions of high viscosity and for immobilized enzymes with relatively low activity. The disadvantage if this bioreactor is that some metabolites tend to decompose upon physical stirring. The batch system is generally suitable for the production of rather small amounts.

Continuous stirred tank

A continuous stirred tank reactor is composed of the same parts as the batch tank as well as a substrate inlet and a product outlet for continuous harvesting. The continuous stirred tank reactor is more efficient than a batch stirred tank reactor, but the equipment is slightly more complicated (Fig. 7.2). The continuous stirred tank is run at steady state with continuous flow

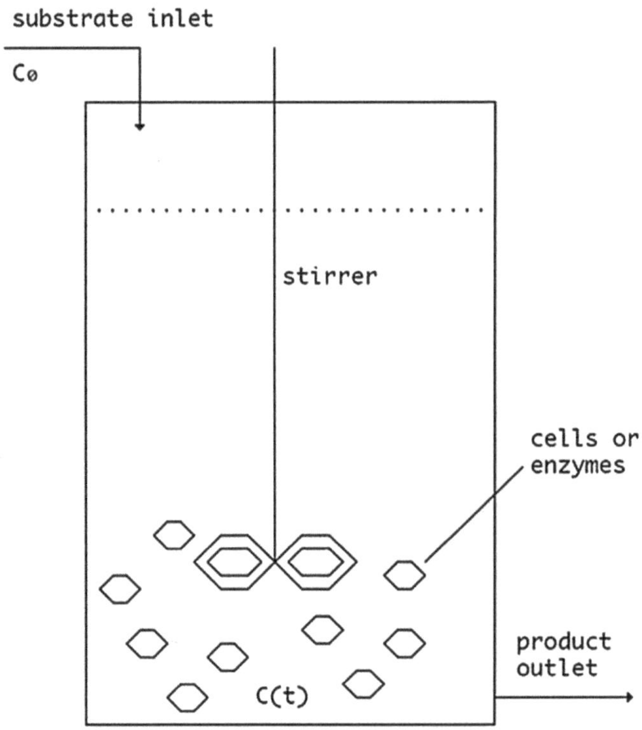

Figure 7.2
Principle of a continuous stirred tank reactor.

of reactants and products; the feed assumes a uniform composition throughout the reactor, and the product outlet stream has the same composition as in the tank.

Fed-batch culture

A fed-batch culture is another technique for operating a stirred tank where one or more nutrients are fed to the bioreactor during cultivation and in which the product(s) remain in the bioreactor until the end of the cultivation. The advantage of the fed-batch culture is that the concentration of fed-substrate in the culture liquid can be adapted to the desired level, so extremely high densities can be reached (Fig. 7.3).

Advantages: It is easy to construct, control, and clean.
Disadvantage: Low conversion rates per volume, relatively high shear forces.

Bubble columns

Bubble columns are cylindrical vessels with a gas inlet at the bottom. The gas is sparred in the form of bubbles into a liquid or a liquid–solid suspension. These reactors are generally referred to as slurry bubble column reactors when a solid phase exists (Fig. 7.4).

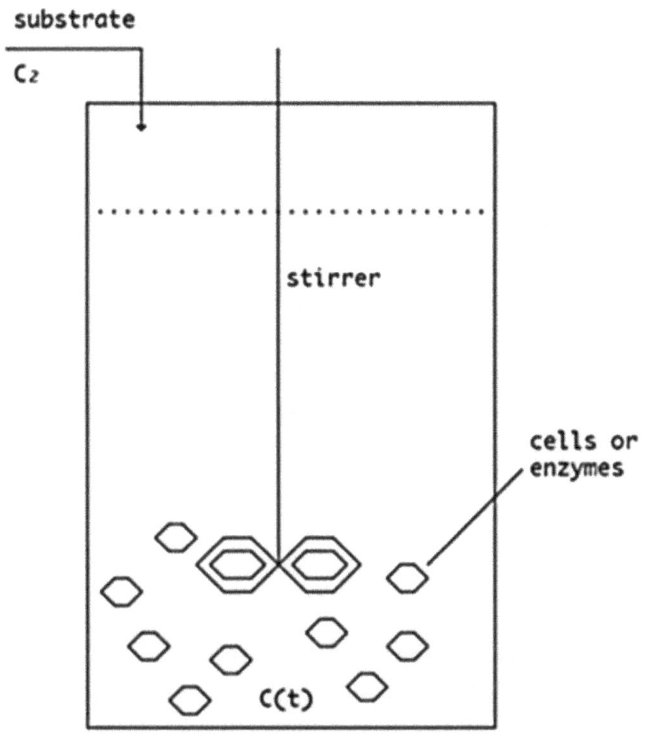

Figure 7.3
Principle of a fed-batch stirred tank reactor.

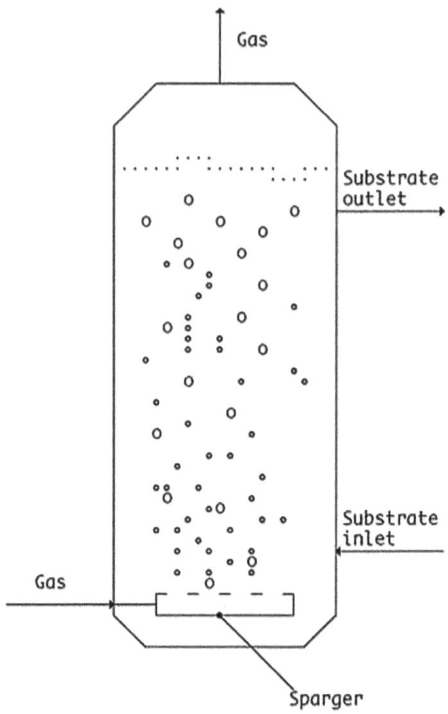

Figure 7.4
Principle of a bubble column reactor.

The gas inlet is placed at the bottom of the column and causes a turbulent stream to enable an optimum gas exchange. The mixing is done by the gas sparging, and it requires less energy than mechanical stirring. The liquid can be in parallel flow or countercurrent. The bubble size has a strong influence on hydrodynamic parameters such as bubble rise velocity, gas residence time, gas–liquid interfacial area, and gas–liquid mass transfer coefficient. An important application area of bubble columns is their use as bioreactors for the production of enzymes, proteins, antibiotics, etc., but also within anaerobic wastewater treatment.

Advantages: High heat and mass transfer, low investment costs, easy to construct, low energy costs, absence of moving parts (low shear forces).
Disadvantage: Undefined flow patterns.

Airlift

Airlift bioreactors are tower reactors for large-scale aerobic cultures where the mixing of the culture broth is done by the inserted gas via an airlift pump (Fig. 7.5). This pump injects compressed air at the bottom of the discharge pipe, which is immersed in the liquid. The compressed air mixes with the liquid, causing the gas–liquid mixture to be less dense than the rest of the liquid around it and therefore is displaced upwards through the discharge pipe by the surrounding

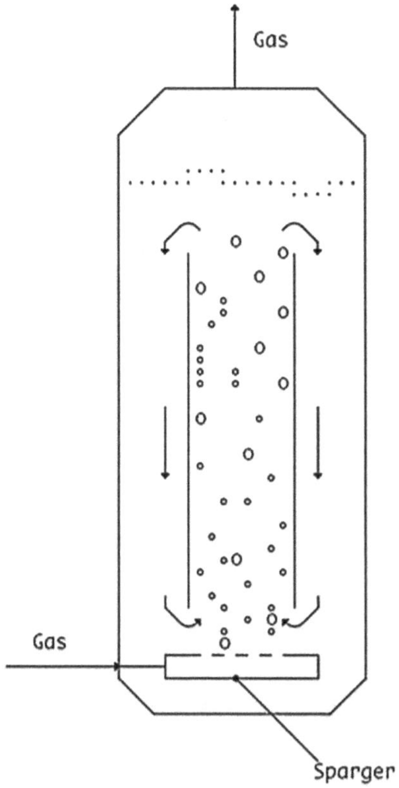

Figure 7.5
Principle of an airlift bioreactor.

liquid of higher density. Solids may be entrained in the flow and, if small enough to fit through the pipe, will be discharged with the rest of the flow at a shallower depth or above the surface. The only energy required is provided by compressed gas, which is prepared by a compressor.

Advantage: Only compressed air (or other gas) required, low shear forces.
Disadvantage: Relatively high costs for compressing flow rate, difficult to adjust.

Fluidized bed

In this type of bioreactor, substrate is passed through the granular or pelleted biomass or encapsulated enzymes at high enough velocities to suspend the solid and is mixed by a gaseous phase inserted at the bottom of the reactor. A porous plate is separating the gas inlet from the liquid–solid dispersion. As soon as the optimal parameters are reached, the mixing performance is comparable to an agitated tank, but without shear forces (Fig. 7.6).

Advantages: Homogenous particle mixing, uniform temperature gradients.
Disadvantages: High space requirement, entrainment of solids.

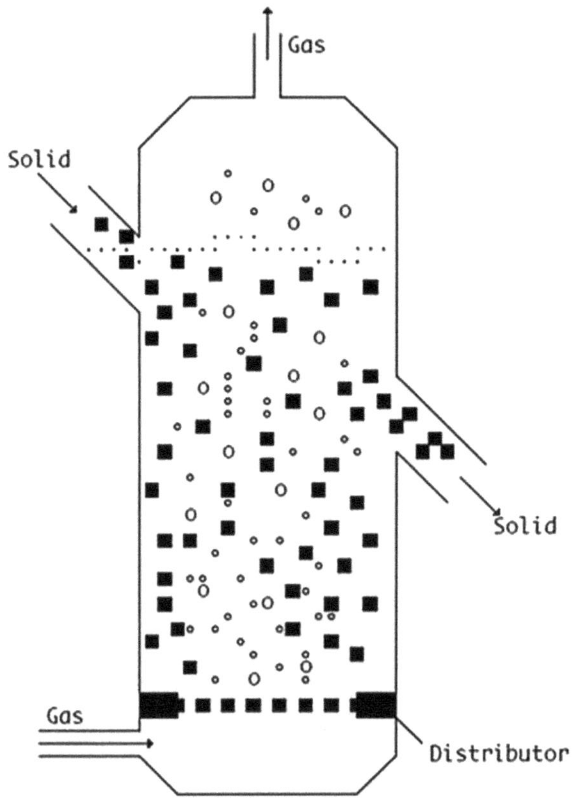

Figure 7.6
Principle of a fluidized bed bioreactor.

Trickle bed

A trickle bed reactor uses the downward movement of the liquid substrate and usually a gas over a packed bed of the granular or pelleted biomass or encapsulated enzymes. This reactor type is relatively easy to construct with just a liquid and a gas in- and outlet as well a double jacket for temperature control. This reactor type is applied in refineries as wastewater treatment as well as for catalytic reactions within the food industry (e.g. vinegar production) (Fig. 7.7).

Advantage: Easy to construct.
Disadvantage: Fluid dynamics not completely understood, limited mixing performance.

7.2.3 Process Control

The aim of process control is safeguarding constant process parameters that enable mass production of consistent products from operated fermentation processes.

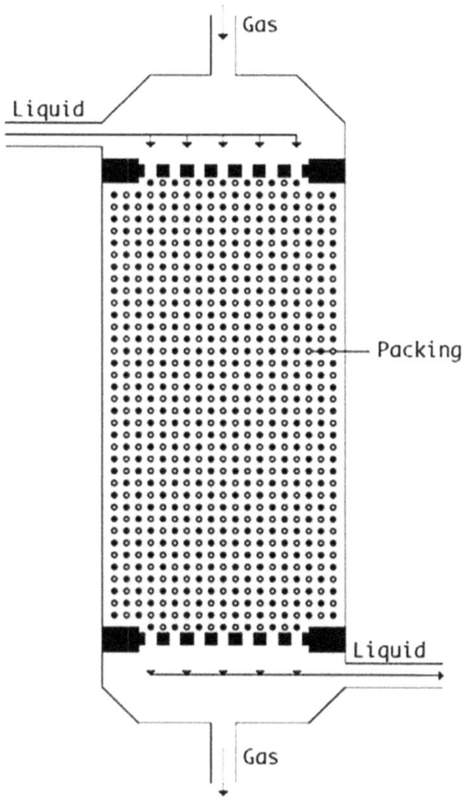

Figure 7.7
Principle of a trickle bed bioreactor.

During fermentation process, monitoring of the variables pH, temperature, dissolved oxygen, off gas, substrate concentration in the reaction tank, level, and turbidity can be carried out and the parameters can either be manually adjusted or automatically controlled. The input variables are those that independently stimulate the system and can thereby induce change in the internal conditions of the process. The output variables are those by which one obtains information about the internal state of the system. The input variables can be manipulated and are thus control variables. Additional disturbance variables cannot be controlled themselves but are considered for controlling the input variables for the optimal process conditions.

The aim of industrial-scale submerged fermentation processes is to get a pure product with a high concentration, which is achieved by regulating:

- oxygen;
- temperature;
- pH; and
- other measurable and variable parameters at optimal levels.

Temperature and heat transfer

In order to maintain the optimal temperature for the cultivated strain, temperature control is of utmost importance within submerged fermentation processes. The control unit constantly assesses the current temperature of the nutrient broth and controls a heater to increase or decrease the temperature of the double jacket according to user-defined setting.

Heat generation is directly related to metabolic activity of the microorganism, particularly respiration during growth, which is related to oxygen consumption and CO_2 formation. The generated heat must be departed since microbial growth is very sensitive to temperature rise, as this will be affecting spore germination, growth, product formation, and sporulation.

Biomass and growth kinetics/time and yield

The growth of bacteria, yeasts, or fungi is the cell division of one cell into two genetically identical cells. The growth of microorganisms in a liquid nutrient broth is related to the concentration of a limiting nutrient. The microbial growth in a batch culture in batch culture can be modeled with four different phases: (1) lag phase; (2) exponential phase; (3) stationary phase; and (4) death phase (Fig. 7.8).

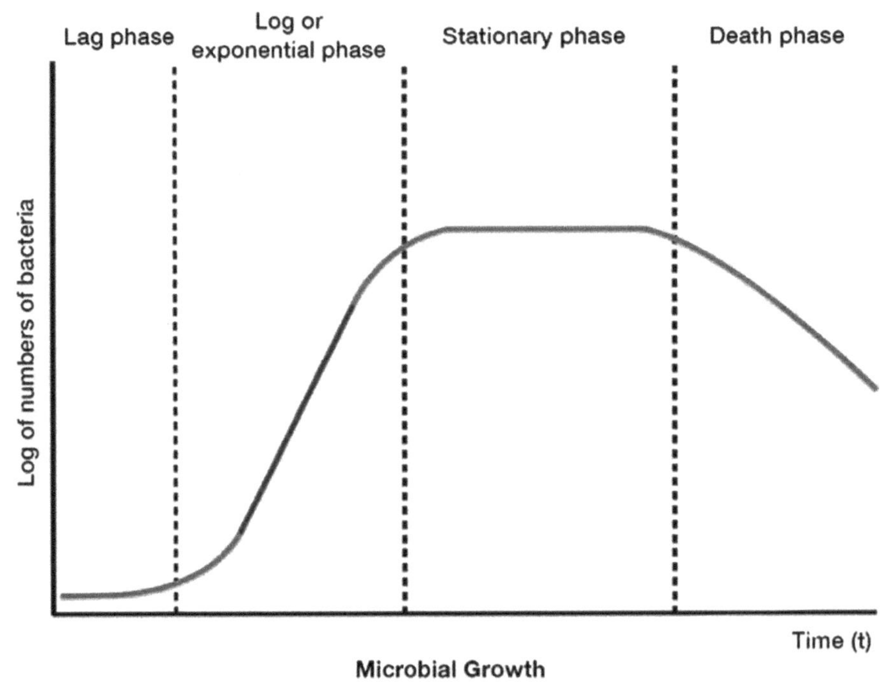

Figure 7.8
Microbial growth dynamics.

1. During lag phase, the microorganisms adapt themselves to growth conditions. They are not yet able to divide.
2. During the exponential phase, the cells grow by doubling. The number of new organisms appearing per time is proportional to the present population. If growth is not limited, doubling will continue at a constant rate, so both the number of cells and the rate of population increase doubles with each consecutive time period. The growth rate within this phase depends on the growth conditions (nutrient concentration, aeration).
   The cell concentration during the exponential phase can be described with the following equation:

   $$c\,(t) = c_0 \cdot e^{\mu(t) \cdot t}$$

   Whereas in the stationary phase:

   $$c_{\max} = c_0 \cdot e^{\mu_{\max} \cdot t}$$
3. As soon as the nutrient concentration is limited or an inhibitory substance is produced, the stationary phase is reached in which growth rate and death rate is equal.
4. At death phase, the population dies because of inappropriate living conditions (exhausted nutrients, oxygen, wrong temperature, etc.).

The biomass and growth kinetics are described by the Monod and the Michaelis–Menten equations, respectively. The Monod equation is a mathematical model for the growth of microorganisms. It is named for Jacques Monod, who proposed using an equation of this form to relate microbial growth rates in an aqueous environment to the concentration of a limiting nutrient. The Monod equation has the same form as the Michaelis–Menten equation but differs in that it is empirical, while the latter is based on theoretical considerations.

The specific growth rate depends on the concentration of the growth-limiting substrate until the concentration is increased to a nonlimiting level. The dependence of the growth rate on substrate concentration follows the Monod kinetics, which are provided in the following equation:

$$\mu = \mu_{\max} \cdot \frac{C\,(S)}{K_s \cdot C\,(S)}$$

where K_s is the saturation constant.

Oxygen

Dissolved oxygen levels in fermentation systems are affected by the gas flow rate and stirrer speed, as well as the oxygen partial pressure in the gas. Since the first two variables cannot easily be separated, a method for dissolved oxygen control has been devised based on the variation of oxygen partial pressure in the gas.

7.2.4 Microorganisms

There is an increasing demand for enzymes and other biomolecules on the market for various industrial applications; among detergent development, enzymes for food applications as well as the bioconversion of lignocellulosic biomass for ethanol production are the major application fields.

Although these molecules can also be extracted from plant material, the advantage of using microorganisms like bacteria, fungi, and yeasts for their production is the economic bulk production capacity. Microorganisms are also easy to manipulate in order to optimize the products' properties.

Depending on the strain and the culture conditions, the produced biomolecules can be tailored for the desired properties.

7.2.5 Substrates

All kinds of (micro)organisms require nutrients for growth, metabolism, and activity of cells. Microorganisms can grow on variable growth substrates and possess versatile metabolic activities. Nutritional versatility makes bacteria, fungi, and yeasts adaptable to different types of environments.

In order to maintain the optimal pH for the cultivated strain, the pH control unit assesses the pH of the nutrient broth and controls a pump, which adjusts the pH by adding acid or base, respectively, according to user-defined setting.

If cultivated, microorganisms require a sterile nutrient broth (substrate) containing a dedicated carbon source, minerals, and trace elements.

Catabolic response

Bacteria are exposed to a constantly changing physical and chemical environment. Within limits, bacteria can react to changes in their environment through changes in patterns of structural proteins, transport proteins, toxins, enzymes, etc., which adapt them to a particular ecological situation.

Bacteria that are fed two different carbon sources together will use one source first because it takes less enzymes to use this one than it does to use the second carbon source. Bacteria are able to develop a sophisticated gathering and transport system if it senses that one nutrient is in short supply in its environment.

Generally, bacteria do not synthesize degradative (catabolic) enzymes unless the substrates for these enzymes are present in their environment.

Bacteria have developed diverse mechanisms for the control of biosynthetic (anabolic) pathways. Bacteria stop biosynthesis when the end product of the pathway is not needed or already obtained from the environment. For example, if a bacterium could find a preformed amino acid in its environment, it would make sense to shut down its own pathway of amino acid biosynthesis and thereby conserve energy. These control mechanisms for all these metabolic pathways must be reversible, since the environment can change quickly and drastically.

Agroindustrial by-products transformation

The use of these lower-cost carbon sources, such as by-products from fruit and vegetable processing industry, like processing residue, press cake, and pomace, are supposed to lead to reductions in production cost.

Renewable by-products from the fruit and vegetable processing industry not only have the advantage of being more environmentally sustainable, since they are derived from fruit and vegetable sources that capture CO_2 from the atmosphere, but they have also recently proven to have more stable price and supply than raw materials derived from fossil sources.

Nowadays, submerged fermentation processes have implemented the use of purified carbohydrates derived from corn starch, beet, and sugarcane as main raw materials, thus competing with their food uses. Accordingly, different types of renewable waste resources have been used for the production of biomolecules by means of fermentation: industrial by-products as well as organic matter in municipal wastewater.

7.2.6 Biomolecules Production

During the past decades, considerable progress has been made toward solving problems within using crude substrates such as food by-products for biomolecule production by submerged fermentation. Chemicals and fuel can be produced that way as well as succinic acid, bioplastics, ethanol, or even enzymes to be applied as detergent or as food ingredients.

Currently, people prefer foodstuff that is not only nutritive, safe, and nice in sight, but also offers health benefits like prevention of diseases. The food industry has identified biologically active compounds like peptides, carbohydrates, polyphenols, carotenoids, phytosterols, and fatty acids, which might comprise special features when applied to food products in a certain amount.

Submerged fermentation of specific microorganisms (fungi, bacteria, yeasts) proved to be a promising alternative for production of various biomolecules in industrial scale, as it is faster, easily controlled, more efficient, and of greater interest in the food and drug industry process than the established field cultivation of the plants.

Advantages and application of submerged fermentation

As said before, bacteria, fungi and yeasts are excellent sources of active biomolecules. Especially when these molecules are produced in the extracellular medium, they are easy to cultivate by submerged fermentation, and the product can be obtained by dedicated downstream processing.

As fruiting body formation is a time-consuming process, scientists are more interested in SmF, which has been exploited for production of enzymes, exopolysaccharides, and other bioactive molecules of innumerable nutraceutical, pharmaceutical, and cosmeceutical properties.

7.3 Solid-State Fermentation

SSF may be defined as the culture of aerobic microorganisms, mostly filamentous mold or single, embedded on moistened substrate particles, where interstitial cavity allows the passive gas diffusion, employing either a natural support or an inert support as a solid material (Pandey et al., 2000).

7.3.1 Biotechnological Potential

SSF is a bioprocess carried out in absence or near absence of free water, but with the moisture necessary to allow growth and metabolic activity of microorganisms. These conditions resemble the natural environment in which they exist and from which microorganisms are isolated (Thomas et al., 2013).

In recent years, SSF has gained attention from the academic and industrial sectors because of its potential for producing industrially important products. SSF has some advantages over submerged fermentation for the production of some bioproducts such as enzymes, spores, and secondary metabolites at laboratory level. However, technical difficulties in the operation and control of fermentation in solid mediums have discouraged the use of this technique for industrial production (Hölker et al., 2004). This has motivated the study of biochemical engineering aspects of SSF (Mitchell et al., 2000, 2003, 2004; Raghavarao et al., 2003).

The design of bioreactors and a bioprocess capable of controlling the operating variables in SSF is necessary to successfully implement this system at the industrial level. Next, we will describe some of the bioreactors available to carry out SSF processes, as well as strategies for monitoring and control of operating variables that have been used.

7.3.2 Types of Bioreactors

Most of the research on SSF at small scale is carried out using standard laboratory glassware as bioreactors, such as Petri dishes, Erlenmeyer flasks, and Roux bottles. These systems are simple, inexpensive, and allow one to explore a large number of experimental conditions simultaneously. However, they provide little information on the operating variables and are not suitable for scaling up the process.

Several bioreactors had been designed to monitor and control the process variables in small- and large-scale SSF. These bioreactors can be static or mixed, forcefully aerated, or with

Table 7.1: Main Characteristics of SSF Bioreactors.

Bioreactor	Mixing	Aeration	Scale
Column	Unmixed	Forced	Lab scale
Try	Unmixed	Unforced	Lab scale to industrial scale
Rotating drum	Mixed	Unforced	Lab scale to industrial scale
Continuous screw	Mixed/unmixed	Forced/unforced	Pilot scale

unforced aeration (Table 7.1). Most of SSF bioreactors operate in batch mode. However, there have been some attempts to design SSF systems capable of continuous operation. Below, we describe some of the bioreactors most used. Main characteristics of SSF bioreactors are shown in Table 7.1.

Column

Packed bed column reactor, also known as Raimbault column, is an unmixed forcefully aerated reactor. It usually consists of a relatively thin (1–2 cm ID) and short (10–20 cm) glass column, although larger columns have been constructed mainly to study heat transfer phenomena. In its most typical form, the column stands vertically in a water bath at a controlled temperature. The column is perforated at the bottom and at the top. The fermented material is packed inside the column. With the aid of an air pump, a valve system, and a humidifier, saturated air is passed through the column from the bottom. The top is usually covered with a perforated stopper to allow passage of the air stream. In most cases, the exhaust air is dried and analyzed by either gas chromatography or by electronic sensors.

Packed column bioreactors are particularly useful to optimize culture conditions and for kinetic experiments. Because of its compact design, column reactors allow to test a large number of experimental conditions simultaneously. Unlike Erlenmeyer flasks, column reactors allow online monitoring of the exhaust gases, which can be used as an indirect measure of microbial growth (Mitchell and Krieger, 2006).

However, column bioreactors have a major drawback for their use at commercial scale. They are difficult to scale up, mainly because of problems to dissipate heat in larger scales. For example, Saucedo-Castañeda et al. (1990) studied heat transfer in 1 L bioreactor (6 cm ID) packed with cassava wet meal and inoculated with *Aspergillus niger*. Under experimental conditions studied, they found a maximum radial gradient of nearly 5°C/cm.

A scheme for a column reactor (Durand, 2003) is shown in Fig. 7.9.

Tray

Tray bioreactors are the simplest reactors used for SSF. These bioreactors have been used since ancient times for the production of traditional Asian foods such as koji and soy sauce. Tray bioreactors consist of wood, plastic, or metal trays, which may be perforated at the bottom. Trays are filled with the moistened and inoculated solid substrate to a bed height of

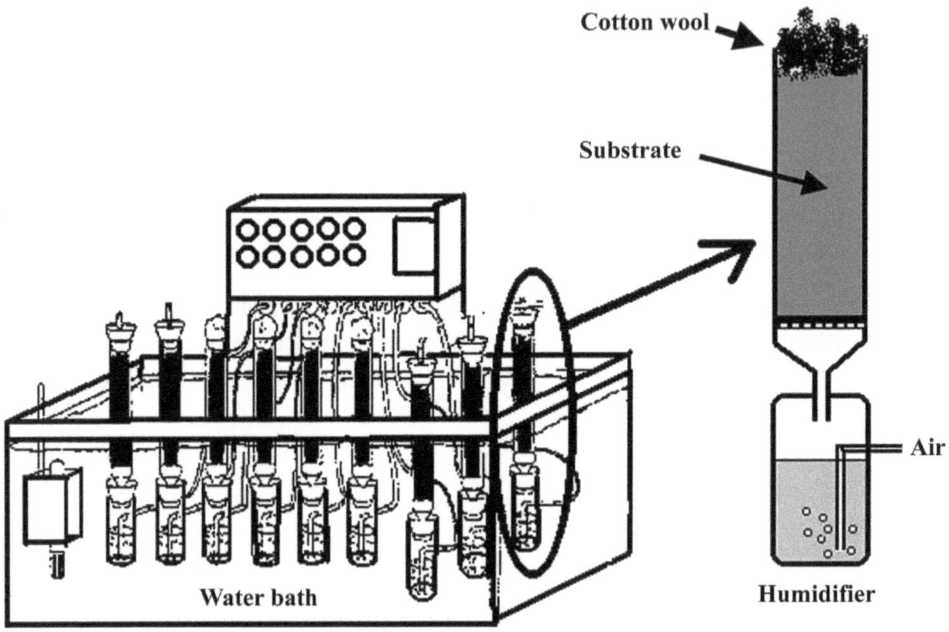

Figure 7.9
Column reactor (Durand, 2003).

5–15 cm. Trays are stacked one above other in trolleys and moved to a room with controlled temperature and humidity (Krishna, 2005).

This technology can be scaled up easily; it is only required to increase the number of trays. Therefore, it has been successfully used at industrial level, mainly in Asian countries. However, this technology requires large areas and is labor intensive. Furthermore, it is difficult to operate tray reactors aseptically (Durand, 2003).

Another drawback of tray bioreactors is mass and heat transfer. In tray bioreactors, O_2 and CO_2 can only move within the bed and the headspace by diffusion. Temperature of the bed surface depends on the heat transfer coefficient of the bed-to-air component, which is affected by the velocity of air between the tray surfaces (Ali and Zulkali, 2011). Chen et al. (2005) studied the effect of the aeration on the temperature gradient within the bed height using a novel tray bioreactor. They improved heat transfer using internal air circulation and air pressure pulsation. Under optimized conditions, the maximum temperature gradient was 0.12°C/cm. This is much lower than the 3°C/cm gradient obtained for a conventional static tray bioreactor. A scheme for a tray reactor (Durand, 2003) is shown in Fig. 7.10.

Rotating drum

The rotary drum reactor consists of a horizontal or near horizontal drum, which may or may not have baffles. Substrate bed occupies generally 10–40% of the reactor volume, and air is

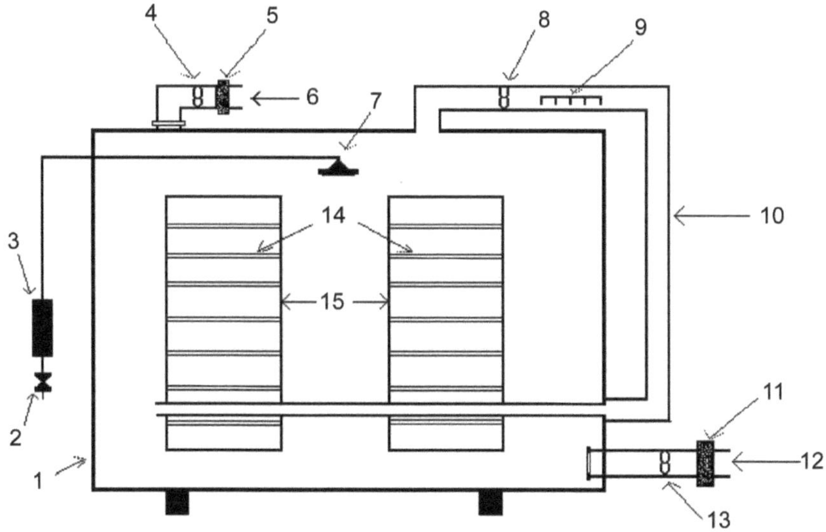

Figure 7.10
Tray reactor (Durand, 2003).

supplied at the top of the reactor. This reactor can operate continuously or be intermittently rotating. The frequency of mixing and rotational speed of the reactor speed can be controlled. (Mitchell et al., 2000).

This type of bioreactor is efficient to dissipate heat generated during microbial growth. However, mixing can damage the mycelium in fungal cultures. Therefore, generally low rotational speeds (1–15 rpm) are used. Intermittent agitation is also less harmful to the mycelium (Krishna, 2005).

Rotating drums bioreactors were first utilized for the production of amylase in SSF by *Aspergillus oryzae* using wheat bran as a substrate in the early 1900s. Then, in the 1940s, rotating drum bioreactors were improved and applied in the commercial-scale production of penicillin. Since then, few technological improvements have been made to this kind of bioreactors (Mitchell et al., 2000). A scheme for a rotary drum reactor (Durand, 2003) is shown in Fig. 7.11.

Continuous screw

Continuous screw reactors are based on screw conveyors. They can move solids with almost zero mixing in the direction of flow. When mixing is desired to be static or dynamic, mixers can be added. If there is no back mixing, external inoculation is necessary. Inoculation might be done by recycling a part of the fermented product, avoiding the need for a separate process for inoculum production (Sánchez et al., 2006).

Continuous screw reactors can be operated in batch or continuous mode. Continuous operation mode can reduce the time required to load, unload, and clean the reactor. In addition,

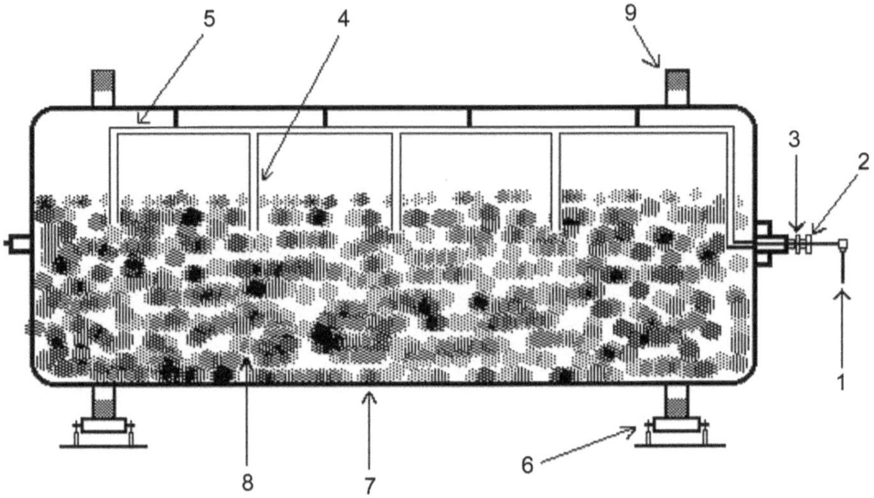

Figure 7.11
Rotary drum reactor (Durand, 2003).

continuous processes allow the use of smaller equipment in upstream and downstream processing (Khanahmadi et al., 2006). Unfortunately, there is limited experimental data on the use of continuous screw SSF bioreactors.

Gibbons et al. (1984, 1986) studied the production of ethanol from fodder beet and sweet sorghum in a continuous screw SSF bioreactor at intermediate scale (about 85 L). This reactor was not aerated, since the alcoholic fermentation is favored by anaerobic conditions. Under these conditions, continuous production of ethanol was maintained for 400 h. In this process, high final ethanol concentration (6–8%) and high yield (about 85% of theory) were reached.

More recently, Varzakas et al. (2008) studied the production of glucoamylases by *A. niger* in SSF using a continuous countercurrent reactor. The reactor consisted of a stainless steel tube of 3.4 m divided into nine compartments. The fermenting material was moved with a continuous stainless steel screw, and aeration was accomplished by natural convection. They observed a continuous production of glucoamylase (40 U/g). However, they also observed problems of heat accumulation and drying in the last compartment. These problems could probably be solved with the forced aeration with moist air. A scheme for a continuous screw reactor (Durand, 2003) is shown in Fig. 7.12.

7.3.3 Process Control

SSF, like any other bioprocess, is subject to disturbances that may affect performance. To ensure proper microbial growth, obtaining the desired products is necessary to measure and control different process variables. These variables include moisture, water activity (A_w),

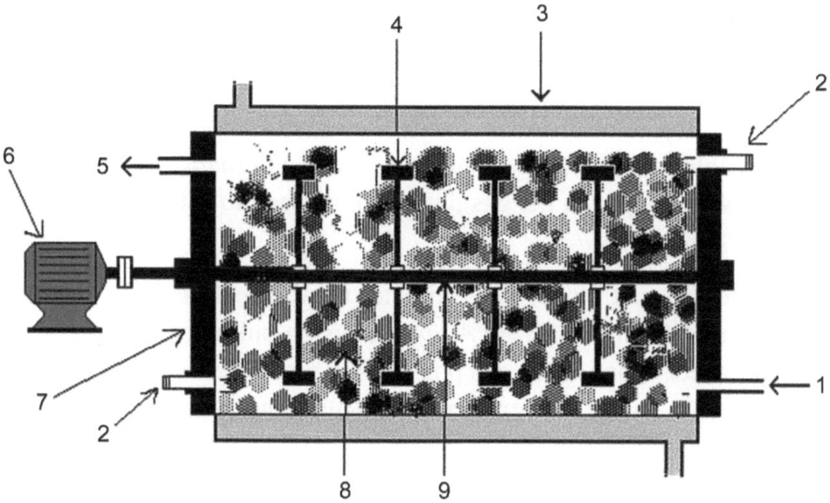

Figure 7.12
Continuous screw reactor (Durand, 2003).

temperature, and pH of the solid bed, as well as composition of the gas phase. In addition, knowledge of transport phenomena and growth kinetics are needed for the design, operation, and scaling up of SSF bioreactors.

Unlike submerged fermentation bioreactors, where there is a large number of standard tools for monitoring and control of process, measurement and control of operating variables in SSF bioreactors remains a challenge. The heterogeneity of the solid bed and the absence of a free aqueous phase hinder these tasks. Description of the most important process variables in SSF and strategies used for measurement and control of those variables are described.

Substrate moisture and water activity

One of the main characteristics of SSF is the absence of freely flowing water. Therefore, the water content (moisture) in SSF depends on the water retention capacity of the carrier used. Under these conditions, any loss of water can affect microbial growth. Water is necessary for microbial growth and has important biological functions. Water acts as a solvent, providing nutrients in soluble form and allowing the removal of waste metabolites. Water also has a structural function, which is implicated in the stability and the function of the biological structures organized at the molecular and cellular levels (Gervais and Molin, 2003).

Water requirements should be defined in terms of A_w, a thermodynamic parameter related to the chemical potential of water. A reduction in A_w normally extends the lag phase of the microbial growth, causing a low biomass production (Sermanni and Tiso, 2008). The water activity of the solid substrate can decrease during SSF as a result of dehydration of the solid bed or by accumulation of solutes in the substrate (Nagel et al., 2001).

In lab-scale SSF bioreactors, water losses can be avoided by forced aeration with humidified air. However, in large-scale SSF bioreactors, evaporative cooling results in large moisture losses and should therefore be combined with water addition to assure moisture-content control. Online monitoring of moisture and water activity of the solid bed is complicated, as there are no devices able to perform these measurements automatically. Instead, continuous measurement of the water vapor in the gas phase can be used to estimate the moisture loss from the solid bed.

Nagel et al. (2001) described a model that estimates the extracellular and overall water contents of wheat grains during SSF with *A. oryzae*, using online measurements of oxygen, carbon dioxide, and water vapor in the gas phase. This model can be used to calculate the water addition required to control the extracellular water content in a 35-L horizontal paddle mixer SSF bioreactor.

Temperature and heat transfer

Temperature control and heat dissipation are major concerns in the design, operation, and scaling up of SSF bioreactors. The heat generation is directly proportional to the metabolic activity. However, microbial growth and product formation are usually sensitive to temperature (Krishna, 2005).

Heat removal in SSF is difficult to achieve because of to the low thermal conductivity of most solid substrates. Evaporative cooling is the most important mechanism to remove metabolic heat in SSF. However, evaporative cooling leads to loss of moisture. Thus, supplemental water needs to be added to ensure the normal microbial growth (Chen, 2013). Forced aeration improves heat transfer in two ways: transferring heat directly and influencing the evaporation rate (Long, 2013). Nevertheless, in static bioreactors, forced aeration can lead to the formation of moisture and temperature gradients along the axis. The continuous or discontinuous mixing of the solid bed facilitates heat dissipation. It also prevents or reduces the formation of gradients temperature and moisture. However, mixing can also affect the mycelial growth in fungal cultures. In large-scale SSF bioreactors, a combination of aeration, water addition, and mixing is used to control temperature and moisture in solid bed.

Biomass and growth kinetics

Biomass is a fundamental parameter in the characterization of SSF, even if the biomass is not the final product. Biomass estimation is needed to analyze the progress of the fermentation, to study the growth kinetics, and develop mathematical models to calculate specific yields, etc. (Krishna, 2005).

Direct measurement of biomass is complicated in SSF due to the difficulty to separate the biomass and the solid substrate. Therefore, indirect measurement of the biomass is commonly used. There are several biochemical methods based on measurement of a component of the biomass, such as protein, glucosamine, ergosterol, and DNA (Durand et al., 1997; Mazaheri and Shojaosadati, 2013). These methods are tedious and time-consuming, and they are not suitable for process control. Moreover, the variation of the composition of the biomass during fermentation can lead to significant errors (Desgranges et al., 1991a,b). There are also several methods for online estimation of biomass in SSF. These methods are based on measurement of exhaust gases (CO_2 and O_2) (Figueroa-Montero et al., 2011; Saucedo-Castañeda et al., 1990), the electrical permittivity of biomass (Davey et al., 1991), and indirect measurement of cellular components by infrared spectroscopy (Desgranges et al., 1991a). Among these methods, the analysis of the exhaust gases is by far the most used technique.

SSF, like other bioprocesses, requires mathematical models for optimization and scale up. To develop a mathematical model for describing the SSF processes, a growth kinetic model should be developed first. Microbial growth kinetics can be described by empirical equations or by mechanistic models. In recent years, stoichiometric models considering the microbial metabolism, metabolic pathways, and metabolism regulation have been developed (Mazaheri and Shojaosadati, 2013).

Most of the growth kinetic profiles reported for SSF systems are based on simple empirical equations. Linear, exponential, logistic, and fast-acceleration/slow-deceleration are commonly used empirical growth rate models in SSF. Empirical equations do not include the effect of nutrient concentration on growth. Empirical equations are simply fitted to experimental growth profiles by nonlinear regression (Mitchell et al., 2004).

The logistic model, first used by Okazaki et al. (1980) is the most widely used model to describe the growth in SSF. The logistic model is symmetric around the inflection point. That means it considers that the acceleration and deceleration rates are the same. The logistic model is very convenient since it describes the entire growth curve with one simple equation. However, the logistic model cannot describe the entire growth profile accurately, especially during early stages of growth. In SSF, the real growth profile consists of two periods: a short period of rapid acceleration followed by a long period of slow growth deceleration. This behavior may be described accurately using a two-phase kinetic model (Mazaheri and Shojaosadati, 2013).

Mechanistic models attempt to describe how the growth can be affected by intraparticle diffusion of O_2, enzymes, hydrolysis products, other nutrients, and the role in the fermentation of other phenomena such as particle shrinkage and spatial microbial biomass distribution (Mitchell et al., 2004). Mechanistic models have been used to a lesser extent because they are more complicated and require more computing power. In addition, the acquisition of reliable experimental data on diffusion, substrate consumption, particle size, etc., is difficult.

Stoichiometric models are the most advanced approaches. Stoichiometric models focus on metabolic pathways and metabolic regulations. In these models, intracellular interactions and extracellular phenomena are considered (Mazaheri and Shojaosadati, 2013). Stoichiometric models have been used for studying phenomena that occur mainly in submerged fermentations, but there are a few reports on the application of this method for modeling SSF processes. For example, Mazutti et al. (2010) developed a mathematical model based on an artificial neural network to predict the cell growth and inulinase production by *Kluyveromyces marxianus* as a function of fermentation time, initial total reducing sugar concentration, and inlet and outlet temperatures in SSF.

pH value

pH is an important factor in any fermentation process. Each microorganism possesses a pH range for its growth and activity and an optimum value within the range. pH value may change in response to metabolic activities. The secretion of organic acids will cause the pH to drop; the assimilation of organic acids, present in some media, will lead to an increase in pH, and urea hydrolysis will result in alkalization (Krishna, 2005).

pH monitoring and control in SSF is difficult due to the lack of proper equipment and electrodes to determine the pH in solid materials and the existence of pH gradients due to the heterogeneous characteristics of SSF (Rodriguez-Leon et al., 2008). The direct measurement of pH in SSF is not reliable due to poor contact between the solid and the sensitive part of the electrode. There are flat electrodes that adapt better to solid samples, but their applicability is limited to static beds because agitation may damage the electrode. The measurement cannot be extrapolated to the rest of the bed due to its heterogeneity (Fernández and Pérez-Correa, 2006). Thus, pH cannot be measured online accurately in SSF, and discontinuous offline measurements are preferred.

An attempt to overcome the problem of pH variability during SSF process is obtained by substrate formulation, considering the buffering capacity of the different components employed or by the use of buffer formulation with components that have no deleterious influence on the biological activity (Cuadra et al., 2008). The use of urea as a nitrogen source is one way of controlling the pH (Krishna, 2005; Saucedo-Castañeda et al., 1992a).

7.3.4 Microorganisms

Bacteria, fungi, and yeasts can grow on solid substrates; however, filamentous fungi are the best adapted for SSF and dominate in research works. While bacteria are mainly involved in composting, ensiling, and some food processes (Doelle et al., 1992), yeasts have been used for ethanol and food or feed production (Saucedo-Castañeda et al., 1992a). However, filamentous fungi are the most important group of microorganisms used in SSF process, owing to their physiological, enzymatic, and biochemical properties. The hyphal mode of fungal growth and their good tolerance to low A_w and high osmotic pressure conditions make fungi efficient and competitive in natural microflora for bioconversion of solid substrates. Koji and tempeh are the two most important applications of SSF with filamentous fungi. *Aspergillus oryzae* is grown on wheat bran and soybean for koji production, which is the first step of soy sauce or citric acid fermentation. Koji is a concentrated hydrolytic enzyme medium required in further steps of the fermentation process. The hyphal mode of growth gives a major advantage to filamentous fungi over unicellular microorganisms in the colonization of solid substrates and for the utilization of available nutrients. The basic mode of fungal growth is a combination of apical extension of hyphal tips and the generation of new hyphal tips through branching (Viniegra-González et al., 2003). An important feature is that although extension occurs only at the tip at a linear and constant rate, the frequency of branching makes the kinetic growth pattern of biomass exponential, mainly in the first steps of the vegetative stage (Aguilar et al., 2001). The hyphal mode of growth gives the filamentous fungi the power to penetrate into the solid substrates. The cell wall structure attached to the tip and the branching of the mycelia makes the fungi the microorganism most adapted to the SSF culture (Viniegra-González and Favela-Torres, 2006).

A large number of microorganisms, including bacteria, yeast, and fungi that produce different groups of enzymes have been enumerated by Soccol and Vandenberghe (2002); they enumerate the variety of microbial cultures employed for enzyme production in SSF. The selection of a suitable strain for the required purpose depends upon a number of factors, in particular upon the nature of the substrate and environmental conditions. Generally, hydrolytic enzymes are produced by fungal cultures, since such enzymes are used in nature by fungi for their growth. Although commercial production of amylases is carried out using both fungal and bacterial cultures, bacterial α-amylase is generally preferred for starch liquefaction due to its high temperature stability. However, in a high productivity with less production cost, genetically modified strains would hold the present microbe selection for both SmF and SSF.

7.3.5 Substrates

One of the advantages of the SSF process depends on the low cost and availability of the substrate. A number of agroindustrial residues have been employed for the production of a wide number of biologic molecules by SSF. Wheat bran has the most commonly been used for enzyme production; however, a long list of low-cost substrates can be enumerated including apple, pomace, aspen pulp, banana waste, cassava flour, cassava waste, coconut coir pith, coconut oil cake, corn flour, corncobs, gram bran, grape orujo, maize bran, mustard oil cake, palm oil mill waste, peanut meal, rapeseed cake, rice bran, rice husk, rice straw, sago humps, sawdust, soy hull, starch, steam pretreated willow, steamed rice, sugar beet pulp, sugarcane bagasse, sweet sorghum pulp, tea waste, wheat flour, and wheat straw. In SSF, the solid substrate has different functions; the substrate supplies the nutrients to the microbial culture growing and also serves as an anchorage for microbial growth. In SSF, microbial growth and product formation occurs at and between the solid mass to be fermented. Also, a rigid substrate structure to keep the porosity through the process is needed. Most agroindustrial residues will cover the required texture through the particle-sized selection; however, in case of a very soft material, a mixing of different texture residues can be the solution to achieve the best water-holding capacity and moisture critical point.

Catabolite repression

Most microorganisms may grow in several carbon sources, but glucose and fructose are the preferred. Microorganism produces the enzymes required for the use of alternative carbon sources, but enzyme production rate can be slow or null (Gancedo, 1998). This phenomenon is known as catabolite repression, used to describe the inhibition of microbial synthesis of many enzymes that require a chemical signal, known as the inducer, and are repressed when there is abundance in the medium of a ready fermentable substrate such as glucose, galactose, maltose, glycerol, or other carbon source (Aguilar et al., 2001; Hogema et al., 1997; Viniegra-González and Favela-Torres, 2006).

Resistance to catabolite repression is one of the most important differences between SSF and SmF and therefore might contribute to the observed differences in physiology (Barrios-González, 2012). Catabolite repression of enzyme synthesis in the presence of glucose and other carbon sources in SmF has been reported, while SSF could reduce significantly the catabolite repression on several enzymes production in some microorganisms (Kriaa and Kammoun, 2015). Aguilar et al. (2001) reported the induction and repression patterns (using tannic acid and glucose, respectively) of tannase production by *A. niger* Aa20 in SSF and SmF. Authors observed that the tolerance to high concentrations of tannic acid was higher in SSF than SmF, while the addition of glucose (>20 g/L) produced a strong catabolite repression in SmF, demonstrating the capacity of SSF to reduces the catabolite repression. In addition, SSF provides less degradation on products by proteases (Hansen et al., 2015).

Agroindustrial by-products transformation

A large amount of agroindustrial residues are generated from the industrial processing of agricultural products, including straws, stems, stalks, leaves, seeds, husk, peels from fruits, legumes or cereals, sugarcane bagasse, wheat bran, rice bran, maize bran, gram bran, wheat straw, rice straw, rice husk, soy hull, sago hampas, grapevine trimmings dust, sawdust, corncobs, coconut coir pith, banana waste, tea waste, cassava waste, palm oil mill waste, aspen pulp, sugar beet pulp, sweet sorghum pulp, apple pomace, peanut meal, rapeseed cake, coconut oil cake, mustard oil cake, cassava flour, wheat flour, corn flour, steamed rice, steam pretreated willow, and starch (Flores-Maltos et al., 2014). Those by-products are mainly composed by sugars, proteins, fibers (cellulose, hemicellulose, and lignin) and minerals, which are considered as abundantly available and cheap of nutrients. The interest on the valorization of agroindustrial by-products is based on economic and environmental aspects. The agroindustrial by-products could be used as low-cost substrates to produce several value-added compounds (Mussatto et al., 2012).

Several studies have been reported the use of agroindustrial by-products transformation by SSF, obtaining value-added compounds such as valuable enzymes peptides, antibiotics, mycotoxins, alkaloids, food grade pigments, plant growth factor, and phenolic compounds, among many others (Martins et al., 2011).

7.3.6 Biomolecules Produced by Solid-State Fermentation

SSF has been used in two potential areas; one of them is the environmental issues, and it has been used on processes such composting and ensiling animal feed from solid wastes. Recently has been implemented in the bioremediation and biodegradation of hazardous compounds and in detoxification of agroindustrial wastes. However, in the last three decades, SSF started to be used to obtain value-added compounds such as enzymes, mushrooms, amino acids, biopesticides, biofuels, biosurfactants, organic acids, flavors, colorants, aromatic compounds, biologically active secondary metabolites, and other substances of industrial interest (Mussatto et al., 2012).

7.4 Solid-State Fermentation Versus Submerged Fermentation: General Considerations

From ancient times, the fermentation process has been extensively used for the biological conversion of complex substrates into simple compounds by several microorganisms, such as fungi, bacteria, and yeast. Fermentation techniques have great importance due to their economic and environmental advantages, as previously discussed. The compounds produced by fermentation process are highly beneficial to individuals and industry. Researchers have focused on the development of fermentation techniques to improve and maximize the productivity. SSF and SmF have been refined based on several parameters and the microorganism used in the process. However, there are still many problems that need to be overcome. Some differences between SSF and SmF are discussed in Table 7.2.

Established differences among the two systems should not be seen as a replace one to the other. Few of the many products on the market are produced by SSF. However, there are products in which SSF process is better than SmF, and now they are produced commercially. Despite this, it is necessary to develop technology that allows obtaining parameters to achieve the full potential of the SSF. Notwithstanding the increasing number of publications dealing with both SSF and SmF, it is not easy to underline general considerations from the literature presented. This is due mostly to the nonadjustment on the process variables and products units reported that would allow impartial comparison between fermentative methods. Both have advantages and disadvantages, and the way to solve the second makes it better one fermentative process over the other one.

Research is focused on the general applicability of SmF and SSF for the production of microbial metabolites; for instance, on a gram bench scale, SSF appears to be higher bioactivity than SmF in several aspects. SSF appears to possess several biotechnological advantages, though at a laboratory scale, than higher fermentation productivity, higher end-concentration of products, higher product stability, lower catabolic repression, water-insoluble substrates, or mixed cultivation of various fungi and some bacteria and yeast, and nonneed of sterility due to the low water activity. However, the SSF upscaling for industrial scale raises severe

Table 7.2: Differences Between Solid-State (SSF) and Submerged Liquid Fermentation (SmF).

Characteristic	SSF	SmF
Cost of culture medium	Low (use of agroindustrial by-products as carbon source)	High (use of chemical reagents)
Water amount	Low	High (main component of the culture)
Nutrients' availability	Easy (absorbed from wet solid substrate, there is a nutrient concentration gradient)	Difficult (absorbed from the liquid culture, there is no nutrient concentration gradient)
Culture system	Three phases (gas, liquid, solid), where gas is the continuous phase	Two phases (water and gas), water is the continuous phase
Inoculum size	Large, more than 10%	Small, less than 10%
Required oxygen	Obtained from gas phase, low energy consumption	Obtained from dissolved oxygen, there is a larger amount of dissolved oxygen
Microbial growth	Microorganisms adsorb on or penetrate into the solid matrix	Microorganisms growth uniformly distributed in the culture system
Products concentration	High, due to the medium is a wet state substrate	Low, due to the medium is liquid
Yield	Greater	Smaller
Fermentation process (temperature control, pH regulation, agitation control)	Difficult (hard to detect and control online)	Easy (can be detected and controlled online)
System	Heterogeneity	Homogeneity
Extraction process	Simple and controllable, little waste organic water	Usually complex, there is a large amount of waste organic water
Bacterial contamination	Low risk	High risk
Morphology of fungal growth	Mycelial form, where both aerial and substrate-penetrating hyphae are produced	Varies from freely dispersed mycelium to mycelial clumps and dense pellets
Catabolic response	Low catabolic repression	High catabolic repression
Reactor design	Hard to scale up, simple fermentation container	Easy to work in any scale, sealed fermentation container
Energy consumption	Low	High
Energy requirement	Low	High
Equipment investment	Low	High
Raw material cost	Low	High

Data from Chen, H., 2013. Modern Solid State Fermentation: Theory and Practice. Springer Netherlands, New Delhi, India; Hansen, G.H., Lübeck, M., Frisvad, J.C., Lübeck, P.S., Andersen, B., 2015. Production of cellulolytic enzymes from ascomycetes: comparison of solid state and submerged fermentation. Process Biochemistry 50, 1327–1341.

engineering problems due to the temperature, pH, O_2, substrate, and moisture gradients along the length and wide of the solid fermenters; then comparison works have to also focus on progress toward industrial engineering. The fact that physiological adaptation of the microorganisms during growth on SSF compared with SmF solutions has been neglected, despite the fact that it may make SSF advantageous against the SmF (Viniegra-González et al., 2003). Then, as it was said before, it is not a matter of "better that" statement; it is rather what can be achieved with the available chain of recourses.

7.5 Perspectives and Conclusions

SSF and SmF systems are both very promising for the production of value-added products. In addition, they offer potential advantages. However, several aspects related to bioreactor design have to be studied in detail. Higher development on automation of the SSF process is needed for increasing its industrial exploitation; while lowering the catabolic inhibition and reduction in water consumption are drown backs for SmF. No method can be called the better, as there are many factors to be considered to be so, and it is still difficult to make a fair comparison due to the limitation still on instrumental devices for the proper variable control of both fermentation methods. Work on the technological facilities for the matter has to be done before one fermentation method may claimed as the best; however, it is not to be forgotten that the best fermentation method is the one can be performed with the substrate, stains, and fermenters available for the matter.

References

Aguilar, C., Augur, C., Favela-Torres, E., Viniegra-González, G., 2001. Induction and repression patterns of fungal tannase in solid-state and submerged cultures. Process Biochemistry 36, 565–570.

Ali, H.K.Q., Zulkali, M.M.D., 2011. Design aspects of bioreactors for solid-state fermentation: a review. Chemical and Biochemical Engineering Quarterly 25, 255–266.

Barrios-González, J., 2012. Solid-state fermentation: physiology of solid medium, its molecular basis and applications. Process Biochemistry 47, 175–185.

Chen, H., 2013. Modern Solid State Fermentation: Theory and Practice. Springer Netherlands, New Delhi, India.

Chen, H.-Z., Xu, J., Li, Z.-H., 2005. Temperature control at different bed depths in a novel solid-state fermentation system with two dynamic changes of air. Biochemical Engineering Journal 23, 117–122.

Cuadra, T., Fernández, F.J., Tomasini, A., Barrios-González, J., 2008. Influence of pH regulation and nutrient content on cephalosporin C production in solid-state fermentation by *Acremonium chrysogenum* C10. Letters in Applied Microbiology 46, 216–220.

Davey, C.L., Peñaloza, W., Kell, D.B., Hedger, J.N., 1991. Real-time monitoring of the accretion of *Rhizopus oligosporus* biomass during the solid-substrate tempe fermentation. World Journal of Microbiology and Biotechnology 7, 248–259.

Desgranges, C., Georges, M., Vergoignan, C., Durand, A., 1991a. Biomass estimation in solid state fermentation II. On-line measurements. Applied Microbiology and Biotechnology 35, 206–209.

Desgranges, C., Vergoignan, C., Georges, M., Durand, A., 1991b. Biomass estimation in solid state fermentation I. Manual biochemical methods. Applied Microbiology and Biotechnology 35, 200–205.

Doelle, H.W., Mitchell, D.A., Rolz, C.E., 1992. Solid Substrate Cultivation. Elsiever Science Publication Ltd, London & New York, 466 p.

Durand, A., 2003. Bioreactor designs for solid state fermentation. Biochemical Engineering Journal 13, 113–125.

Durand, A., Vergoignan, C., Desgranges, C., 1997. Biomass estimation in solid state fermentation. In: Roussos, S., Lonsane, B.K., Raimbault, M., Viniegra-Gonzalez, G. (Eds.), Advances in Solid State Fermentation. Springer Netherlands, pp. 23–37.

Fernández, M., Pérez-Correa, J.R., 2006. Instrumentation for monitoring SSF bioreactors. In: Mitchell, D., Berovič, M., Krieger, N. (Eds.), Solid-State Fermentation Bioreactors. Springer Berlin Heidelberg, pp. 363–374.

Figueroa-Montero, A., Esparza-Isunza, T., Saucedo-Castañeda, G., Huerta-Ochoa, S., Gutiérrez-Rojas, M., Favela-Torres, E., 2011. Improvement of heat removal in solid-state fermentation tray bioreactors by forced air convection. Journal of Chemical Technology and Biotechnology 86, 1321–1331.

Flores-Maltos, D.A., Mussatto, S.I., Contreras, J.C., Buenrostro, J.J., Rodríguez, R., Teixeira, J.A., Aguilar, C.N., 2014. Typical Mexican agroindustrial residues as supports for solid-state fermentation. American Journal of Agricultural and Biological Sciences 9, 289–293.

Gancedo, J., 1998. Yeast carbon catabolite repression. Microbiology and Molecular Biology Reviews 62, 334–361.

Gervais, P., Molin, P., 2003. The role of water in solid-state fermentation. Biochemical Engineering Journal 13, 85–101.

Gibbons, W.R., Westby, C.A., Dobbs, T.L., 1984. A continuous, farm-scale, solid-phase fermentation process for fuel ethanol and protein feed production from fodder beets. Biotechnology and Bioengineering 26, 1098–1107.

Gibbons, W.R., Westby, C.A., Dobbs, T.L., 1986. Intermediate-scale, semicontinuous solid-phase fermentation process for production of fuel ethanol from sweet sorghum. Applied and Environmental Microbiology 51, 115–122.

Hansen, G.H., Lübeck, M., Frisvad, J.C., Lübeck, P.S., Andersen, B., 2015. Production of cellulolytic enzymes from ascomycetes: comparison of solid state and submerged fermentation. Process Biochemistry 50, 1327–1341.

Hogema, B.M., Arents, J.C., Inada, T., Aiba, H., Van Dam, K., Postma, P.W., 1997. Catabolite repression by glucose 6-phosphate, gluconate and lactose in *Escherichia coli*. Molecular Microbiology 24, 857–867.

Hölker, U., Höfer, M., Lenz, J., 2004. Biotechnological advantages of laboratory-scale solid-state fermentation with fungi. Applied Microbiology and Biotechnology 64, 175–186.

Khanahmadi, M., Mitchell, D.A., Beheshti, M., Roostaazad, R., Sánchez, L.R., 2006. Continuous solid-state fermentation as affected by substrate flow pattern. Chemical Engineering Science 61, 2675–2687.

Kriaa, M., Kammoun, R., 2015. Producing *Aspergillus tubingensis* CTM507 glucose oxidase by solid state fermentation versus submerged fermentation: process optimization and enzyme stability by an intermediary metabolite in relation with diauxic growth. Journal of Chemical Technology and Biotechnology. http://dx.doi.org/10.1002/jctb.4753.

Krishna, C., 2005. Solid-state fermentation systems—an overview. Critical Reviews in Biotechnology 25, 1–30.

Long, L., 2013. Process engineering of solid-state fermentation. In: Solid State Fermentation for Foods and Beverages. CRC Press, pp. 55–94.

Martins, S., Mussatto, S.I., Martínez-Avila, G., Montañez-Saenz, J., Aguilar, C.N., Teixeira, J.A., 2011. Bioactive phenolic compounds: production and extraction by solid-state fermentation. A review. Biotechnology Advances 29, 365–373.

Mazaheri, D., Shojaosadati, S.A., 2013. Mathematical models for microbial kinetics in solid-state fermentation: a review. Iranian Journal of Biotechnology 11, 156–167.

Mazutti, M.A., Zabot, G., Boni, G., Skovronski, A., Oliveira, D.d., Luccio, M.D., Rodrigues, M.I., Treichel, H., Maugeri, F., 2010. Kinetics of inulinase production by solid-state fermentation in a packed-bed bioreactor. Food Chemistry 120, 163–173.

Mitchell, D., Berovic, M., Krieger, N., 2000. Biochemical engineering aspects of solid state bioprocessing. In: New Products and New Areas of Bioprocess Engineering. Springer Berlin Heidelberg, pp. 61–138.

Mitchell, D.A., Krieger, N., 2006. Growth kinetics in SSF systems: experimental approaches. In: Mitchell, D., Berovič, M., Krieger, N. (Eds.), Solid-State Fermentation Bioreactors. Springer Berlin Heidelberg, pp. 207–218.

Mitchell, D.A., von Meien, O.F., Krieger, N., 2003. Recent developments in modeling of solid-state fermentation: heat and mass transfer in bioreactors. Biochemical Engineering Journal 13, 137–147.

Mitchell, D.A., von Meien, O.F., Krieger, N., Dalsenter, F.D.H., 2004. A review of recent developments in modeling of microbial growth kinetics and intraparticle phenomena in solid-state fermentation. Biochemical Engineering Journal 17, 15–26.

Mussatto, S.I., Ballesteros, L.F., Martins, S., Teixeira, J.A., 2012. Use of agro-industrial wastes in solid-state fermentation processes. In: Kuan-Yeow, S., Xinxin, G. (Eds.), Industrial Waste. InTech, Croatia, pp. 121–140.

Nagel, F.-J.J.I., Tramper, J., Bakker, M.S.N., Rinzema, A., 2001. Model for on-line moisture-content control during solid-state fermentation. Biotechnology and Bioengineering 72, 231–243.

Okazaki, N., Sugama, S., Tanaka, T., 1980. Mathematical model for surface culture of koji mold: growth of koji mold on the surface of steamed rice grains (ix). Journal of Fermentation Technology 58, 471–476.

Pandey, A., Soccol, C.R., Mitchell, D., 2000. New development in solid state fermentation. I. Bioprocecess and products. Process Biochemistry 35, 1153–1169.

Raghavarao, K.S.M.S., Ranganathan, T.V., Karanth, N.G., 2003. Some engineering aspects of solid-state fermentation. Biochemical Engineering Journal 13, 127–135.

Rodriguez-Leon, J., Soccol, C., Pandey, A., Rodriguez, D., 2008. Factors affecting solid-state fermentation. In: Pandey, A., Soccol, C., Larroche, C. (Eds.), Current Developments in Solid-State Fermentation. Springer, New York, pp. 26–47.

Sánchez, L.R., Khanahmadi, M., Mitchell, D., 2006. Continuous solid-state fermentation bioreactors. In: Mitchell, D., Berovič, M., Krieger, N. (Eds.), Solid-State Fermentation Bioreactors. Springer Berlin Heidelberg, pp. 141–158.

Saucedo-Castañeda, G., Gutiérrez-Rojas, M., Bacquet, G., Raimbault, M., Viniegra-González, G., 1990. Heat transfer simulation in solid substrate fermentation. Biotechnology and Bioengineering 35, 802–808.

Saucedo-Castañeda, G., Lonsane, B.K., Navarro, J.M., Roussos, S., Raimbault, M., 1992a. Importance of medium pH in solid state fermentation for growth of *Schwanniomyces castellii*. Letters in Applied Microbiology 15, 164–167.

Sermanni, G., Tiso, N., 2008. Aspects of design of bioreactors in SSF. In: Pandey, A., Soccol, C., Larroche, C. (Eds.), Current Developments in Solid-State Fermentation. Springer, New York, pp. 117–144.

Soccol, C.R., Vandenberghe, L.P.S., 2002. Overview of applied solid state fermentation in Brazil. Biochemical Engineering Journal 3648, 1–14.

Thomas, L., Larroche, C., Pandey, A., 2013. Current developments in solid-state fermentation. Biochemical Engineering Journal 81, 146–161.

Varzakas, T.H., Roussos, S., Arvanitoyannis, I.S., 2008. Glucoamylases production of *Aspergillus nige*r in solid state fermentation using a continuous counter-current reactor. International Journal of Food Science and Technology 43, 1159–1168.

Viniegra-González, G., Favela-Torres, E., Aguilar, C.N., Romero-Gómez, S.J., Diaz-Godinez, G., Augur, C., 2003. Advantages on fungal enzyme production in solid state over liquid fermentation systems. Culture Biochemical Engineering Journal 13, 157–167.

Viniegra-González, G., Favela-Torres, E., 2006. Why solid-state fermentation seems to be resistant to catabolite repression? Food Technology and Biotechnology 44, 397–406.

CHAPTER 8

Composition of Plant Biomass Biotech Engineering of Cell Wall to Optimize Biofuel Production

F. Francocci, I.B. Reca
CNR-ISPA, National Research Council of Italy, Lecce, Italy

8.1 Introduction

Plant cell walls represent the most abundant renewable source of energy of the Earth. For thousands of years, their use mainly concerned animal feed and heat power as well as the large employ in the paper industry. This abundant resource is today considered a promising source of energy that can be converted into biofuels for transport and products for electricity and heat production, as well as biogas International Energy Agency (IEA) (IEA, 2008). Cellulose, that is the main component of the plant cell wall, is embedded in a matrix of other polysaccharides, and together they form the skeleton of the plant tissue (Somerville, 2006) (Keegstra, 2010). This structure is naturally recalcitrant to the deconstruction carried out by microbes and enzymes (Himmel et al., 2007; Kalluri et al., 2014), so breakthrough technologies are needed to overcome barriers to developing cost-effective processes for the conversion of biomass into fuels and chemicals. For all these reasons, the interest in deeply understanding cell wall structure and composition of plants is becoming even more essential.

The progress made in understanding plant cell wall biosynthesis and remodeling and the biotechnological approaches to reducing biomass recalcitrance are here described.

8.2 The Plant Cell Wall: An Overview

The cell wall is a dynamic interface that changes during the growth and differentiation of the cell. It plays crucial roles for a number of essential functions of a plant, such as providing shape to cells, giving sustainment to plants tissues and organs, and forming the interface between adjacent cells. The cell wall is also involved in intercellular communication (Tameshige et al., 2015) and cellular response to external stimuli, including pathogens (Bellincampi et al., 2014). The plant cell wall is a highly complex structure formed by a mixture of polysaccharides, proteins, and other polymers assembled to form a rigid network with a fine and dynamical

Biotransformation of Agricultural Waste and By-Products. http://dx.doi.org/10.1016/B978-0-12-803622-8.00008-2

organization (Alberts et al., 2002). The primary wall surrounds growing or dividing cells; these walls provide mechanical strength but must also expand to allow the cell to grow and divide; secondary, more thickened, cell walls instead surround specialized cells such as vessel elements or fiber cells (Cosgrove and Jarvis, 2012) and account for most of the carbohydrates in biomass. The division in two cell wall-type categories is indicative because all differentiated cells contain their proper, distinct wall compositions, generating a variability in which primary and secondary walls are considered at the two opposite extremes. Both primary and secondary walls contain cellulose and hemicelluloses, but while primary walls contain pectins, enzymes, and structural proteins, secondary walls normally contain lignin and a small amount of protein and pectin (Somerville et al., 2004; Cosgrove and Jarvis, 2012).

Most plant biomass, consisting primarily of cellulose, hemicellulose, and lignin, is comprised respectively of about 40–50%, 20–30%, and 5–25% by weight of these cell wall polymers (Pattathil et al., 2015; Anwara et al., 2014). Primary cell walls of grasses contain high levels of cellulose and of cell wall-bound hydroxycinnamic acids but negligible amounts of pectin (1–2%) (Vogel, 2008; Molinari et al., 2013; Pauly and Keegstra, 2008). In dicotyledonous plants, instead, the primary wall consists of around 30% cellulose, 30% hemicelluloses, 35% pectin, and 1–5% structural proteins on a dry weight basis (Vorwerk et al., 2004; Pattathil et al., 2015). The structures of the individual polymers are well known, but their specific organizations in the cell wall are not completely understood yet.

While cellulose is synthesized at the plasma membrane, the remaining matrix polysaccharides are first made in the Golgi and then exported to the cell wall by exocytosis, where they intercalate among cellulose microfibrils to form a composite wall (Somerville et al., 2004).

Although many progresses have been made in defining the structures of the various polysaccharides that make up plant cell walls, still there are holes in our understanding of the biosynthesis of all wall polysaccharides. The processes and players involved in assembling functional cell walls from components synthesized at different cellular locations are largely unknown.

One reason for the lack of information regarding the biosynthesis of cell wall polysaccharides is that the application of the traditional biochemical techniques is not useful; by the contrary, the application of genetic and genomic strategies, has led to significant progress in identifying the genes and proteins required for cell wall biosynthesis (Keegstra, 2010). As described in this chapter, however, many important questions regarding the biochemistry, cell biology, and regulation of polysaccharide biosynthesis remain unanswered.

8.3 Organization and Biosynthesis of Wall Components

8.3.1 Cellulose

Cellulose is the most abundant polysaccharide that represents the core of plant cell walls where it functions as a scaffold for the binding of other wall components. It consists of β-1,4-linked glucose chains that are then packed into highly ordered microfibrils in which all chains lie in parallel direction; those microfibrils are embedded in a matrix that contains other polysaccharides, glycoproteins, and proteins. Cellulose is made at the plasma membrane by cellulose synthase complexes that contain multiple cellulose synthase subunits, which form a rosette structure and are deposited directly into the cell wall in a directional manner (Somerville, 2006). The plasma membrane rosettes contain the cellulose synthase catalytic subunit (CESA) proteins that are encoded by the CESA genes. It is thought that each hexameric rosette comprises six rosette subunits and that each rosette subunit contains six CESA proteins, providing a total of 36 CESA proteins per rosette. In Arabidopsis, CESA1, CESA3, and CESA6 are required for cellulose biosynthesis in primary cell walls, whereas CESA4, CESA7, and CESA8 are required for cellulose biosynthesis during secondary wall deposition. Similar conclusions have been reached in other plants, although the numbering system for CESA proteins varies among plant species (Guerriero et al., 2010; Hill et al., 2014; Kumar and Turner, 2015).

Other important cellulose biosynthesis machinery components were identified as associated wall proteins (KORRIGAN, KOBITO, COBRA) whose molecular functions are still unknown (Pagant et al., 2002; Zuo et al., 2000; Roudier et al., 2005; Gu et al., 2010). There are some detailed review articles to consult on the subject of cellulose biosynthesis, here we report Guerriero et al. (2010), Somerville (2006), Somerville et al. (2004), and Kumar and Turner (2015).

8.3.2 Hemicellulose

Hemicellulosic polysaccharides are complex molecules that, associated with cellulose microfibrils, provide a cross-linked matrix. The hemicelluloses are more abundant in secondary walls than in the primary walls of both dicots and monocot species. Monocot species have significantly more hemicellulose and less pectin than dicots, and also have mixed-linkage glucans (MLG) that make up a major proportion of monocot hemicellulose polysaccharides (Pauly and Keegstra, 2008). Hemicelluloses can be divided into four main classes: xyloglucans (XyG), which contain a heavily substituted β-1,4-glucan backbone; heteromannans, containing a variably substituted backbone that includes β-1,4-linked mannose (glucose and mannose) residues; heteroxylans (GAX), containing a substituted β-1,4-linked xylan backbone; and MLG, which involve an unsubstituted backbone of glucosyl residues containing both β-1,3- and β-1,4-linkages (Somerville et al., 2004).

XyG is likely to interact with cellulose microfibrils as they are synthesized into the primary wall matrix, causing microfibrils of smaller diameter (less chains per fiber) than those found in secondary walls. The binding of XyG to cellulose is also known to weaken cellulose networks, but increases the expansibility of such networks (Chanliaud et al., 2002). For a detailed list and function of all known hemicellulose biosynthetic enzymes, consult Pauly et al. (2013).

8.3.3 Pectins

Pectins of higher plants form the structurally most complex family of polysaccharides in nature that make up 35% of the primary cell wall in dicotyledonous plants and nongraminaceous (nongrass) monocots, while they are present only in a percentage of 2–10% of grass primary walls and up to 5% of wood tissues (Mohnen, 2008).

The simplest and most abundant of pectic substructures is homogalacturonan (HG), a linear polymer of (1–4)-α-D-GalA that is synthesized in the Golgi and transported to the wall in membrane vesicles (Harholt et al., 2010). Carboxylic functions of GalA residues are partly methylesterified by wall-localized pectin methylesterases (PMEs) (Pelloux et al., 2007). The negatively charged forms of HG in the presence of positively charged ions such as Ca++ results in the formation of HG:HG salt bridges that contribute to cell–cell adhesion (Daher and Braybrook, 2015). HG can be substituted at O-3 by xylose to form xylogalacturonan, in apple, watermelon (Zablackis et al., 1995), and other plant species (Zandleven et al., 2007). The second major substructure, rhamnogalacturonan I (RG-I), is composed of a repeating disaccharide [–2)-α-L -Rha -(1–4)-α-D-GalA-(1-]n; the Rha residues in the backbone can be substituted and may vary among plants (Harholt et al., 2010). Rhamnogalacturonan II (RG-II) is a minor substructure where the HG backbone carries four different and highly conserved side chains containing different sugars and derivatives (O'Neill et al., 2004). Many are the reports about the models hypothesized for pectin assembly in the wall (Talmadge et al., 1973; Vincken et al., 2003; Voragen, 2009), but there is no general consensus about that. However, data from literature support a model with HG, RG-I, and RG-II linked via their backbone. Pectins are also supposed to be associated or covalently linked to other types of wall polysaccharides as xyloglucans and xylans, indicating the possibility that pectins may help to hold at least some hemicelluloses in the wall (Caffall and Mohnen, 2009). Approximately 67 glycosyltransferase, methyltransferase, and acetyltransferase activities are predicted to be required for pectin synthesis as described in a detailed review (Mohnen, 2008). What is well demonstrated, however, is the key role of HG in maintaining a correct cell adhesion in vivo. Two mutants, *quasimodo1* and *quasimodo2*, carrying mutations respectively in a galacturonosyltransferase (a clade of the glycosyltransferase 8 (GT8) family (Hao et al., 2014; Yin et al., 2010), and in a putative pectin methyltransferase, were shown to be affected in cellular adhesion (Bouton et al., 2002; Mouille et al., 2007). In *quasimodo2*, the cell adhesion defect was related to a 25 percent decrease in the absolute amount of HG in the cell wall without modification of the methyl esterification rate of the HG. This strong phenotype suggests that the HG content of the cell wall must be timely and temporally regulated during plant development. For a detailed review on pectin biosynthesis, consult Harholt et al. (2010), Mohnen (2008), and Caffall and Mohnen (2009).

8.3.4 Lignin

Lignin, a major component of plant cell walls, is an abundant aromatic biopolymer that occurs via oxidative coupling of monolignols, which are synthesized from the phenylpropanoid pathway. Lignin biosynthesis is well characterized, and all the enzymes required for the synthesis of its three major building blocks (monolignols), *p*-coumaryl, coniferyl, and sinapyl alcohols, are well known and highly conserved in all vascular plants. At least 10 enzymes are involved in catalyzing monolignol biosynthesis; between these, there are phenylalanine ammonia lyase, cinnamate 4-hydroxylase (C4H), caffeoyl CoA O-methyltransferase, or cinnamoyl CoA reductase (Bonawitz and Chapple, 2010). Monolignols are synthesized in the cytosol, transported into cell walls, and here polymerized into lignin via oxidative reactions catalyzed by oxidases, as laccases and peroxidases (Vanholme et al., 2010). Lignin composition, that is, the proportion of each of three polymerized monolignols, differs significantly among different plant species but also among different cell types of the same species. Lignin fills the spaces in the cell wall between cellulose, hemicellulose, and pectin components, especially in tracheid, sclereid, and xylem cells. It also links hemicellulose depositing on the cellulose surface, thus conferring strength properties to the cell wall.

Several review articles are available for a detailed view of the lignin biosynthesis pathway (Bonawitz and Chapple, 2010; Vanholme et al., 2010; Zhao and Dixon, 2011).

8.4 Enzymes and Their Role in Cell Wall Modification (and Deconstruction)

8.4.1 Cell Wall Modification/Deconstruction: Learning From the Plant

Plants have endogenous mechanisms to modify their own cell wall (Rose and Bennett, 1999; Roberts et al., 2002) that are required in a broad spectrum of plant processes as in fruit ripening, aerenchyma formation, leaf abscission, mobilization of cell wall storage polysaccharides, and so on. Recent studies proposed a model for cell wall disassembly as a repetition of modules that include cell separation, cell expansion, hemicellulose, and cellulose hydrolysis. For more details, see (Tavares et al., 2015). During endogenous cell wall hydrolysis, the plant glycosyl hydrolases (GHs) are produced and secreted to the wall in a temporal sequence that follows specific cell and tissue processes. These can be seen in the cell separation and expansion modules where expansins and pectinases are recruited first, and only afterward in the process, xyloglucanases and cellulases participate to loosen the walls.

Pectins

In nature, the attack on pectins is an early step in different endogenous cell wall degradation events prior to gain access to other cell wall polysaccharides. Particularly in dicots, the cohesive pectin matrix is critical for tissue integrity and accessibility. Intermolecular bonds of pectin, mediated by acidic HG (HGA), are the major strength-conferring elements and influence wall plasticity (Ezaki et al., 2005) and cell adhesion (Willats et al., 2001). HGA, as widely described before, is synthesized and secreted as a highly methylesterified polymer (Zhang and Staehelin, 1992), and its structure can be modified by the activity of cell wall enzymes (Willats et al., 2001; Mohnen, 2008; Micheli, 2001). HG-modifying enzymes belong to large multigenic families (CAZy, http://cazy.org/); HG degree of methyl-esterification is modified by pectin methylesterases (EC 3.1.1.11), a modifying enzyme whose activity is regulated by endogenous pectin methylesterase inhibitors (Pelloux et al., 2007; Wolf et al., 2009) and by pectin acetylesterases (EC 3.1.1.6) that hydrolyze the O-acetylated bonds. Overall, the partially demethylesterified HGs (HGA) not only is involved in the formation of the so-called "egg-box" structures, but in lignified tissues, could enhance the formation of benzyl-uronate crosslinks (Grabber et al., 2000). HGA may also become a target for pectin-degrading enzymes such as polygalacturonases (endo-PGs, EC 3.2.1.15; and exo-PGs, EC 3.2.1.67) and pectate lyases (endo-PLs, EC 4.2.2.2). The HG-modifying enzymes action results in a range of natural processes such as fruit ripening or lysigenous aerenchyma formation (Yamauchi et al., 2014; Brummell, 2006). Indeed, there has been observed a change in the pattern of pectin esterification in the middle lamella during the early stages of aerenchyma development, possibly related to the increase in polygalacturonase and pectin lyase expression during aerenchyma formation (Gunawardena et al., 2001a,b).

For a detailed review, consult Frankova and Fry (2013) and Tavares et al. (2015).

Hemicellulose

In dicots, hemicelluloses such as XyGs, mannans, glucomannans, and galactomannans are involved in exceptionally efficient hydrolytic systems (Buckeridge, 2010; Meier, 1982; De Mason DA, 1983; Reid, 1985; Buckeridge and Tiné, 2000). Natural mechanisms that involve mannan deconstruction are related mainly to seed germination. In lettuce, for instance, β mannanase enzyme activity starts in the radicle tip and endosperm cap, and may be involved in cell wall extensibility (Karssen et al., 1989). In storage tissues, XyG transglycosylase/hydrolase (XTH) has been demonstrated to play a key role in XyG mobilization (Fanutti et al., 1993). Decreasing levels of storage XyG have been reported during seedling development in cotyledons of the tropical tree Hymenaea courbaril (Tiné et al., 2000) and fruit ripening in avocado (O'Donoghue et al., 1994). In maize, increased expression of a gene coding for an XTH was observed during aerenchyma formation (Saab and Sachs, 1996; Rajhi et al., 2011). The XyG depolymerization by endo-β-1,4-glucanases can be increased further by exoenzymes that will accomplish XyG side-chain removal, making more sites available to the attack of endoenzymes (Rose and Bennett, 1999; Brummell, 2006). The complete degradation of XyG requires indeed α-xylosidases, β-galactosidases, and β-glucosidases to remove the XyG side chains (Tiné et al., 2000). Additionally, the presence of expansins is critical in the process of detaching XyG bound to the surface of cellulose micro/macrofibrils. This is due mainly to the reduced XyG accessibility by XTH action when it is bound to cellulose (Pauly et al., 1999).

Cellulose

Reports on cellulase activity are either a consequence of other polymer disassembly or overall modifications, but it rarely takes place as an isolated event. The detection of high cellulase activity in expanded cells during bean leaf abscission indicates that cell expansion must be coupled with cellulose hydrolysis. Both features are related to cell wall loosening required for accomplishment of the abscission process (Ueda et al., 1996). Cellulase action does occur in parallel with endopolygalacturonase activity in the abscission of orange calyx (Burns et al., 1998) and cotton leaves (Du et al., 2014). Pectin modifications also seem to be relevant to cellulase action in lysigenous aerenchyma formation in maize roots (Gunawardena et al., 2001a,b).

8.4.2 Cell Wall Deconstruction/Modification: Learning From the Microbes

In nature, there are organisms that have evolved the ability to overcome the wall barrier of a plant cell and release nutrients from it. Many fungal pathogens gain access to plant tissues by directly penetrating plant cell walls (Guerriero et al., 2015). As the cell wall is the first line of defense against the outside environment, it is not surprising that plants have evolved means to perceive and respond to the physical or chemical events associated with such attack (Malinovsky et al., 2014). The various polymers that made up the plant cell wall are substrates to the numerous enzymes secreted by microbial pathogens, and their degradation provides them with nutrients (Walton, 1994). Cellulose and also hemicellulose in plant cell walls are limitedly accessible to the rumen microbes because of the direct (covalent) or indirect (ester or ether) linkage to lignin and pectins (Ding et al., 2012; Vanholme et al., 2010). Because of the complexity of the plant cell wall, microorganism counter with a set of highly specific enzymes. The deconstruction of lignocellulosic biomass requires the synergistic action of numerous oxidative, hydrolytic, and nonhydrolytic enzymes (King et al., 2011; Guerriero et al., 2015). The very first secreted enzymes by microbial pathogens are an array of pectinases, most notably polygalacturonases, pectin and pectate lyases, and pectin esterases directed against the homogalacturonan domain, as well as rhamnogalacturonases, thereby indicating a central role in pathogenesis. Filamentous ascomycetes are able to produce oxidative enzymes that include laccases (oxygen oxidoreductase, EC 1.10.3.2) and class II peroxidases, such as lignin peroxidase (EC 1.11.1.14), manganese peroxidase (EC 1.11.1.13) and hybrid lignin/manganese versatile peroxidase (EC 1.11.1.16), which catalyze the oxidative cleavage of C-C, C-O-C bonds in a wide variety of organic compounds, including polyphenols and lignin, by generating free radicals (Siddiqui et al., 2014). Fungi can also release hydrolytic enzymes that degrade both cellulose and hemicellulose polymers and include endocellulases (endo-1,4-β-d-glucanase, EC 3.2.1.4), which act randomly on internal cellulose bonds and generate oligosaccharides, and exocellulases (exo-1,4--d-glucanase, cellobiohydrolase EC 3.2.1.91), which only attack cellulose polymer from nonreducing ends to produce disaccharide cellobiose. The disaccharide is finally hydrolyzed by β-glucosidase (EC 3.2.1.21) to glucose units. Hydrolysis of hemicelluloses, complex heteropolysaccharides, involves many enzymes acting in a cooperative manner, e.g. endo-*β*-1,4-xylanases (EC 3.2.1.8), β-d-xylosidase (EC 3.2.1.37), that cleave bonds in xylan, 1,4-β-d-endo-mannanases (EC 3.2.1.78) and 1,4-β-d-mannosidases (EC 3.2.1.25) that hydrolyze mannan, as well as β-l-arabinosidases, β-glucosidases, β-galactosidases, β-glucuronidases, acetyl xylan esterases, arabinofuranosidases, acetylmannanesterases, feruloylesterases, galacto- and glucomannanases, *p*-coumaric acid-, and acetyl xylan esterases (Shallom and Shoham, 2003). Another example of microbial organism is Basidiomycetes, involved in wood decay and divided into two main classes: brown- and white-rot fungi. While white-rot fungi can degrade all the components of secondary walls, including lignin, differently, brown-rot fungi can degrade cellulose and hemicelluloses but can only modify lignin. This difference is due to the absence of class II peroxidases from the genomes of brown-rot fungi. The main lignin modification carried out by brown-rot fungi is a demethylation reaction, where O-methyl groups are removed. For a detailed story, please refer to Guerriero et al. (2015).

Considering key modifications and/or degradation of cell wall polysaccharides found in plant natural processes, as well as in plant pathogen interaction, in the following sections, we will discuss how they can drive improvements in the technologies for 2G bioethanol production.

8.5 Manipulating Polysaccharides and Lignin Content or Composition to Optimize Biofuel Production

8.5.1 Plants Cell Wall as Source of Energy

The current world's growing demand for energy and the concern about consume of fossil oil has prompted the scientific efforts in plant biology to establish a cost-effective production of alternative fuels. Bioethanol is the most common biofuel, accounting for more than 90 percent of total biofuel usage in 2013 (IEA-ETSAP). In 2014, ethanol reduced CO_2-equivalent greenhouse gas (GHG) emissions from transportation by 39.6 million metric tons (RFA, 2015); up to now, the ethanol-fuel economy is based on first-generation bioethanol (1G) that is produced from sucrose or starch, mainly in Brazil and in the USA (Table 8.1).

Table 8.1: Worldwide Production of Bioethanol.

World Fuel Ethanol Production by Country or Region (in Million Gallons)								
Country	2007	2008	2009	2010	2011	2012	2013	2014
USA	6521	9309	10,938	13,298	13,948	13,300	13,300	14,300
Brazil	5019	6472	6578	6922	5573	5577	6267	6190
Europe	570	734	1040	1209	1168	1179	1371	1445
China	486	502	542	542	555	555	696	635
Canada	211	238	291	357	462	449	523	510
Rest of world	315	389	914	985	698	752	1272	1490
WORLD	**13,123**	**17,644**	**20,303**	**23,311**	**22,404**	**21,812**	**23,429**	**24,570**

Worksheet available at www.afdc.energy.gov/data/.

There are serious concerns about the sustainability with first-generation bioenergy and, as an example, FAO set up a Bioenergy and Food Security project (BEFS) to assess how bioenergy developments could be implemented without hindering food security.

Most authorities agree that selected first-generation biofuels have contributed to the recent increases in world prices for food and animal feeds, and they do not exclude the possibility of creating undue competition for land and water used for food and fiber production. Many of the problems associated with first-generation biofuels can be addressed by the production of biofuels manufactured from agricultural and forest residues and from nonfood crop feedstocks.

Plant biomass is considered today the most abundant and highly wasted organic raw material available (Pauly and Keegstra, 2008), and indeed, crop residues are considered to be a major biomass resource for biofuel production. Grasses such as sugarcane, sorghum, switchgrass, maize, and miscanthus have an appealing biotechnological potential to be used as feedstocks for second-generation bioethanol production (Carroll and Somerville, 2009). However, none of them fulfill all the needs for an ideal biofuel crop.

The conventional production of bioethanol is a well-known process based on enzymatic conversion of starchy biomass into sugars and/or fermentation of six-carbon sugars with final distillation of ethanol to fuel (Canilha et al., 2012). While conventional processes use only the sugar and starch biomass components, in second-generation processes, all the available lignocellulosic materials are used to increase the variety and quantity of suitable feedstock. Many attempts have been made for different conversion routes, trying to develop the most promising future option. However, the two main conversion routes are:

1. Biochemical: the process is based on enzymatic hydrolysis of the lignocellulosic material through a range of enzymes that break the cellulosic material into sugars; those sugars are fermented, in a second step, into alcohol, which is then distilled into ethanol.
2. Thermochemical: the first step is the gasification of the feedstock under high temperature into a synthesis gas. This gas can then be transformed into different types of liquid or gaseous fuel, the so-called "synthetic fuels" (i.e. bio-SNG).

More detailed information can be found in the IEA publication "From 1st-to 2nd-Generation Biofuel Technologies" (IEA, 2008) or on the http://ethanolrfa.org website.

8.5.2 Cell Wall Recalcitrance, the Major Obstacle to Bioconversion

In order to convert plant biomass into fermentable sugars (a process named **saccharification**) for bioethanol production, it is necessary to access the polysaccharides (e.g. cellulose, hemicelluloses, and pectins) that form plant cell walls. On the other hand, plant cell wall is naturally recalcitrant to hydrolysis as described in detail and represents the major bottleneck for the industrial scale up of the overall process (Himmel et al., 2007) to be circumvented. The structural integrity of plant cell walls, which most probably causes cell wall recalcitrance, is the net result of physicochemical features of the various wall polymers and the complex interactions/associations among those components. As an alternative to harsh thermochemical pretreatments, enzymatic hydrolysis is the most promising and environmentally friendly technology available for saccharification (Yu and Zhang, 2004). In industry, the breakdown of these polymers is usually accomplished by chemical biomass pretreatment followed by hydrolysis with glycosyl hydrolase cocktails (IEA, 2013). Although considerable improvements in microbial fermentation and enzymatic cocktails efficiencies have been made (Li et al., 2015), cell wall recalcitrance to hydrolysis remains a major barrier. Furthermore, the large amount of enzyme required for polysaccharide hydrolysis corresponds to 9% of the total cost for biomass conversion, which can make second-generation bioethanol unfeasible in a larger scale (Banerjee et al., 2010). Moreover, for the overall process, yield should be also considered that the use of common pretreatments might represent a loss of polysaccharide of about 12–33% that could potentially be used to produce second-generation bioethanol (Tavares et al., 2015).

In this sense, any scientific information that can help decrease cell wall recalcitrance, and consequently, the number of pretreatments and the amount of enzyme used by the industry will probably have an important impact on bioethanol production as a renewable and economically sustainable biofuel (Lynd et al., 2011).

8.5.3 Modification of the Cell Wall Structure to Improve the Saccharification Process

A brief description of cell wall bioengineering strategies and emerging technologies that could be used to generate tomorrow's bioenergy and biochemical crops is here reported. New strategies are undertaken to reduce cell wall recalcitrance without altering plant development or causing undesirable effects. Classical cell wall modification methods typically repress the expression or activity of cell wall biosynthetic genes but often have failed because they drastically reduce the content of cell wall components in a nonselective way. Otherwise, there are examples in which genetic manipulations have been used to moderately reduce lignin content or modify its composition in biomass, partially improving saccharification efficiency (Eudes et al., 2014). Lignin fills the spaces in the cell wall between cellulose, hemicellulose, and pectin components, especially in tracheid, sclereid, and xylem cells (Vanholme et al., 2010). It also links hemicellulose depositing on the cellulose surface, thus constituting an obstruction for effective hydrolysis. With cell wall polysaccharides, lignin forms cross-linkages that confer strength properties to cell wall, and therefore the plant that results in the recalcitrance of biomass to biofuels conversion processes.

Lignin, pectin, and xylan are therefore the cell wall components to be modified in terms of manipulation of their content, composition, and distribution to facilitate cellulose extraction and hydrolyzation (Bonawitz and Chapple, 2010; Chen and Dixon, 2007). A detailed and updated description on new approaches to alter plant cell walls and reduce biomass recalcitrance could be found on Tavares et al. (2015).

8.5.4 Lignin Modification to Improve Cellulose Accessibility

Two main goals of engineering plants with an altered cell wall composition, in order to lower costs and improve efficiency of biofuel production, is to decrease recalcitrance by decreasing the lignin content or altering the lignin composition (Kalluri et al., 2014). Unfortunately, it is also the primary contributor to the high cost of lignocellulosic sugar production because cell wall polysaccharides are encrusted with lignin, which make them highly resistant to extraction and enzymatic hydrolysis. Moreover, lignin has almost no commercial value aside from its role as a source of heat, and it is generally treated as a waste product (Eudes et al., 2014). Ways to reduce pretreatments and improve the overall saccharification process by modifying lignin biosynthetic pathways are described (Fu et al., 2011; Kalluri et al., 2014). As an example, it has been shown that the reduction of lignin content by downregulation of *tobacco peroxidase 60* brought to a 30% improvement in sugar released per milligram of cell wall material compared to the wild type (Blee et al., 2003; Cook et al., 2012). That the reduction of lignin content improves the saccharification efficiency was demonstrated also in alfalfa, although it can reduce biomass yield (Chen and Dixon, 2007). Lignin is the primary material responsible for biomass recalcitrance but cannot be simply removed from growing plants without causing serious developmental defects. Unfortunately, the use of natural mutants or silencing strategies for genetic manipulation resulted in a nonselective reduction of lignin content (Eudes et al., 2014). Anyway, studies starting from 2012 up to 2015 report that lignin composition and distribution can be manipulated by using tissue-specific promoters to reduce its recalcitrance by changing its biophysical properties. A recent work showed that expressing the bacterial hydroxycinnamoyl-CoA hydratase-lyase (HCHL) in *Arabidopsis* allowed the overproduction of such C6C1 aromatics, which incorporate into the lignin and reduce its molecular weight (Eudes et al., 2012).

In conclusion, the increasing knowledge on lignin pathways may allow the manipulation of lignin at the various levels of flux control and the incorporation of monolignols, and therefore, the type of lignin structure and regulation of polymerization. To expand the use of lignin modification in many processes, novel and multidisciplinary analytical approaches aimed to investigate the consequence of soft transgenic manipulation are needed also at the molecular level. Important aspects in the engineering of lignin are extensively reviewed by Eudes et al. (2014).

On the other hand, it has been reported that high levels of delignification do not result in the expected digestibility (Rollin et al., 2011). This is explained by the fact that not only lignin but also many different factors influence tissue digestibility, and that increasing cellulose accessibility is more important than removing lignin.

8.5.5 Hemicellulose and Pectin Modification to Overcome Cell Wall Recalcitrance

As lignin, pectin, and xylan are effective barriers to cellulose extraction and hydrolysis, reducing their presence in cell walls may influence saccharification (Biswal et al., 2015).

Hemicelluloses are the main domain present in cell walls of grass bioenergy feedstocks. In grasses, cellulose microfibrils can form bundles, possibly tethered by hemicelluloses (xylans and/or XyGs), generating macrofibrils (Ding and Himmel, 2006; Kozlova et al., 2014). Particularly in dicots, pectin is critical for tissue integrity and accessibility to cell wall-degrading enzymes, and its role in influencing the degradation of other cell wall components has been recognized also in grasses for bioenergy (Chung et al., 2014). Thus, the degradation of these polysaccharides is crucial to achieve efficient cell wall disassembly.

It has been reported that removal of hemicellulose, especially xylose, xylooligomers, and xylan from lignocellulosic materials prior to enzymatic hydrolysis has a positive effect on enzyme accessibility, conversion rates, and yield of plant biomass hydrolysis (Bindschedler et al., 2007; Kalluri et al., 2014). By the modification of hemicellulose biosynthesis pathway, for instance, by a xylan downregulated tobacco line (uxs) showed a significant improvement in saccharification efficiency (Cook et al., 2012).

In a similar fashion, changes in pectin content and/or composition pattern led to increased saccharification (Francocci et al., 2013). For instance, the downregulation of a putative glycosyltransferase results in reduced recalcitrance, increased growth, and reduced xylan and pectin in a woody biofuel feedstock (Biswal et al., 2015).

The deconstruction of cell wall polysaccharides to fermentable sugars by enzymes is a crucial step in lignocellulose conversion. Fungi, which degrade complex and recalcitrant plant polymers, secrete different enzymes that hydrolyze plant cell wall polysaccharides (Guerriero et al., 2015). It has been shown that the expression of fungal acetyl xylan esterase in *Arabidopsis thaliana* improves saccharification of stem lignocellulose (Pawar et al., 2015). Similarly in poplar, tobacco, wheat, and *Arabidopsis*, the overexpression of plant pectinases such as endopolygalacturonase, pectate lyase, or pectin methylesterase led to increased release of simple sugars (Biswal et al., 2015; Lionetti et al., 2010; Tomassetti et al., 2015). In terms of bioenergy, the synergistic action between hemicellulose and pectin-degrading enzymes may reduce the requirement for chemical pretreatments by coordination of wall disassembly.

8.6 Conclusions and Outlook

The large amount of knowledge acquired on the basis of recent sequencing, omics profiling, and genome engineering is expected to further improve saccharification efficiency, limiting the negative growth effects. Here we have highlighted those processes that involve a series of cell wall modifications, resulting in obtaining plants able to change their own cell walls in order to produce endogenous biological pretreatments to improve the saccharification process. All the efforts advised here contribute to the construction of a sustainable bioethanol production, both from an environmental and economic point of view. Beyond showing appropriate cell wall characteristics, the most promising bioenergy crop will also require improved plant performance, resulting in high productivity even in the absence of optimal soil conditions. Adverse effects on plant growth and development can be avoided with a controlled activation of cell wall modification modules by using promoters with the right spatio-temporal expression profile. Furthermore, transcription factors controlling GH expression could be relevant targets to understand how to activate cell wall degradation processes.

As an alternative to the use of transgenic plants in bioenergy, some important points should be reconsidered. New pretreatment technologies need to be developed; for instance, this may include pretreatment of biomass with lignin-degrading fungi that may remove lignin prior to saccharification. Preliminary and promising data are available (Cook et al., 2014; Balasubramaniam and Rajarathinam, 2013), but more studies need to be made to optimize the overall ethanol production process.

References

Alberts, B.J.A., Lewis, J., Raff, M., Roberts, K., Walter, P., 2002. Molecular Biology of the Cell, fourth ed. New York.

Anwara, Z., Gulfraz, M., Irshada, M., 2014. Agro-industrial lignocellulosic biomass a key to unlock the future bio-energy: A brief review. Journal of Radiation Research and Applied Sciences 7 (2), 163–173.

Balasubramaniam, M.K., Rajarathinam, R., 2013. Implementation of white rot fungal pretreated rice straw for sustainable bioethanol production by *Saccharomyces cerevisiae*. International Journal of Engineering Research and Technology 2 (11).

Banerjee, G., Car, S., Scott-Craig, J.S., Borrusch, M.S., Walton, J.D., 2010. Rapid optimization of enzyme mixtures for deconstruction of diverse pretreatment/biomass feedstock combinations. Biotechnology for Biofuels 3, 22. http://dx.doi.org/10.1186/1754-6834-3-22.

Bellincampi, D., Cervone, F., Lionetti, V., 2014. Plant cell wall dynamics and wall-related susceptibility in plant-pathogen interactions. Frontiers in Plant Science 5, 228. http://dx.doi.org/10.3389/fpls.2014.00228.

Bindschedler, L.V., Tuerck, J., Maunders, M., Ruel, K., Petit-Conil, M., Danoun, S., Boudet, A.M., Joseleau, J.P., Bolwell, G.P., 2007. Modification of hemicellulose content by antisense down-regulation of UDP-glucuronate decarboxylase in tobacco and its consequences for cellulose extractability. Phytochemistry 68 (21), 2635–2648. http://dx.doi.org/10.1016/j.phytochem.2007.08.029.

Biswal, A.K., Hao, Z., Pattathil, S., Yang, X., Winkeler, K., Collins, C., Mohanty, S.S., Richardson, E.A., Gelineo-Albersheim, I., Hunt, K., Ryno, D., Sykes, R.W., Turner, G.B., Ziebell, A., Gjersing, E., Lukowitz, W., Davis, M.F., Decker, S.R., Hahn, M.G., Mohnen, D., 2015. Downregulation of GAUT12 in *Populus deltoides* by RNA silencing results in reduced recalcitrance, increased growth and reduced xylan and pectin in a woody biofuel feedstock. Biotechnology for Biofuels 8, 41. http://dx.doi.org/10.1186/s13068-015-0218-y.

Blee, K.A., Choi, J.W., O'Connell, A.P., Schuch, W., Lewis, N.G., Bolwell, G.P., 2003. A lignin-specific peroxidase in tobacco whose antisense suppression leads to vascular tissue modification. Phytochemistry 64 (1), 163–176.

Bonawitz, N.D., Chapple, C., 2010. The genetics of lignin biosynthesis: connecting genotype to phenotype. Annual Review of Genetics 44, 337–363. http://dx.doi.org/10.1146/annurev-genet-102209-163508.

Bouton, S., Leboeuf, E., Mouille, G., Leydecker, M.T., Talbotec, J., Granier, F., Lahaye, M., Hofte, H., Truong, H.N., 2002. QUASIMODO1 encodes a putative membrane-bound glycosyltransferase required for normal pectin synthesis and cell adhesion in Arabidopsis. The Plant Cell 14 (10), 2577–2590.

Brummell, D.A., 2006. Cell wall disassembly in ripening fruit. Functional Plant Biology 33 (2), 103–119.

Buckeridge, M.S., 2010. Seed cell wall storage polysaccharides: models to understand cell wall biosynthesis and degradation. Plant Physiology 154 (3), 1017–1023. http://dx.doi.org/10.1104/pp.110.158642.

Buckeridge, M.S.S.H., Tiné, M.A.S., 2000. Mobilisation of storage cell wall polysaccharides in seeds. Plant Physiology and Biochemistry 38, 141–156.

Burns, J.K., Lewandowski, D.J., Nairn, C.J., Brown, G.E., 1998. Endo-1,4- β-glucanase gene expression and cell wall hydrolase activities during abscission in Valencia orange. Physiologia Plantarum 102, 217–225.

Caffall, K.H., Mohnen, D., 2009. The structure, function, and biosynthesis of plant cell wall pectic polysaccharides. Carbohydrate Research 344 (14), 1879–1900. http://dx.doi.org/10.1016/j.carres.2009.05.021.

Canilha, L., Kumar Chandel, A., dos Santos Milessi, T.S., Fernandes Antunes, F.A., da Costa Freitas, W.L., das Gracas Almeida Felipe, M., da Silva, S.S., 2012. Bioconversion of sugarcane biomass into ethanol: an overview about composition, pretreatment methods, detoxification of hydrolysates, enzymatic saccharification, and ethanol fermentation. Journal of Biomedicine and Biotechnology 2012, 989572. http://dx.doi.org/10.1155/2012/989572.

Carroll, A., Somerville, C., 2009. Cellulosic biofuels. Annual Review of Plant Biology 60, 165–182. http://dx.doi.org/10.1146/annurev.arplant.043008.092125.

Chanliaud, E., Burrows, K.M., Jeronimidis, G., Gidley, M.J., 2002. Mechanical properties of primary plant cell wall analogues. Planta 215 (6), 989–996. http://dx.doi.org/10.1007/s00425-002-0783-8.

Chen, F., Dixon, R.A., 2007. Lignin modification improves fermentable sugar yields for biofuel production. Nature Biotechnology 25 (7), 759–761. http://dx.doi.org/10.1038/nbt1316.

Chung, D., Pattathil, S., Biswal, A.K., Hahn, M.G., Mohnen, D., Westpheling, J., 2014. Deletion of a gene cluster encoding pectin degrading enzymes in *Caldicellulosiruptor bescii* reveals an important role for pectin in plant biomass recalcitrance. Biotechnology for Biofuels 7 (1), 147. http://dx.doi.org/10.1186/s13068-014-0147-1.

Cook, C., Francocci, F., Cervone, F., Bellincampi, D., Bolwell, P.G., Ferrari, S., Devoto, A., 2014. Combination of pretreatment with White rot fungi and modification of primary and secondary cell Walls improves saccharification. Bioenergy Research 8, 175–186.

Cook, C.M., Daudi, A., Millar, D.J., Bindschedler, L.V., Khan, S., Bolwell, G.P., Devoto, A., 2012. Transcriptional changes related to secondary wall formation in xylem of transgenic lines of tobacco altered for lignin or xylan content which show improved saccharification. Phytochemistry 74, 79–89. http://dx.doi.org/10.1016/j.phytochem.2011.10.009.

Cosgrove, D.J., Jarvis, M.C., 2012. Comparative structure and biomechanics of plant primary and secondary cell walls. Frontiers in Plant Science 3, 204. http://dx.doi.org/10.3389/fpls.2012.00204.

Daher, F.B., Braybrook, S.A., 2015. How to let go: pectin and plant cell adhesion. Frontiers in Plant Science 6, 523. http://dx.doi.org/10.3389/fpls.2015.00523.

De Mason, D.A., Sexton, R., Grant Reid, J.S., 1983. Structure, composition and physiological state of the endosperm of *Phoenix dactylifera*. Annals of Botany 52, 71–80.

Ding, S.Y., Himmel, M.E., 2006. The maize primary cell wall microfibril: a new model derived from direct visualization. Journal of Agricultural and Food Chemistry 54 (3), 597–606. http://dx.doi.org/10.1021/jf051851z.

Ding, S.Y., Liu, Y.S., Zeng, Y., Himmel, M.E., Baker, J.O., Bayer, E.A., 2012. How does plant cell wall nanoscale architecture correlate with enzymatic digestibility? Science 338 (6110), 1055–1060. http://dx.doi.org/10.1126/science.1227491.

Du, Q., Wand, L., Zhou, D., Yang, H., Gong, C., Pan, W., Zhang, D., 2014. Allelic variation within the *S*-adenosyl-L-homocysteine hydrolase gene family is associated with wood properties in Chinese white poplar (*Populus tomentosa*). BMC Genetics 15 (Suppl. 1), S4.

Eudes, A., George, A., Mukerjee, P., Kim, J.S., Pollet, B., Benke, P.I., Yang, F., Mitra, P., Sun, L., Cetinkol, O.P., Chabout, S., Mouille, G., Soubigou-Taconnat, L., Balzergue, S., Singh, S., Holmes, B.M., Mukhopadhyay, A., Keasling, J.D., Simmons, B.A., Lapierre, C., Ralph, J., Loque, D., 2012. Biosynthesis and incorporation of side-chain-truncated lignin monomers to reduce lignin polymerization and enhance saccharification. Plant Biotechnology Journal 10 (5), 609–620. http://dx.doi.org/10.1111/j.1467-7652.2012.00692.x.

Eudes, A., Liang, Y., Mitra, P., Loque, D., 2014. Lignin bioengineering. Current Opinion in Biotechnology 26, 189–198. http://dx.doi.org/10.1016/j.copbio.2014.01.002.

Ezaki, N., Kido, N., Takahashi, K., Katou, K., 2005. The role of wall Ca2+ in the regulation of wall extensibility during the acid-induced extension of soybean hypocotyl cell walls. Plant and Cell Physiology 46 (11), 1831–1838. http://dx.doi.org/10.1093/pcp/pci199.

Fanutti, C., Gidley, M.J., Reid, J.S., 1993. Action of a pure xyloglucan endo-transglycosylase (formerly called xyloglucan-specific endo-(1→4)-beta-D-glucanase) from the cotyledons of germinated nasturtium seeds. The Plant Journal 3 (5), 691–700.

Francocci, F., Bastianelli, E., Lionetti, V., Ferrari, S., De Lorenzo, G., Bellincampi, D., Cervone, F., 2013. Analysis of pectin mutants and natural accessions of Arabidopsis highlights the impact of de-methyl-esterified homogalacturonan on tissue saccharification. Biotechnology for Biofuels 6 (1), 163. http://dx.doi.org/10.1186/1754-6834-6-163.

Frankova, L., Fry, S.C., 2013. Biochemistry and physiological roles of enzymes that "cut and paste" plant cell-wall polysaccharides. Journal of Experimental Botany 64 (12), 3519–3550. http://dx.doi.org/10.1093/jxb/ert201.

Fu, C., Mielenz, J.R., Xiao, X., Ge, Y., Hamilton, C.Y., Rodriguez Jr., M., Chen, F., Foston, M., Ragauskas, A., Bouton, J., Dixon, R.A., Wang, Z.Y., 2011. Genetic manipulation of lignin reduces recalcitrance and improves ethanol production from switchgrass. Proceedings of the National Academy of Sciences United States of America 108 (9), 3803–3808. http://dx.doi.org/10.1073/pnas.1100310108.

Grabber, J.H., Ralph, J., Hatfield, R.D., 2000. Cross-linking of maize walls by ferulate dimerization and incorporation into lignin. Journal of Agricultural and Food Chemistry 48 (12), 6106–6113.

Gu, Y., Kaplinsky, N., Bringmann, M., Cobb, A., Carroll, A., Sampathkumar, A., Baskin, T.I., Persson, S., Somerville, C.R., 2010. Identification of a cellulose synthase-associated protein required for cellulose biosynthesis. Proceedings of the National Academy of Sciences United States of America 107 (29), 12866–12871. http://dx.doi.org/10.1073/pnas.1007092107.

Guerriero, G., Fugelstad, J., Bulone, V., 2010. What do we really know about cellulose biosynthesis in higher plants? Journal of Integrative Plant Biology 52 (2), 161–175. http://dx.doi.org/10.1111/j.1744-7909.2010.00935.x.

Guerriero, G., Hausman, J.F., Strauss, J., Ertan, H., Siddiqui, K.S., 2015. Destructuring plant biomass: focus on fungal and extremophilic cell wall hydrolases. Plant Science 234, 180–193. http://dx.doi.org/10.1016/j.plantsci.2015.02.010.

Gunawardena, A.H., Pearce, D.M., Jackson, M.B., Hawes, C.R., Evans, D.E., 2001a. Characterisation of programmed cell death during aerenchyma formation induced by ethylene or hypoxia in roots of maize (*Zea mays* L.). Planta 212 (2), 205–214.

Gunawardena, A.H.L.A.N., Pearce, D.M.E., Jackson, M.B., Hawes, C.R., Evans, D.E., 2001b. Rapid changes in cell wall pectic polysaccharides are closely associated with early stages of aerenchyma formation, a spatially localized form of programmed cell death in roots of maize (*Zea mays* L.) promoted by ethylene. Plant, Cell and Environment 24, 1369–1375.

Hao, Z., Avci, U., Tan, L., Zhu, X., Glushka, J., Pattathil, S., Eberhard, S., Sholes, T., Rothstein, G.E., Lukowitz, W., Orlando, R., Hahn, M.G., Mohnen, D., 2014. Loss of Arabidopsis GAUT12/IRX8 causes anther indehiscence and leads to reduced G lignin associated with altered matrix polysaccharide deposition. Frontiers in Plant Science 5, 357. http://dx.doi.org/10.3389/fpls.2014.00357.

Harholt, J., Suttangkakul, A., Vibe Scheller, H., 2010. Biosynthesis of pectin. Plant Physiology 153 (2), 384–395. http://dx.doi.org/10.1104/pp.110.156588.

Hill Jr., J.L., Hammudi, M.B., Tien, M., 2014. The Arabidopsis cellulose synthase complex: a proposed hexamer of CESA trimers in an equimolar stoichiometry. The Plant Cell 26 (12), 4834–4842. http://dx.doi.org/10.1105/tpc.114.131193.

Himmel, M.E., Ding, S.Y., Johnson, D.K., Adney, W.S., Nimlos, M.R., Brady, J.W., Foust, T.D., 2007. Biomass recalcitrance: engineering plants and enzymes for biofuels production. Science 315 (5813), 804–807. http://dx.doi.org/10.1126/science.1137016.

IEA, 2008. From 1st- to Second-generation Biofuel Technologies: An Overview of Current Industry and RD&D Activities. OECD/IEA, Paris. http://www.iea.org/textbase/papers/2008/2nd_Biofuel_Gen.pdf.

IEA, 2013. Large Industrial Users of Energy Biomass. Paris. Available from: http://www.bioenergytrade.org/downloads/t40-large-industrial-biomass-users.pdf.

IEA-ETSAP, 2013. Liquid Biofuels. Available from: http://iea-etsap.org/web/ThanksDl.asp?file=P10.

Kalluri, U.C., Yin, H., Yang, X., Davison, B.H., 2014. Systems and synthetic biology approaches to alter plant cell walls and reduce biomass recalcitrance. Plant Biotechnology Journal 12 (9), 1207–1216. http://dx.doi.org/10.1111/pbi.12283.

Karssen, C.M., Zagorski, S., Kepczynski, J., Groot, S.P.C., 1989. Key role for endogenous gibberellins in the control of seed germination. Annals of Botany 63, 71–80.

Keegstra, K., 2010. Plant cell walls. Plant Physiology 154 (2), 483–486. http://dx.doi.org/10.1104/pp.110.161240.

King, B.C., Waxman, K.D., Nenni, N.V., Walker, L.P., Bergstrom, G.C., Gibson, D.M., 2011. Arsenal of plant cell wall degrading enzymes reflects host preference among plant pathogenic fungi. Biotechnology for Biofuels 4, 4. http://dx.doi.org/10.1186/1754-6834-4-4.

Kozlova, L.V., Ageeva, M.V., Ibragimova, N.N., Gorshkova, T.A., 2014. Arrangement of mixed-linkage glucan and glucuronoarabinoxylan in the cell walls of growing maize roots. Annals of Botany 114 (6), 1135–1145. http://dx.doi.org/10.1093/aob/mcu125.

Kumar, M., Turner, S., 2015. Plant cellulose synthesis: CESA proteins crossing kingdoms. Phytochemistry 112, 91–99. http://dx.doi.org/10.1016/j.phytochem.2014.07.009.

Li, X., Yu, V.Y., Lin, Y., Chomvong, K., Estrela, R., Park, A., Liang, J.M., Znameroski, E.A., Feehan, J., Kim, S.R., Jin, Y.S., Glass, N.L., Cate, J.H., 2015. Expanding xylose metabolism in yeast for plant cell wall conversion to biofuels. Elife 4. http://dx.doi.org/10.7554/eLife.05896.

Lionetti, V., Francocci, F., Ferrari, S., Volpi, C., Bellincampi, D., Galletti, R., D'Ovidio, R., De Lorenzo, G., Cervone, F., 2010. Engineering the cell wall by reducing de-methyl-esterified homogalacturonan improves saccharification of plant tissues for bioconversion. Proceedings of the National Academy of Sciences United States of America 107 (2), 616–621. http://dx.doi.org/10.1073/pnas.0907549107.

Lynd, L.R., Aziz, R.A., de Brito Cruz, C.H., Chimphango, A.F., Cortez, L.A., Faaij, A., Greene, N., Keller, M., Osseweijer, P., Richard, T.L., Sheehan, J., Chugh, A., van der Wielen, L., Woods, J., van Zyl, W.H., 2011. A global conversation about energy from biomass: the continental conventions of the global sustainable bioenergy project. Interface Focus 1 (2), 271–279. http://dx.doi.org/10.1098/rsfs.2010.0047.

Malinovsky, F.G., Fangel, J.U., Willats, W.G., 2014. The role of the cell wall in plant immunity. Frontiers in Plant Science 5, 178. http://dx.doi.org/10.3389/fpls.2014.00178.

Meier, H.R.J., 1982. Reserve polysaccharides other than starch in higher plants. In: Loewus, F.A., Tanner, W. (Eds.), Encyclopedia of plant physiology, vol. 13A. Springer, Berlin, pp. 418–471.

Micheli, F., 2001. Pectin methylesterases: cell wall enzymes with important roles in plant physiology. Trends in Plant Science 6 (9), 414–419.

Mohnen, D., 2008. Pectin structure and biosynthesis. Current Opinion in Plant Biology 11 (3), 266–277. http://dx.doi.org/10.1016/j.pbi.2008.03.006.

Molinari, H.B., Pellny, T.K., Freeman, J., Shewry, P.R., Mitchell, R.A., 2013. Grass cell wall feruloylation: distribution of bound ferulate and candidate gene expression in *Brachypodium distachyon*. Frontiers in Plant Science 4, 50. http://dx.doi.org/10.3389/fpls.2013.00050.

Mouille, G., Ralet, M.C., Cavelier, C., Eland, C., Effroy, D., Hematy, K., McCartney, L., Truong, H.N., Gaudon, V., Thibault, J.F., Marchant, A., Hofte, H., 2007. Homogalacturonan synthesis in *Arabidopsis thaliana* requires a golgi-localized protein with a putative methyltransferase domain. The Plant Journal 50 (4), 605–614. http://dx.doi.org/10.1111/j.1365-313X.2007.03086.x.

O'Neill, M.A., Ishii, T., Albersheim, P., Darvill, A.G., 2004. Rhamnogalacturonan II: structure and function of a borate cross-linked cell wall pectic polysaccharide. Annual Review of Plant Biology 55, 109–139. http://dx.doi.org/10.1146/annurev.arplant.55.031903.141750.

O'Donoghue, E.M., Huber, D.J., Timpa, J.D., Erdos, G.W., Brecht, J.K., 1994. Influence of avocado (*Persea americana*) C*x*-cellulase on the structural features of avocado cellulose. Planta 194, 573–584.

Pagant, S., Bichet, A., Sugimoto, K., Lerouxel, O., Desprez, T., McCann, M., Lerouge, P., Vernhettes, S., Hofte, H., 2002. KOBITO1 encodes a novel plasma membrane protein necessary for normal synthesis of cellulose during cell expansion in Arabidopsis. The Plant Cell 14 (9), 2001–2013.

Pattathil, S., Hahn, M.G., Dale, B.E., Chundawat, S.P., 2015. Insights into plant cell wall structure, architecture, and integrity using glycome profiling of native and AFEXTM-pre-treated biomass. Journal of Experimental Botany 66 (14), 4279–4294. http://dx.doi.org/10.1093/jxb/erv107.

Pauly, M., Albersheim, P., Darvill, A., York, W.S., 1999. Molecular domains of the cellulose/xyloglucan network in the cell walls of higher plants. Plant Journal 20 (6), 629–639.

Pauly, M., Gille, S., Liu, L., Mansoori, N., de Souza, A., Schultink, A., Xiong, G., 2013. Hemicellulose biosynthesis. Planta 238 (4), 627–642. http://dx.doi.org/10.1007/s00425-013-1921-1.

Pauly, M., Keegstra, K., 2008. Cell-wall carbohydrates and their modification as a resource for biofuels. The Plant Journal 54 (4), 559–568. http://dx.doi.org/10.1111/j.1365-313X.2008.03463.x.

Pawar, P.M., Derba-Maceluch, M., Chong, S.L., Gomez, L.D., Miedes, E., Banasiak, A., Ratke, C., Gaertner, C., Mouille, G., McQueen-Mason, S.J., Molina, A., Sellstedt, A., Tenkanen, M., Mellerowicz, E.J., 2015. Expression of fungal acetyl xylan esterase in *Arabidopsis thaliana* improves saccharification of stem lignocellulose. Plant Biotechnology Journal. http://dx.doi.org/10.1111/pbi.12393.

Pelloux, J., Rusterucci, C., Mellerowicz, E.J., 2007. New insights into pectin methylesterase structure and function. Trends in Plant Science 12 (6), 267–277. http://dx.doi.org/10.1016/j.tplants.2007.04.001.

Rajhi, I., Yamauchi, T., Takahashi, H., Nishiuchi, S., Shiono, K., Watanabe, R., Mliki, A., Nagamura, Y., Tsutsumi, N., Nishizawa, N.K., Nakazono, M., 2011. Identification of genes expressed in maize root cortical cells during lysigenous aerenchyma formation using laser microdissection and microarray analyses. The New Phytologist 190 (2), 351–368. http://dx.doi.org/10.1111/j.1469-8137.2010.03535.x.

Reid, J.S.G., 1985. Cell-wall storage carbohydrates in seeds – biochemistry of the seed 'gums' and 'hemicelluloses'. Advances in Botanical Research 11, 125–155.

RFA, 2015. Renewable Fuel Association (RFA), 2015 Ethanol Industry Outlook. Available from: https://ethanolrfa.3cdn.net/c5088b8e8e6b427bb3_cwm626ws2.pdf.

Roberts, J.A., Elliott, K.A., Gonzalez-Carranza, Z.H., 2002. Abscission, dehiscence, and other cell separation processes. Annual Review of Plant Biology 53, 131–158. http://dx.doi.org/10.1146/annurev.arplant.53.092701.180236.

Rollin, J.A., Zhu, Z., Sathitsuksanoh, N., Zhang, Y.H.P., 2011. Increasing cellulose accessibility is more important than removing lignin: a comparison of cellulose solvent-based lignocellulose fractionation and soaking in aqueous ammonia. Biotechnology and Bioengineering 108, 22–30.

Rose, J.K., Bennett, A.B., 1999. Cooperative disassembly of the cellulose-xyloglucan network of plant cell walls: parallels between cell expansion and fruit ripening. Trends in Plant Science 4 (5), 176–183.

Roudier, F., Fernandez, A.G., Fujita, M., Himmelspach, R., Borner, G.H., Schindelman, G., Song, S., Baskin, T.I., Dupree, P., Wasteneys, G.O., Benfey, P.N., 2005. COBRA, an Arabidopsis extracellular glycosyl-phosphatidyl inositol-anchored protein, specifically controls highly anisotropic expansion through its involvement in cellulose microfibril orientation. The Plant Cell 17 (6), 1749–1763. http://dx.doi.org/10.1105/tpc.105.031732.

Saab, I.N., Sachs, M.M., 1996. A flooding-induced xyloglucan endo-transglycosylase homolog in maize is responsive to ethylene and associated with aerenchyma. Plant Physiology 112 (1), 385–391.

Shallom, D., Shoham, Y., 2003. Microbial hemicellulases. Current Opinion in Microbiology 6 (3), 219–228.

Siddiqui, K.S., Ertan, H., Charlton, T., Poljak, A., Daud Khaled, A.K., Yang, X., Marshall, G., Cavicchioli, R., 2014. Versatile peroxidase degradation of humic substances: use of isothermal titration calorimetry to assess kinetics, and applications to industrial wastes. Journal of Biotechnology 178, 1–11. http://dx.doi.org/10.1016/j.jbiotec.2014.03.002.

Somerville, C., 2006. Cellulose synthesis in higher plants. Annual Review of Cell and Developmental Biology 22, 53–78. http://dx.doi.org/10.1146/annurev.cellbio.22.022206.160206.

Somerville, C., Bauer, S., Brininstool, G., Facette, M., Hamann, T., Milne, J., Osborne, E., Paredez, A., Persson, S., Raab, T., Vorwerk, S., Youngs, H., 2004. Toward a systems approach to understanding plant cell walls. Science 306 (5705), 2206–2211. http://dx.doi.org/10.1126/science.1102765.

Talmadge, K.W., Keegstra, K., Bauer, W.D., Albersheim, P., 1973. The structure of plant cell Walls: I. The macromolecular components of the walls of suspension-cultured Sycamore cells with a detailed analysis of the pectic polysaccharides. Plant Physiology 51 (1), 158–173.

Tameshige, T., Hirakawa, Y., Torii, K.U., Uchida, N., 2015. Cell walls as a stage for intercellular communication regulating shoot meristem development. Frontiers in Plant Science 6, 324. http://dx.doi.org/10.3389/fpls.2015.00324.

Tavares, E.Q., De Souza, A.P., Buckeridge, M.S., 2015. How endogenous plant cell-wall degradation mechanisms can help achieve higher efficiency in saccharification of biomass. Journal of Experimental Botany 66 (14), 4133–4143. http://dx.doi.org/10.1093/jxb/erv171.

Tiné, M.A., Cortelazzo, A., Buckeridge, M.S., 2000. Xyloglucan mobilisation in cotyledons of developing plantlets of *Hymenaea courbaril* L. (Leguminosae-Caesalpinoideae). Plant Science 154, 117–126.

Tomassetti, S., Pontiggia, D., Verrascina, I., Reca, I.B., Francocci, F., Salvi, G., Cervone, F., Ferrari, S., 2015. Controlled expression of pectic enzymes in *Arabidopsis thaliana* enhances biomass conversion without adverse effects on growth. Phytochemistry 112, 221–230. http://dx.doi.org/10.1016/j.phytochem.2014.08.026.

Ueda, J., Miyamoto, K., Hashimoto, M., 1996. Jasmonates promote abscission in bean petiole expiants: its relationship to the metabolism of cell wall polysaccharides and cellulase activity. Journal of Plant Growth Regulation 15, 189–195.

Vanholme, R., Demedts, B., Morreel, K., Ralph, J., Boerjan, W., 2010. Lignin biosynthesis and structure. Plant Physiology 153 (3), 895–905. http://dx.doi.org/10.1104/pp.110.155119.

Vincken, J.P., Schols, H.A., Oomen, R.J., McCann, M.C., Ulvskov, P., Voragen, A.G., Visser, R.G., 2003. If homogalacturonan were a side chain of rhamnogalacturonan I. Implications for cell wall architecture. Plant Physiology 132 (4), 1781–1789.

Vogel, J., 2008. Unique aspects of the grass cell wall. Current Opinion in Plant Biology 11 (3), 301–307. http://dx.doi.org/10.1016/j.pbi.2008.03.002.

Voragen, A.G., 2009. Pectin: structure and function - a special issue dedicated to Professor Walter Pilnik. Editorial. Carbohydrate Research 344 (14), 1786–1787. http://dx.doi.org/10.1016/j.carres.2009.08.023.

Vorwerk, S., Somerville, S., Somerville, C., 2004. The role of plant cell wall polysaccharide composition in disease resistance. Trends in Plant Science 9 (4), 203–209. http://dx.doi.org/10.1016/j.tplants.2004.02.005.

Walton, J.D., 1994. Deconstructing the cell wall. Plant Physiology 104 (4), 1113–1118.

Willats, W.G., Orfila, C., Limberg, G., Buchholt, H.C., van Alebeek, G.J., Voragen, A.G., Marcus, S.E., Christensen, T.M., Mikkelsen, J.D., Murray, B.S., Knox, J.P., 2001. Modulation of the degree and pattern of methyl-esterification of pectic homogalacturonan in plant cell walls. Implications for pectin methyl esterase action, matrix properties, and cell adhesion. Journal of Biological Chemistry 276 (22), 19404–19413. http://dx.doi.org/10.1074/jbc.M011242200.

Wolf, S., Mouille, G., Pelloux, J., 2009. Homogalacturonan methyl-esterification and plant development. Molecular Plant 2 (5), 851–860. http://dx.doi.org/10.1093/mp/ssp066.

Yamauchi, T., Watanabe, K., Fukazawa, A., Mori, H., Abe, F., Kawaguchi, K., Oyanagi, A., Nakazono, M., 2014. Ethylene and reactive oxygen species are involved in root aerenchyma formation and adaptation of wheat seedlings to oxygen-deficient conditions. Journal of Experimental Botany 65 (1), 261–273. http://dx.doi.org/10.1093/jxb/ert371.

Yin, Y., Chen, H., Hahn, M.G., Mohnen, D., Xu, Y., 2010. Evolution and function of the plant cell wall synthesis-related glycosyltransferase family 8. Plant Physiology 153 (4), 1729–1746. http://dx.doi.org/10.1104/pp.110.154229.

Yu, Z., Zhang, H., 2004. Ethanol fermentation of acid-hydrolyzed cellulosic pyrolysate with *Saccharomyces cerevisiae*. Bioresource Technology 93 (2), 199–204. http://dx.doi.org/10.1016/j.biortech.2003.09.016.

Zablackis, E., Huang, J., Muller, B., Darvill, A.G., Albersheim, P., 1995. Characterization of the cell-wall polysaccharides of *Arabidopsis thaliana* leaves. Plant Physiology 107 (4), 1129–1138.

Zandleven, J., Sorensen, S.O., Harholt, J., Beldman, G., Schols, H.A., Scheller, H.V., Voragen, A.J., 2007. Xylogalacturonan exists in cell walls from various tissues of *Arabidopsis thaliana*. Phytochemistry 68 (8), 1219–1226. http://dx.doi.org/10.1016/j.phytochem.2007.01.016.

Zhang, G.F., Staehelin, L.A., 1992. Functional compartmentation of the golgi apparatus of plant cells: immunocytochemical analysis of high-pressure frozen- and freeze-substituted sycamore maple suspension culture cells. Plant Physiology 99 (3), 1070–1083.

Zhao, Q., Dixon, R.A., 2011. Transcriptional networks for lignin biosynthesis: more complex than we thought? Trends in Plant Science 16 (4), 227–233. http://dx.doi.org/10.1016/j.tplants.2010.12.005.

Zuo, J., Niu, Q.W., Nishizawa, N., Wu, Y., Kost, B., Chua, N.H., 2000. KORRIGAN, an Arabidopsis endo-1,4-beta-glucanase, localizes to the cell plate by polarized targeting and is essential for cytokinesis. The Plant Cell 12 (7), 1137–1152.

CHAPTER 9

Production of Dicarboxylic Acid Platform Chemicals Using Yeasts: Focus on Succinic Acid

C. Pais, R. Franco-Duarte, P. Sampaio
Centre of Molecular and Environmental Biology, Universidade do Minho, Braga, Portugal

J. Wildner
ttz Bremerhaven, Bremerhaven Institute for Food Technology and Bioprocess Engineering, Bremerhaven, Germany

A. Carolas, D. Figueira, B.S. Ferreira
Biotrend S.A., Cantanhede, Portugal

9.1 Markets and Applications

9.1.1 Drivers for Biobased Chemicals

The continuously expanding human population and the rising average income of the population cause an increasing demand on energy and raw materials. The steady depletion in fossil feedstocks, the impact of environmental and geopolitical disturbances on price and security of their supply, ecological problems due to emissions of CO_2 and other greenhouse gases, and the consequent impact on climate change are forcing advanced industrialized economies to search and develop alternative and renewable sources for energy, transport fuels, and chemicals. Since 2004, governments, international organizations, and industry associations have issued strategic vision documents and initiated R&D programs for a major transition from the current economy that relies on fossil resources such as crude oil and natural gas to a so-called bioeconomy, based on renewable biomass from agriculture, forestry, and agro-industrial waste streams. Biological raw material from plants and trees, or waste, is renewable in the short term (less than 10 years), as opposed to fossil material renewable in 10 million years. Biobased products can thus make a sizable contribution to reduction in CO_2 emissions and in the dependence on fossil resources, such as crude oil, natural gas, and coal. Moreover, industrial biotechnology frequently shows significant performance benefits compared with conventional chemical technology, such as a higher reaction rate, increased conversion efficiency, improved product purity, lowered energy consumption, and significant decrease in

Biotransformation of Agricultural Waste and By-Products. http://dx.doi.org/10.1016/B978-0-12-803622-8.00009-4

chemical waste generation. So far, the use of biomass for production of transportation biofuels has attracted most political and public attention, although industrial biotechnology allows for the coproduction of a wide range of biobased chemicals from renewable biomass rather than fossil resources, without requiring the massive amounts of raw materials of the fuels market.

The diversity of biobased products that can be produced covers a broad range of intermediate products, product components, and ready-made products, e.g. biobased plastics, biolubricants, biofibers for textiles, composite materials for construction and automotive, chemical and pharmaceutical building blocks, organic acids, amino acids, and enzymes. In addition, biobased products may offer specific innovative properties that have advantages over other products. For example, in sensitive environments, biodegradable lubricants that are nontoxic to soil and water can be used. The addressable market is thus extremely large. It has been estimated by the OECD that the economic contribution of biotechnology in 2030 will be greatest in industrial applications, with 39% of the total potential gross value added from biotechnology, followed by primary production with 36% of the total and health applications at 25% of the total (OECD, 2009). This estimate conflicts sharply with an OECD estimate of the distribution of R&D expenditures by businesses in 2003, where of the vast majority of private sector R&D investment, 87%, went to health applications, with only 2% of biotechnology R&D expenditures spent on industrial applications (OECD, 2009). Since then, the R&D efforts and the development of new technology have experienced an unprecedented dynamism fueled not only by the need to meet the goals, but also by benefiting from the advances in the biological sciences, including a wealth of advanced genomic and molecular tools that are now available at reasonable cost as a result of the investments already made in the advancement of pharmaceutical biotechnology and biomedicine. This chapter illustrates the multidisciplinary endeavor required to develop sustainable processes that respect the environment, improve quality of life, and are competitive in the marketplace. The ultimate goal is the development of a clean (bio)chemical technology, starting from renewable raw materials and energy, with minimal waste generation, and maximal productivity and competitiveness. It encompasses the integrated application of disciplines such as biochemistry, bioinformatics, genetics, systems biology, and bioprocess engineering to develop useful processes and products, based on the catalytic action of microbial cells. The transformation of renewable carbon sources to a variety of biotechnological end products and precursors for chemical processes has been generally limited to products of natural metabolic pathways. Even when highly sophisticated biocatalysts have been developed, the genetic changes were targeted to enhancing the productivity and yields. The emerging field of synthetic biology is opening new opportunities by reassembling existing metabolic pathways and even introducing synthetic metabolic pathway modules into living systems.

Among the possible entry points of biobased products in the value chain from feedstocks to end products, platform chemicals seem to provide a particularly promising opportunity. These biobased intermediate compounds can be then transformed in the same diversity of end products currently produced from oil-derived building blocks. Dicarboxylic acids are an example of such platform chemicals.

9.1.2 Current Uses and Production of Dicarboxylic Acids

Succinic acid

Succinic acid is a key building block for a wide range of secondary chemicals and finds use in chemical, pharmaceutical, food, and agricultural industries. Succinic acid is a C4-dicarboxylic acid, presently produced primarily from petroleum feedstock. Four carbon dicarboxylic acids have the potential to be key building blocks for deriving both commodity and specialty chemicals. The basic chemistry of succinic acid is similar to that of the petrochemically derived maleic acid/anhydride, which has a market of over 1.6 million metric tons and is used in the production of other chemicals such as succinic anhydride, 1,4-butanediol (BDO), fumaric acid, diethyl maleate, glyoxylic acid, and other plastics. Succinic acid has remained a largely underexploited market, despite its potential as a platform molecule, primarily due to the high production cost associated with the petroleum-derived process. Its average market price typically fluctuates between $3–8 per kilo, significantly linked to the fluctuations of oil prices and more than double the typical price of maleic anhydride. The current market for succinic acid is around $240 million (TMR, 2013). Biobased succinic acid is estimated to expand the succinic acid marketplace with emerging applications in the production of biobased 1,4-BDO, gamma-butyrolactone and tetrahydrofuran (THF) and their derivatives, and use as building block of polyesters through the polymerization of succinic acid with diols. Indeed, in 2012 the market of succinic acid was around 60 kton/year and was expected to grow to 665 kton/year until 2020. For example, succinic acid together with 1,4-BDO forms polybutylene succinate (PBS), which is a biodegradable polyester with similar properties to polyethylene terephthalate, but could also substitute polyethylene, polylactic acid and polypropylene in some applications. The market for PBS is currently small, around 15 kton/year; it is expected to increase quickly as a 100% biobased product, using biobased succinic acid and biobased 1,4-BDO, which can also be derived from biobased succinic acid.

Many different microorganisms have been screened and studied for succinic acid production from various carbon sources. Among them, *Anaerobiospirillum succiniciproducens*, *Actinobacillus succinogenes*, *Mannheimia succiniciproducens*, and *Basfia succiniciproducens* have been most intensively studied due to their ability to produce a relatively large amount of succinic acid. Usually, these natural succinate-producing strains require complex and rich nutrient media, which increase substrate requirements and cost, and the productivity rarely reaches the desired target for feasible operation (Beauprez et al., 2010). As conventional strain improvement is often not sufficient, metabolic engineering is also applied to improve strain productivity, but this approach may be delayed by the lack of molecular genetic tools for these particular organisms (BREW, 2006). There has been much effort in developing recombinant *Escherichia coli* strains, which are capable of enhanced succinic acid production under aerobic and anaerobic conditions. In contrast to natural succinic acid producers, which employ only one major pathway for succinic acid production, *E. coli* uses a total of six biosynthetic routes. Pathway engineering has been performed in order to direct carbon flow predominantly to succinic acid, assisted by *in silico* metabolic pathway analysis of engineered strains. The microbe had been optimized to produce succinate salts from sugar and CO_2, with 25% of the carbon in the product coming from CO_2. Overall, the biosynthesis consumes CO_2 rather than generating it. This process has already been validated using wheat-derived glucose at 350,000L (Ritter, 2011). In an ideal situation, the fermentation would be run at low pH, most preferably without requiring any neutralization and avoiding the need to convert the resulting sodium succinate to the free acid required for subsequent derivatization. The conversion of the salt to the free acid adds significant costs, and undesired salts are produced. As such, low pH-tolerant yeast strains have been successfully engineered to produce succinic acid with less by-products (van den Tweel, 2010; Ritter, 2011). The use of yeast could also allow using retrofit bioethanol facilities for the production of succinic acid. The major fermentation feedstock is still glucose, but ongoing efforts are aiming at introducing cheaper, less-processed feedstocks including sorghum grits and hydrolyzed agricultural and forest residues, such as corn stover, bagasse, rice straw, and wood chips. Several companies have announced the launch of demonstration facilities (Table 9.1), which are supplying succinic acid to the market, particularly to strategic partners with which development of specific applications downstream the value chain is being performed.

Fumaric acid

Fumaric acid had an estimated annual production volume of 230kton in 2013 (GVR, 2015) and is particularly used in food and animal feed for pH adjustment, preservation, and flavor enhancement. Fumaric acid derivatives have also been tested in a pharmaceutical application for treatment of psoriasis. Addition of fumaric acid to animal feed can lead to large reduction of the methane emissions by cattle (up to 70%). Fumaric acid contains a double bond and two carboxylic groups, which makes it very suitable for production of polyester and alkyd resins.

Table 9.1: Demonstration-Scale Facilities for the Production of Biobased Succinic Acid.

Company	Technology	Capacity (ton/year)
BioAmber	Genetically modified *E. coli* using glucose; currently developing an alternative yeast platform	30,000
Myriant	Genetically modified *E. coli* using glucose; planned future use of lignocellulosic sugars	10,000
Reverdia (joint venture between DSM and Roquette)	Genetically modified yeast using glucose; planned future use of lignocellulosic sugars	10,000
Succinity (joint venture between BASF and Corbion)	*Basfia succiniciproducens*	10,000

In the early 20th century, fumaric acid was biotechnologically produced at industrial scale, but after the Second World War, the production process was switched to a petrochemical route. Today, fumaric acid is produced in an amount exceeding 100 kton/year by the chemical route from maleic anhydride. The cost of fumaric acid ranges between US $2.0–2.2/kg, which is about 10% higher than the price of maleic anhydride. Fumaric acid is attractive for polymer production due to its nontoxic nature as well as its ability to give special properties as hardness to the polymer structure. The currently available native fumaric acid fungal producers represent a challenge for industrial fermentations due to filamentous morphology. The presently known recombinant hosts produce fumaric acid at low titer, rate, and yield. Fumaric acid is cited in the development pipeline of various companies currently focusing on succinic acid, including DSM and Myriant.

Glutaric acid

The applications of glutaric acid remain fairly limited to niche applications. It is produced as a by-product of the production process of adipic acid (about 2% of the output of an adipic acid plant is glutaric); however, it does not usually reach the market because it is converted into dibasic esters and sold as environmentally friendly solvents. The bulk price of glutaric acid ranges between US $10 and US $17/kg. Glutaric acid can be produced through various chemical routes, for example, from cyclopentane by oxidation with molecular oxygen and cobalt (III) catalysts or by ozonolysis; and from cyclopentanol–cyclopentanone by oxidation with oxygen and $Co(CH_3CO_2)_2$, with potassium peroxide in benzene, or with N_2O_4 or nitric acid. Together with succinic acid, glutaric acid is formed as a by-product during oxidation of cyclohexanol–cyclohexanone in the adipic acid production process. None of these routes yield a biobased product.

The company Rennovia (USA) has submitted a patent that discloses a method using chemical catalysis for the production of glutaric acid from carbohydrate-containing materials (WO2010144871). Briefly, pentoses, for example those sourced from hemicellulose-containing raw materials, are oxidized into the corresponding pentaric acid in the presence of an oxidation catalyst. Thereafter, the pentaric acid is reacted with hydrogen in the presence of a hydrodeoxygenation catalyst and a halogen source (for example, hydrogen bromide or iodide) to yield glutaric acid. No biotechnological processes for the production of glutaric acid have been developed so far, despite ongoing early-stage research using *E. coli* or *Saccharomyces cerevisiae* as host. In the case of the *E. coli* platform, lysine and α-ketoglutarate have been used as co-substrates, with impact on the economic feasibility of the process.

Itaconic acid

Itaconic acid is an important building block in the chemical industry. It is a white crystalline powder and readily biodegrades in soil. Hence, it is an optimum substitute for petro-derived chemicals such as acrylic acid, maleic anhydride, or acetone cyanohydrin in various end-user industries. The demand for itaconic acid is high in the manufacturing of superabsorbent polymers, mainly used in diapers, adult incontinence, and feminine hygiene products. Itaconic acid is used as a cross-linking agent due to its ability to efficiently take part in addition polymerization. It also finds large application in seed coating, root dipping, ornamental gardens, food packaging, and artificial snow. Moreover, increasing demand for unsaturated polyester resins in pipes, artificial stones, electrical cabinets, and laminating resins is expected to increase the demand for itaconic acid. High price of itaconic acid is the major factor hampering the growth of itaconic acid market. Polyitaconic acid (a derivative of itaconic acid) has the potential to replace sodium tripolyphosphate in detergents. However, strong establishment of other phosphate-free builders impedes the growth of itaconic acid in detergents application. Other application segments include lubricant oil, adhesives, paints and coatings, pharmaceuticals, emulsifiers, herbicides, printing chemicals, and acrylic fiber. The global market is estimated to be worth around US $126 million/kg (TMR, 2015). The production in China has boomed, and as a result, the market price decreased to around US $2/kg or even lower (Boy and Lappe, 2012).

Itaconic acid is a biobased product mainly produced by fermentation using certain filamentous fungi (e.g. *Ustilago*, *Helicobasidium*, and *Aspergillus*). A mixture of itaconic acid, citraconic acid, and citraconic anhydride is also obtained by reaction of succinic anhydride with formaldehyde at 200–500°C in the presence of alkali or alkaline earth hydroxides (could at least partially be biobased if biobased succinate is used as raw material for the production of succinic anhydride). Other methods involve carbonylation of propargyl chloride with metal carbonyl catalysts and thermal decomposition of citric acid, which is also a biobased chemical. *Aspergillus terreus* is the strain commonly used for the industrial production of itaconic acid. A significant amount of research has been put into the reduction of the production costs: the replacement of sugar, used as the carbon source, by cheaper alternative substrates such as cellulolytic biomass; optimizing the bioreactor type and configuration; deriving innovations by which the process becomes more energy saving; strain improvement by genetic and metabolic engineering, allowing the effective use of cheap alternative substrates, etc. Recent patent activity has particularly focused on the improvement of the producing strain, mainly by using recombinant DNA techniques, and several patents have been submitted worldwide in the last 10 years. There is a significant market opportunity for the development of biobased products from the C5 building block, itaconic acid. The major challenges are primarily associated with reducing the overall cost of the fermentation. It was estimated that in order to render the products derived from biobased itaconic acid competitive with petrochemical-derived products, the fermentation cost needed to be below US $1/kg, which is a significant technical challenge and should be undertaken with a longer-term perspective.

Other acids

The production of various other dicarboxylic acids using biotechnological routes has been being increasingly discussed. For example, although adipic acid is commercially the most important aliphatic dicarboxylic acid, no biobased product is yet on the market, with a current annual market of 2.5–3 million tons with an estimated value of US $5–7 billion. In large-scale production, the six carbon atoms of the adipic acid backbone are usually derived from benzene, which is hydrogenated to cyclohexane or phenol. Phenol can be hydrogenated to form cyclohexanol, while cyclohexane can be oxidized with air to a ketone–alcohol oil (KA oil, a mixture of cyclohexanol and cyclohexanone). Cyclohexanol, cyclohexanone, or KA oil is then oxidized with nitric acid to form adipic acid. Several processes have been suggested for the production of biobased adipic acid, and many of these proposed by advanced biotechnology companies, including Verdezyne, which achieved proof of concept production and recovery of adipic acid by an engineered *S. cerevisiae* from an alkane feedstock. DSM Venturing, a corporate venture fund from DSM, has invested in Verdezyne, and DSM has recently announced it will invest in biobased adipic acid in the next years, targeting a 100–150 kton/year commercial scale target. Also Genomatica, a synthetic biology company, has developed intellectual property targeting processes to produce over 20 intermediate and basic chemicals, including adipic acid. Building on the success of the development of a feasible biobased succinic acid process, BioAmber and Myriant have also included biobased adipic acid in their pipeline. Other companies active in the biobased adipic acid market include Aemetis, Amyris (using fermentation technology), and Rennovia (through the chemical bioconversion of sugars).

Other dicarboxylic acids with commercial interest include azelaic acid, citraconic acid, maleic acid, and crotonic acid. Biobased fermentation production routes have been proposed for azelaic acid and crotonic acid, but with very low titers and still very far from commercially relevant processes.

9.2 Selection and Improvement of Yeast Strains

The enormous biodiversity of fungi and their essential role as nutrient recyclers, decomposers, mutualists, or pathogens in almost every terrestrial environment are today largely established (Hawksworth, 2001). Yeasts are unicellular fungi that are widely distributed in the natural environment and present the capacity to respond to numerous environmental stresses, such as presence of oxidative, ionic, and osmotic variations; high temperatures; nutrient limitations; starvation; high concentration of ethanol; and the presence of competing organisms. Moreover, their ability to degrade and grow in a wide range of substrates has determined their choice for commercial and industrial processes (Attfield, 1997; Querol et al., 2003; Negi et al., 2013).

Traditionally, yeast refers to *S. cerevisiae* and its close relatives, used for alcoholic fermentation and baking. However, approximately 2000 species of yeasts belonging to 150 genera have been described so far. The genomes of over 100 yeast species have already been sequenced since the completion of the *S. cerevisiae* genome project in 1996. Presently, whole genome sequences for a great number of *S. cerevisiae* isolates are available, revealing considerable variation among strains that might have arisen from artificial selection for novel traits through cross-hybridization (Dunn et al., 2005), horizontal gene transfer (Novo et al., 2009), and genomic mosaicism (Liti et al., 2009). A similar variation can be expected within all the other yeast species described to date, representing a huge reservoir of genetic diversity that may be exploited for the discovery of novel genes, entire metabolic pathways, and potentially valuable end products.

Biotechnologists usually designate all non-*S. cerevisiae* yeasts as nonconventional yeasts. What they have in common is a lower degree of fermentative overflow metabolism and a rather short history of genetic and biological characterization. The yeast *S. cerevisiae* is characterized by flourishing in extremely high sugar concentrations, most of which transformed in the by-product ethanol. The majority of natural habitats, however, do not provide such extreme substrate conditions so that most nonconventional yeasts provide alternative metabolic routes for substrate utilization, product formation, and different regulatory patterns. As an example some species of major interest are *Pichia pastoris* (syn. *Komagataella pastoris*), *Yarrowia lipolytica, Hansenula polymorpha* (syn. *Ogataea parapolymorpha*), *Pichia stipitis* (syn. *Scheffersomyces stipitis*), *Kluyveromyces marxianus*, or *Torulaspora delbrueckii*.

A great number of culture collections has been developed worldwide, both public and in companies. Public collections play an important role since they offer identification facilities to meet an increasing demand for precise characterization of yeast strains of biotechnological interest, they are invaluable in the academic sphere as a source of expertise and fully authenticated material for biological research, and they provide resources to safeguard commercially important genetic variation against market failures. In addition to screening culture collections, access to biodiversity via metagenomic strategies is a promising approach to screen for improved enzymes and/or transport proteins.

9.2.1 Yeasts as Platform Organisms for Dicarboxylic Acid Production

Dicarboxylic acids, namely succinic, malic, and fumaric, are naturally occurring compounds that are found as major metabolic intermediates in most prokaryotic and eukaryotic microorganisms. The wide application of these 1,4-dicarboxylic acids as platform chemicals and the relatively few enzymatic steps required for their production have led to intensive investigation into their microbial synthesis. The yeast *S. cerevisiae* stands out as a platform microorganism for dicarboxylic acid production. Due to its role as a model organism in research, their physiology and genetics are extensively documented, and a well-developed metabolic engineering toolbox is available. In addition, a huge number of genetic mutant strains are already available, such as the Euroscarf collection with around 5100 mutant strains, covering 82% of the ~6200 annotated yeast ORFs (Open Reading Frames). Moreover, wild-type *S. cerevisiae* strains have GRAS (Generally Regarded As Safe) status, so that modified strains are more likely to be allowed in the production of these C4-dicarboxylic acids for food and pharmaceutical applications.

9.2.2 Microorganisms Producing Succinic Acid: Natural Producers Versus Metabolic Engineered Strains

As mentioned above, several bacterial strains are known to produce considerable amounts of succinic acid under anaerobic conditions, but due to their physiological properties, they are not efficient in performing it at low pH, including facultative anaerobes (*A. succinogenes, Mannheimia succiniciproducens*), and strict anaerobe *A. succiniciproducens*. Many fungal species, such as *Paecilomyces variotii* (Ling et al., 1978), *Aspergillus niger* (Papagiani, 2007), and *Penicillium simplicissimum* (Gallmetzer et al., 2002), are also found to be capable of succinate production. They can secrete succinate as a metabolic by-product under aerobic and/or anaerobic conditions, but the productivity is much lower when compared with the bacterial strains. Besides, succinate is mainly synthesized in the mitochondria, and to accumulate as an extracellular by-product, it has to cross the mitochondrial and cellular membrane (Coustou et al., 2005). Yeasts of the genus *Candida* grown on n-alkanes (Kamzolova et al., 2012), *Candida zeylanoides*, and *Candida catenulata*, grown on ethanol (Mandeva et al., 1981; Kamzolova et al., 2009), and *Y. lipolytica*, grown on rape seed oil (Kamzolova et al., 2014), have also been proposed as succinic acid producers. In addition to natural producers, many microorganisms can be metabolically engineered to produce succinate as an end product of fermentation. These are mostly model organisms since they are more prone to be genetically modified, and a completely engineered pathway is required for them to produce succinate, including *E. coli* and *Corynebacterium glutamicum*, the latter one of the few gram-positive bacteria which have been tested for succinate production.

It has been well-documented that *S. cerevisiae* achieves high concentrations of succinate to enhance the quality of wine (Wakai et al., 1980). This yeast species grows well under acidic conditions and is thus an attractive alternative as a biocatalyst for succinic acid production. Although *S. cerevisiae* wild-type strains may be used in succinic acid production, the optimization of strains via metabolic engineering for a more efficient production is currently underway, as mentioned earlier.

9.2.3 Succinate Formation Pathways

The production of organic acids begins in yeast via glycolysis, differing then downstream of pyruvate formation. From pyruvate, three pathways are possible for the succinate formation: via oxidative tricarboxylic acid (TCA) cycle, via reductive branch of TCA cycle, or via oxidative pathway of the glyoxylate cycle.

Oxidative tricarboxylic acid cycle

Under aerobic conditions, succinate is an intermediate metabolite of both the oxidative TCA cycle and the glyoxylate shunt. In the TCA cycle (Fig. 9.1A), acetyl-CoA generated from pyruvate is joined with oxaloacetate, and via the action of citrate synthase (encoded by gene

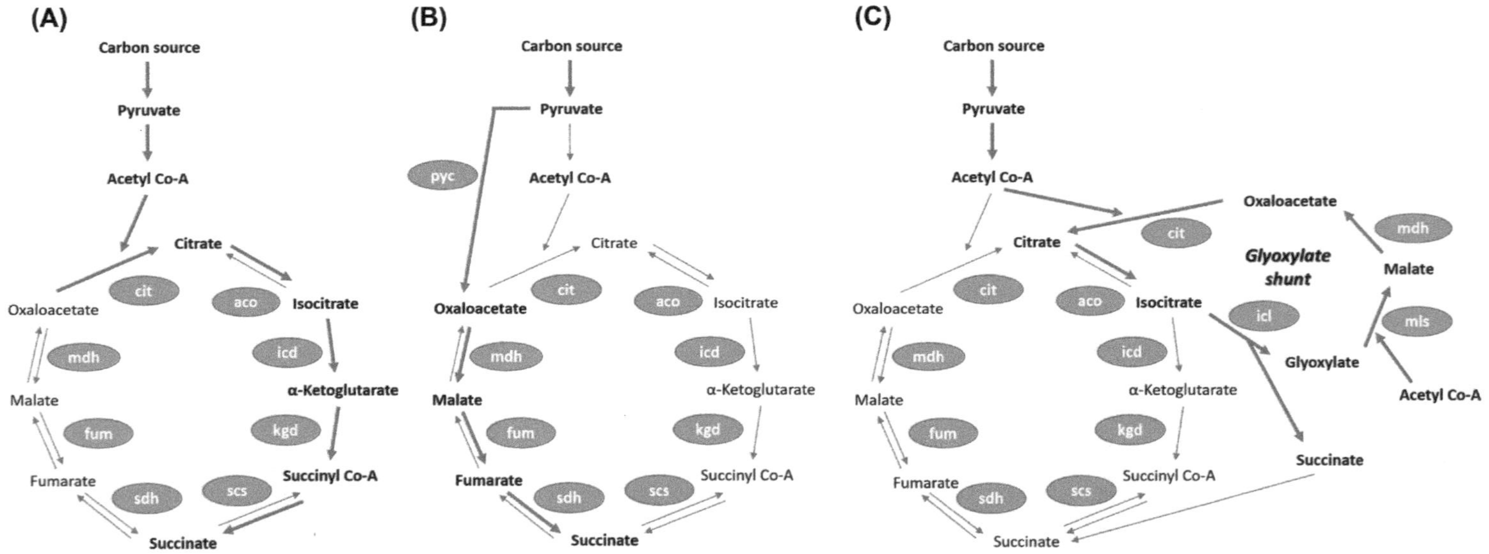

Figure 9.1

Main succinate production metabolic pathways in *Saccharomyces cerevisiae*: (A) oxidative branch of TCA cycle; (B) reductive branch of TCA cycle; (C) glyoxylate pathway. Not all enzymatic steps are shown. Abbreviations used for enzymes are: *aco*, aconitase; *cit*, citrate synthase; *frds*, fumarate reductase; *fum*, fumarase; *icd*, isocitrate dehydrogenase; *icl*, isocitrate lyase; *kgd*, α-ketoglutarate dehydrogenase; *scs*, succinyl-coenzyme A synthetase; *mdh*, malate dehydrogenase; *mls*, malate synthase; *pyc*, phosphoenol pyruvate carboxylase; *sdh*, succinate dehydrogenase.

CIT), will form citrate. Aconitase (ACO1) will then convert citrate to isocitrate, which will be transformed in α-ketoglutarate by isocitrate dehydrogenase (encoded by genes *IDH* and *IDP1*). The molecule of α-ketoglutarate will then be decarboxylated by α-ketoglutarate dehydrogenase and form succinyl-CoA, which is finally converted to succinate by succinyl-CoA synthetase (encoded by gene *LCS1*). If the TCA cycle continues, enzyme succinate dehydrogenase (encoded by genes of the complex *SDH*) will convert succinate to fumarate. Therefore, one key step of metabolic engineering strategies is the inactivation of *SDH* genes in order to stop this conversion to fumarate, and in this way, allowing the accumulation of succinate in the cell.

Reductive branch of tricarboxylic acid cycle

Under anaerobic conditions, the reductive pathway of TCA cycle is activated, since succinate is the H-acceptor instead of oxygen, and pyruvate, originated from glycolysis, is converted to oxaloacetate, malate, fumarate, and then succinate (Fig. 9.1B). This pathway, from phosphoenolpyruvate (which precedes pyruvate in glycolysis) to succinate, requires 2 mol of NADH per mole of succinate produced, which represents a maximum theoretical yield of two molecules of succinate for every glucose molecule, since each molecule of glucose can provide only two molecules of NADH through glycolysis. Therefore, the redirection of the carbon flux only to the anaerobic fermentation pathway is energetically unfavorable, has a theoretical yield higher than the oxidative direction or the glyoxylate shunt (1.71 mol per mole of glucose), and results in net CO_2 fixation. The reductive pathway can be divided in the following steps: (1) pyruvate carboxylation, in which pyruvate in converted in oxaloacetate, a reaction performed by the enzyme pyruvate carboxylase, encoded by gene *PYC*; (2) oxaloacetate reduction to malate, via the action of malate dehydrogenase, encoded by gene *MDH*; (3) translation of malate to fumarate, under the action of fumarase (encoded by *FUMR*); and (4) fumarate reduction performed by fumarate reductase, encoded by genes *FRDS1* and *OSH1*. Yan et al. (2014) reviewed the main obstacles of reductive TCA pathway for succinic acid production: (1) yeast fumarase (*FUM*) only converts fumarate to malate without the possibility to revert the process to fumarate production; (2) fumarate reductase, the key enzyme involved in the reductive production of succinate (Arikawa et al., 1999), coded by genes *FRDS1* (cytosol) and *OSM1* (mitochondria) (Fig. 9.1B), are only produced under anaerobic conditions; and (3) a high amount of NADH is consumed by this branch (2 mol of NADH per each mole of succinic acid formed).

Oxidative pathway of glyoxylate cycle

The glyoxylate shunt is also a possible pathway to produce succinic acid in aerobic conditions. One large part of the glyoxylate cycle is similar to the TCA cycle, sharing three of the five enzymes associated with the TCA cycle and many of its intermediate steps. The main variations are related with the conversion of isocitrate into glyoxylate and succinate by isocitrate lyase (encoded by gene *ICL1*). Then glyoxylate is converted into malate by the action of malate synthase (Fig. 9.1C). Succinate production via glyoxylate cycle instead of

TCA cycle has the advantage of avoiding mitochondrial transport, since it occurs in the cytosol, and also leads to less carbon loss due to only one decarboxylation step. When compared to the reductive cycle, the glyoxylate shunt has a lower theoretical yield but provides ATP for maintenance and active transport processes (Raab and Lang, 2011).

Yet another alternative to produce succinate is the combined use of both the reductive TCA cycle and glyoxylate shunt by joining the oxidative and reductive route, providing in this way an even redox balance, a higher maximum succinate yield, and a fixation of CO_2 instead of its release (reviewed in Raab and Lang, 2011). However, its implementation will be a challenge in the following years, since it requires that oxidative and fermentative metabolism run simultaneously in yeasts. This strategy was already successfully accomplished in *E. coli* strains, with an increase of the theoretical yield (Rezaei et al., 2015; Vemuri et al., 2002a,b).

9.2.4 Metabolic Engineering Strategies for Biotechnological Production of Succinic Acid

Saccharomyces cerevisiae

Since the beggining of the XXI century, research has focused efforts in the biotechnological succinate production process, in particular using metabolic engineering. Succinate is not normally produced at high levels in *S. cerevisiae,* but there are many reasons for choosing this yeast as a microbial cell factory for platform chemicals, including the easiness of genetic manipulation, the extensive library of genetic engineering tools and a very well-annotated genome, their ability to grow both anaerobically and aerobically, and the proven ability of being used at industrial scale.

Different strategies have been used to manipulate yeast strains to an enhanced production of succinic acid, as reviewed in Cheng et al. (2013). The challenge in metabolic engineering of *S. cerevisiae* for the efficient production of succinic acid involves at least three levels: (1) elimination of alcoholic fermentation, which occurs irrespective of the availability of oxygen; (2) engineering fast and efficient metabolic pathways that link the high-capacity glycolytic pathway to the TCA cycle, taking into account NADH and ATP constraints; and (3) engineering of product export.

When increased succinate production is the goal, ethanol production should be avoided, since it will change the carbon flux from pyruvate to ethanol, leading to carbon loss in the process. However, even under fully aerobic conditions, *S. cerevisiae* strains are intrinsically associated to alcoholic fermentation, and it is hard to redirect the flux from pyruvate to organic acids production, instead of ethanol formation (Abbott et al., 2009). The pathway that converts pyruvate into ethanol involves only two targets for metabolic engineering via gene deletion: pyruvate decarboxylase and alcohol dehydrogenase. In the first step to produce ethanol, pyruvate is decarboxylated to acetaldehyde in an irreversible reaction catalyzed by pyruvate decarboxylase. *Saccharomyces cerevisiae* contains three structural genes that encode functional pyruvate decarboxylase isozymes, *PDC1*, *5*, and *6*. Although deletion of all three genes completely eliminates alcoholic fermentation, pyruvate decarboxylase-negative (Pdc^-) strains grow poorly in complex media and fail to grow on glucose as the sole carbon source. This inability is due to the need of producing cytosolic acetyl-CoA, an essential precursor for lysine and lipid synthesis, and the fact that in *S. cerevisiae* the mitochondrial acetyl-CoA, produced via pyruvate dehydrogenase, cannot be transported to the cytosol. Thus, the requirement for C2 compounds (to overcome cytosolic acetyl-CoA need) and the high glucose sensitivity represents major impediments for the use of Pdc^- strains for succinic acid production. However, by applying evolutionary engineering, these two limitations could be overcome, and Pdc^- *S. cerevisiae* isolates are available showing high pyruvate concentrations, up to 135 g/L at a yield of 0.54 g/(g glucose) (van Maris et al., 2004). The second step in the conversion of pyruvate to ethanol is the reduction of acetaldehyde to ethanol through the action of alcohol dehydrogenase. *Saccharomyces cerevisiae* contains four alcohol dehydrogenase (*ADH1* to *ADH4*) structural genes. Strains deleted in these genes were observed to grow poorly, mainly due to intracellular accumulation of glycerol and of toxic acetaldehyde (Drewke et al., 1990). Thus, to block ethanol formation, the best alternative is the elimination of pyruvate decarboxylase activity. Since these *S. cerevisiae* strains produced high titers and yields of pyruvate, Pdc^- strains have a great potential for the production of organic acids, especially those for which pyruvate is an intermediate metabolite (Flikweert et al., 1996; Flikweert and Johannes, 1997).

One of the first attempts to improve succinate production was reported by Arikawa et al. (1999) using sake yeast strains with TCA cycle gene deletions (Table 9.2). The simultaneous deletion of *SDH1* and *FUM1* led to 2.7-fold higher production of succinate in comparison with wild-type strains. However, these enhancements were not observed in the anaerobic

Table 9.2: Main Metabolic Engineering Strategies to Improve Succinate Production in Different Yeast Species.

Strain	Genotype	Succinate Production	References
	Saccharomyces cerevisiae		
Kura	*ΔSDH1ΔFUM1*	2.32 g/L	Arikawaetal (1999)
AH22ura3	*ΔSDH1, ΔSHD2, ΔIDH1, ΔIDP1*	3.62 g/L, 0.11 mol/mol glucose	Raab et al. (2010)
8D	*ΔSDH, ΔSER3/SER33*, overexpression of native *ICL1*	0.9 g/L	Otero et al. (2013)
BY4741	*ΔDIC1*	0.23 g/L	Agren et al. (2013)
PMCFf	*ΔFUM1, ΔPDC, ΔGPD1*	8.09 g/L	Yan et al. (2014)
S149sdh12	*ΔSDH1, ΔSHD2*, expression of *MAE1* gene	2.36 C-mol yield	Ito et al. (2014)
	Yarrowia lipolytica		
Y-3312	*ΔSDH1, ΔSHD2*	45 g/L	Yuzbashev et al. (2010)
	Issachenkia orientalis		
SD108	*ΔPYC, ΔMDH, ΔFUMR, ΔFRD*	11.63 g/L	Xiao et al. (2014)

conditions of sake fermentation. Another metabolic engineering strategy for the oxidative production of succinic acid was the quadruple gene deletion (*SDH1, SDH2, IDH1, IDP1*) described by Raab et al. (2010). In glucose-grown shake-flask cultures, the mutant strain produced a titer of 3.62 g/L (factor 4.8 as compared to wild-type), the highest obtained for yeast until then. Multigene deletion followed by directed evolution was used by Otero et al. (2013) to select a succinate producer mutant. The strategy included deletion of *SDH3* and interruption of glycolysis derived serine by deletion of 3-phosphoglycerate dehydrogenase (*SER3/SER33*) and overexpression of native *ICL1*. The mutant strain presented a 43-fold increase in succinate yield on biomass compared to the reference strain. A modified pathway for succinate production was established by the deletion of genes *FUM1, PDC, GPD1*, and was reported by Yan et al. (2014). The authors obtained a succinate titer of 8.09 g/L, which was further improved to 9.98 g/L through regulation of biotin and urea levels, and to 12.97 g/L through optimal CO_2 conditions in a bioreactor. Another attempt to construct *S. cerevisiae* strains with increased succinate production involved the disruption of *SDH1* and *SDH2* genes, together with the expression of *Schizosaccharomyces pombe* malic acid transporter *MAE1*. The expression of the transporter to export intracellular acid outside of the cell successfully improved succinic acid production (Ito et al., 2014). Table 9.2 summarizes the main metabolic engineering strategies to improve succinate production in *S. cerevisiae* and other yeast species.

Other yeasts

Yuzbashev et al. (2010) proposed a new strategy for construction of an aerobic eukaryotic succinic acid producer, based on the yeast *Y. lipolytica*, with a deletion in the gene coding for one of the succinate dehydrogenase subunits. Firstly, an original *in vitro* mutagenesis-based approach was proposed to construct strains with temperature-sensitive mutations in the *SDH1* gene. These mutants were used to optimize the composition of the media for selection of transformant strains with the deletion in the *Y. lipolytica SDH2* gene. Although the defects of each succinate dehydrogenase subunit prevented the growth on glucose, the mutant strains grew on glycerol and produced succinate in the presence of the buffering agent $CaCO_3$. Subsequent selection of the strain with deleted *SDH2* gene for increased viability allowed the construction of a strain capable of accumulating succinate at the level of more than 45 g/L in shaking flasks with buffering and more than 17 g/L without buffering agent.

Due to its ability to grow at low pH and high tolerance to multiple organic acids, the yeast *Issatchenkia orientalis* has also been used as a microbial platform for succinic acid production (Xiao et al., 2004). Based on the fact that the reductive TCA cycle gives the highest theoretical yield on glucose (1.31 g/g), genes from this metabolic pathway were selected has targets for metabolic engineering. The metabolic strategy designed involved the assembly and integration of four genes (*PYC, MDH, FUMR*, and *FRD*) from the reductive TCA pathway in the genome of a uracil auxotroph strain of *I. orientalis* (SD108). To obtain high transcription levels, the genes were individually cloned to downstream of strong promoters, and to avoid the potential issue with succinic acid being exported across the inner membranes (e.g. from mitochondria to cytoplasm), potential signal peptides were removed so that the enzymes would be expressed in the cytoplasm. In batch cultures using shake flasks, the resulting metabolic engineered strain (*IoΔura3*+SA) was able to consume all glucose (100 g/L) and produce 11.63 g/L of succinic acid, after 110 h of growth, while strain *IoΔura3* + *ura3* consumed 99.29 ± 0.08 g/L glucose and yielded 1.43 ± 0.04 g/L succinic acid during the same period.

9.3 Production of Biobased Dicarboxylic Acids

9.3.1 From Strain Isolation to Initial Physiological Studies

When developing a process for industrial implementation, the central concern is providing specific conditions for the microorganism perform at highest productivity at each stage of the bioprocess. To do so, the process has to be tuned to the specific requirements of the microorganism being used, and thus, the prior knowledge of its physiology and dynamic behavior in culture is crucial. The candidate strains may be well-known strains obtained from culture collections, genetically modified organisms after metabolic engineering optimization rounds or the introduction of heterologous pathways, or new isolates naturally occurring in nature.

Given the diversity of microorganisms naturally occurring in nature, the isolation of a desired population of microorganisms from natural sources is a time-consuming task, which could culminate in the isolation of hundreds of potentially desired strains. When a large number of isolates needs to be tested, high-throughput phenotypic screening can be used to carry out experiments in parallel rather than sequentially. In general, many variables can be tested at once, namely, different strains, carbon and nitrogen sources, nutrient concentrations, pH, and temperature. A simple yet powerful way to do so involves, for example, the use of deep-well microplates coupled with optical density analysis in microplate readers (biomass concentration) and high-performance liquid chromatography (HPLC) analysis with microplate-compatible autosamplers (for analysis of metabolites). The different well geometries and orbital agitation conditions further enable to modulate the mass transfer efficiency and thus aeration (Duetz et al., 2000), which can be varied from microaerobic to fully aerobic conditions. By cultivating the cells in static deep-well microtiter plates, anaerobiosis is achieved. More sophisticated multiplexed systems exist, which allow monitoring online the variation of parameters, such as the optical density, the dissolved oxygen concentration, and the pH, but the high cost of these systems and the time-consuming data analysis of the various time profiles recorded has prevented their widespread use.

Initially, the screening should be performed in a medium known to enable sustained growth of the candidate microorganisms and to favor the production of the target compound. For example, in the case of the production of succinic acid from yeasts, de Klerk (2010) described a culture medium that simulates grape must, used for the isolation of succinic acid producing wine yeasts. According to the author, succinic acid production by fermenting yeasts will be favored by moderate to high fermentation temperatures in must with a nicotinic acid and/or nicotinamide deficiency, high sugar content, moderate amounts of metabolically available nitrogen, presence of flavonoids, and large supplies of unsaturated long-chain fatty acids. Using a number of 96 deep-well microtiter plates, a fast screening is achieved by growing a different isolate in each well, especially if some degree of automation is available. The screening should be focused on parameters relevant for the envisaged larger scale production, such as cell growth kinetics and product concentration. When using the high-throughput system, sampling during the incubation period is limited, since the volume of culture in each well ranges from 1 to 2 mL. For that reason, samples should be only collected in the end of the incubation period or use multiple replicate microtiter plates, which allows characterization of each culture at different incubation periods by simply removing one of the replicas from the incubator at a given point. This array of experiments will typically allow for the selection of a significantly narrowed range of candidate strains, which are then brought to shake-flask experiments, allowing for more detailed physiological studies. Again, the mass transfer during the specific culture conditions will determine the aeration efficiency of the culture. As explained above, this will have major effects on the metabolism and particularly on succinic acid production. The aeration efficiency will depend on the stirring speed, the ratio of volume of liquid versus total flask volume, and on particular features the shake flask may be equipped with (e.g. baffles).

In the case of the production of succinic acid using yeast strains, this screening activity is based on measurements of parameters such as: biomass concentration, residual main carbon source concentration, concentration of succinic acid, concentration of other carboxylic acids, concentration of ethanol. If various conditions are tested, such as different fermentation times and different agitation/aeration regimes, preliminary, but important, information will be derived that will enable designing fermentation strategies to be implemented in fully controlled bioreactors, simulating the large-scale equipment.

9.3.2 Linking Strain Physiology to Bioreaction Conditions

As mentioned above, the way carbon is processed through the central carbon pathway depends not only on the carbon source concentration, but also on the availability of oxygen. When the carbon source and oxygen are available without limitation in *Saccharomyces*, the carbon flux to the TCA cycle will be split at the pyruvate node, due to intrinsic limitation of the respiratory capacity of the yeast (Crabtree effect). As result, part of the carbon source will be diverted into unwanted compounds such as ethanol and eventually acetate, thus diminishing the yield of the product of interest but also inhibiting growth if high concentrations of ethanol or acetate are reached. This will happen not only with *Saccharomyces* strains, but with all Crabtree-positive yeasts. These conditions are typically those occurring when using a batch fermentation strategy in which all components of the fermentation medium are present at the time of inoculation of the bioreactor and the additions to the bioreactor are limited to those required to control the dissolved oxygen concentration (air) and the pH (acid and/or base). The absence of carbon or oxygen limitation results in the most energetically efficient use of carbon source (maximizes the generation of ATP), but mainly targets metabolism to biomass production with generation of CO_2. Regarding succinic acid, although it is produced in the TCA cycle, it is hardly accumulated and readily converted to fumarate and beyond to complete the cycle. Further, the enzymes involved in the glyoxylate cycle are known to be repressed when high concentrations of glucose are present. One way to try to circumvent this diversion of carbon from the target pathway is the adoption of a fed-batch strategy in which a small amount of main carbon source is present in the culture medium at the time of inoculation and the carbon source is gradually fed throughout the culture, e.g. as a concentrated glucose solution, to supply, ideally, the nutrient at the exact rate at which it is being consumed by the culture, thereby keeping a constant concentration in the bioreactor below the threshold of the respiratory capacity of the yeast. The purpose of using a high concentration of sugar is to reduce the dilution of the culture when the solution is added by minimizing the volume of solution that needs to be added to maintain the desired concentration in the bioreactor. This strategy also allows supplying a much higher total amount of nutrient than that possible in a batch fermentation without causing substrate inhibition or the formation of high concentrations of by-products, some of which potentially inhibit growth. Further, when using complex raw materials, particularly complex by-products of agroforestry industries, the growth inhibitors that may be present on those materials will be highly diluted at the initial stages of the fermentation, and their concentrations will only become meaningful when a robust high-cell density culture has developed, thus more able to cope with those inhibitors. In the case of yeast, even if some ethanol or acetic acid are produced, these can be used as substrate as soon as the main carbon source is limiting or absent in aerobic conditions. It has been shown that these are essentially processed through the glyoxylate cycle, which generates succinate and malate. When the main carbon source is mainly processed through a fermentative pathway due to oxygen limitation, ethanol is formed. However, part of it could be directed to the reductive direction of the TCA cycle, although this pathway is much less active than the ethanol production pathway due to the kinetic parameters of the involved enzymes for the reductive route of TCA cycle and due to the resulting redox imbalance. When sufficiently high concentrations of CO_2 are present, the equilibrium of the conversion of isocitrate to α-ketoglutarate and of α-ketoglutarate to succinyl-CoA will be affected, and the conversion of isocitrate to succinate through the glyoxylate cycle will be favored. Additionally, the conversion of pyruvate to oxaloacetate and the cycling of TCA in the reductive direction will also be favored. These carboxylation reactions can be favored by sparging CO_2 or by adding carbonate salts in the fermentation medium.

Since succinic acid is a member of the core carbon metabolism cycle, in normal circumstances, yeasts will not accumulate it, nor any of the other dicarboxylic acids upstream, or downstream succinic acid in that cycle. Another way to foster the accumulation of the intermediate components of the TCA cycle would be to restrict biomass formation. Indeed, the TCA cycle intermediates oxaloacetate and α-ketoglutarate are precursors of biomass components, namely amino acids, and provide carbon diversions from the target product. In order to limit the diversion of carbon and TCA intermediates to biomass formation, restrictions in the supply of, for example, nitrogen, can be imposed. Without nitrogen, those precursors cannot be converted into nitrogen-containing molecules, such as amino acids, the building blocks of proteins, and therefore their diversion from the TCA cycle will be diminished. One very effective way to supply a fermentation process with the required nitrogen for biomass buildup is the control of the pH with ammonium hydroxide, balancing the pH lowering effect of the production and accumulation of organic acids, namely acetic acid. If the pH control agent is shifted from ammonium hydroxide to sodium hydroxide, less nitrogen will be available, eventually creating nitrogen limitation, thus favoring the reactions not involving biomass formation.

9.3.3 Requirements for Scale-Up Toward Commercial Operation

Various technical issues must be tackled in order to deploy a process that can adequately perform at commercial scale. First and foremost, the process must be robust enough in order to be replicated without significant performance deviations between production batches. This implies that the slight variations in raw material quality should not impact significantly on the process, that the implemented controls are able to maintain the process within the desired operating windows, and that the microorganism is stable during the batch, and cultures in the cell banks are adequately standardized. This is particularly important when genetically modified strains are used, since they are normally under a more demanding, stressful, and energy-demanding physiological state than their nongenetically modified counterparts, which, if present, tend to dominate the culture.

Although scale-up of a fermentation-based process may seem a simple multiplication factor of all aspects of the process, it is actually one of the most challenging tasks in bioprocess engineering. Indeed, the larger the scale of a bioreactor, the further will it be from ideal behavior. For example, in larger bioreactors, mass and heat transfer issues become highly relevant since concentration and temperature gradients become significant, which originate diverse microenvironments within the bioreactor that can dramatically affect the metabolism and cell viability. For example, upon feeding a concentrated sugar solution to the bioreactor, the cells on the region in which the sugar solution drops will be subjected to a very high sugar concentration and therefore may trigger the diversion of the metabolism toward the generation of by-products (e.g. ethanol in *S. cerevisiae* or acetic acid in *E. coli*). Similarly, cells that at a certain instant flow past the air sparger will be subjected to a very high dissolved oxygen concentration, while those that at the same instant are flowing at the top of the fermenter will be in an environment with a much lower and eventually limiting oxygen concentration, with evident differences in their metabolic state. These heterogeneities can therefore lead to decreases in productivity and yields as compared to the same process as implemented at small-scale in conditions close to ideal mixing. These scale-up issues are particularly important for aerobic processes (gradients in oxygen concentration) and in fed-batch operation, in which the continuous or periodic addition of solutions, often highly concentrated, is performed.

9.4 Downstream Processing of Biobased Dicarboxylic Acids

9.4.1 General Steps in Downstream Processing

Once produced by fermentation, the dicarboxylic acids in the fermentation broth need to be recovered and purified to meet the specifications of the final product. This process is called downstream processing (DSP) and includes several unit operations. It may comprise all or some of the following steps: cell disruption, separation, recovery, concentration, purification, product stabilization, and formulation of a target molecule (Table 9.3). The DSP also includes consideration of side-streams and reuse of effluents. This section gives an overview of general steps in DSP of a broth produced by submerged fermentation.

The choice of unit operations for a specific DSP sequence has to consider the final product application, as different specifications regarding purity, biological activity, molecular weight, etc., might be required. The understanding of the chemical and physical properties of the target molecule and the specific process is a prerequisite for designing an adequate DSP. As dicarboxylic acids are excreted to the fermentation broth, no cell disruption is required. For separation of cells from the fermentation broth, the common solid/liquid separation technologies are sedimentation or centrifugation, based on the difference in density/gravity, or filtration, based on the difference in particle size, typically microfiltration. After cell removal, the dicarboxylic acids need to be recovered from the fermentation broth. The cells may be recirculated back into the fermentation process or discarded. The following recovery process aims at the crude separation of soluble components (e.g. proteins, sugars), which are different in their physical and chemical properties compared to the target molecule. Here, the separation may be based on differences in molecular size in which filtration is used again, typically through ultrafiltration or nanofiltration membranes; differences in solubility, through precipitation processes; and differences in chemistry, for example, through solvent extraction or reactive extraction. Between steps, it is often desirable to reduce the amount of liquid material, and therefore a concentration step may be used for this purpose. In fact, in submerged fermentations, the fermentation broth may contain up to 95% of water. A simple concentration process is the removal of water or

Table 9.3: General Steps in Downstream Processing, Highlighting Common Principles and Unit Operations as well as the Respective Desired Outcome of Each Step.

Step	Principle	Unit Operation	Aim
Cell disruption	Mechanical	High pressure, homogenization, elongation, shear-stress	Disrupted cells released the target molecule
	Chemical	Addition of lytic agents, e.g. solvents, surfactants, alkalis	
	Physical	Heat, sonication, freezing, osmotic shock	
	Biological	Specific enzymes	
Separation (liquid/solid)	Density/gravity Particle size	Centrifugation, sedimentation Filtration	Particle-free liquid or solid containing the target molecule depending on solubility in fermentation broth
Recovery and concentration	Solubility	Precipitation	Removal of components with different chemical/physical properties compared to the target molecule, e.g. proteins, sugars, water
	Molecular size	Membrane filtration	
	Chemical	Solvent extraction, reactive extraction	
	Physical	Evaporation of water/solvent	
Purification	Solubility Affinity Ionic charge	Crystallization Chromatography, adsorption Electrodialysis	Removal of components with similar chemical/physical properties compared to the target molecule, e.g. salts, other organic acids
Product stabilization & formulation	Drying Agglomeration Sterilization Blending/packing	Spray drying, freeze-drying Fluidized bed drying Heating Ingredients for specific product properties	Final product formulation

solvents via evaporation. For heat-sensitive metabolites (like proteins), evaporation under reduced pressure or a membrane process can also be applied. However, the energy-demanding water removal is often a later-stage step after less demanding operations allowed for a preliminary quantity reduction of processed streams. Subsequent purification results in the removal of components (e.g. salts, other organic acids), which are difficult to separate because of their similar physical and chemical properties compared to the target molecule. Steps for separation of similar molecules usually contribute the most to the cost of DSP due to high equipment and operating costs. Depending on the type of molecule and final product requirements, further purification steps can include crystallization, based on differences of solubility; chromatography and adsorption, based on different affinity toward a solid phase; and electrodialysis or ion exchange chromatography, mainly based on differences in charges of molecules. After sufficient purification, the final product may need to be stabilized for shipping, storage, and sale. The most common principle here is drying the product. Depending on sensitivity of the product, different unit operations are available such as freeze-drying, spray drying, fluidized bed drying, etc. Agglomeration, sterilization, or blending might also be required for some final product formulations.

9.4.2 Factors Influencing Downstream Processing

Downstream processing is an integral part of the entire production process and is influenced by many factors. Not only the bioprocess itself, but also the raw materials used, the specifications of the final product or the operations that may follow directly after DSP, such as formulation or direct *in situ* use, such as polymerization, need to be considered. Further, in some instances, at least part of the DSP sequence may be closely integrated with the fermentation step, whereby *in situ* separation of products can be performed. DSP make up a significant part of the total production costs, ranging between 50% and 90% of the total production process.

The type of microorganism (filamentous fungi, yeast, bacteria), product (intra-/extracellular, chemical and physical properties), fermentation substrate, fermentation process (submerged, solid state) as well as target application (purity specifications) influence the decision for the specific choice of DSP. In most cases, different combinations of steps and different unit operations have to be used. As mentioned earlier, various by-products may arise from succinic acid fermentation, which have similar structures and properties, including malic acid and lactic acid, and their selective separation has a significant impact on DSP costs. The use of recombinant strains may allow for less production of metabolic by-products.

Bacterial fermentation for the production of succinic acid is usually performed at neutral pH. Due to constant release of organic acids, the pH decreases during fermentation, and the effect of product inhibition usually occurs, unless pH control is implemented, normally through the use of calcium, magnesium, sodium or potassium hydroxides, carbonates, or ammonium salts or simply ammonia. This leads to the production of a succinate salt, which then needs to be converted into succinic acid through acidification with sulfuric acid, generating gypsum as a by-product. Yeasts are more tolerant to acidic conditions and may be used in processes at significantly lower pH. At a pH below 3, many dicarboxylic acids are predominantly present in their undissociated form (e.g. succinic acid $pK_{a1} = 4.2$, and $pK_{a2} = 5.6$ at 25°C (Dean, 1999)). In this case, the requirements of pH-controlling elements are smaller, resulting in lower consumption of chemicals, and less waste produced since the undissociated free acid is directly recovered (Abbott et al., 2009).

Generally, carbon sources are typically provided by fermentation substrates containing easily fermentable carbohydrates such as mono- or disaccharides. An increasing diversity of low-cost but complex raw materials, such as agro-industrial wastes, are being considered by the chemical industry as it replaces its petroleum based production processes with industrial biobased fermentation processes. Pretreatment of material (e.g. digestion, removal of inhibitors) becomes necessary, allowing for a cleaner broth, better fermentation results, and consequently lower DSP costs. Yet, fermentation based on these alternative raw materials includes a more diverse initial mix of complex components, which need to be separated after the fermentation in order to meet product specifications.

The choice of unit operation for the DSP is also influenced by the intended application of the final product. For instance, some applications require large amounts of the target product in lower quality, while others require smaller amounts of high quality and biological active material. Further, the nature of contaminants can be determinant. In general, for common applications of succinic acid a high purity is required (>98%). If the product is intended to be used in the production of polymers, such as PBS, it can tolerate small amounts of other dicarboxylic acids, which still have two functional groups sustaining the chain extension during the polymerization process. Conversely, even minute amounts of monocarboxylic acids will be detrimental to the final product, since they only have one functional group and, as such, will act as polymerization terminators.

The type of target metabolite and thus its chemical and physical characteristics have the highest impact on designing the DSP. This mainly includes polarity, acid dissociation constant, sensitivity, solubility, etc. In summary, it can be said that designing an effective DSP is very specific and needs to evaluate the chemical and physical properties of the target molecule, the requirements of desired application, and the reuse of effluents and side products.

9.4.3 State of the Art DSP for Dicarboxylic Acids and New Trends

Precipitation

State of the art recovery processes of dicarboxylic acids take advantage of precipitation technologies that are currently industrially applied, i.e. in the production and purification of lactic, citric, and acetic acids (Kurzrock and Weuster-Botz, 2010; Yang et al., 2007). Thus, equipment, technology, and infrastructure can often be adapted or even directly employed (Cheng et al., 2012a). In the case of succinic acid, biomass is removed from the carboxylate-rich fermentation broth, which is then acidified. The control of pH during fermentation through the addition of $Ca(OH)_2$ has the disadvantage of the lower solubility of the calcium carboxylate, which may lead to the precipitation of calcium salts already during fermentation. This is the reason why alternative pH controlling agents, like sodium and magnesium hydroxide, were investigated (De Haan et al., 2013). The precipitation may be facilitated by cooling the fermentation broth, and the resulting succinates can be separated via filtration or centrifugation and then converted to succinic acid by adding concentrated H_2SO_4. Further purification steps may include active carbon absorption or ion exchange followed by concentration and crystallization via evaporation. Disadvantages of this process are the high dosages of $Ca(OH)_2$, or CaO and H_2SO_4, which cannot be reused. Additionally, the undesirable by-product, $CaSO_4$ (equimolar to succinic acid), is of low economic value

(Datta, 1992; Datta et al., 1992; Berglund et al., 1999). A yield of 91% (Datta et al., 1992) and a purity of 94.2% can be achieved. The precipitation of ammonium carboxylates has the advantage of possible regeneration of reagents (including sulfuric acid and ammonia through thermal cracking). Thus, lower amounts of waste by-products accumulate since the recovered acid and base can be reused during the fermentation and the DSP, while achieving higher recovery rates (yield 93.3%) (Yedur et al., 2001). Disadvantages include the low selectivity of precipitation, high energetic costs, and the need for durable equipment that is able to withstand the severe conditions of temperature and acidity during recycling of used reagents.

Extraction

The cell-free fermentation broth is an aqueous phase from which a specific dicarboxylic acid needs to be selectively extracted into another liquid phase (aqueous–solvent or aqueous–aqueous), immiscible with the fermentation broth. Each target molecule has a special distribution coefficient also depending on the parameters of extraction, such as pH, temperature, ionic strength, and type of solvent (Ottens et al., 2006). The distribution coefficient is defined as the ratio of the concentration of the target acid between two immiscible phases at equilibrium. The distribution coefficient should be high for the targeted product and low for other substances in the fermentation broth, indicating preferential solubility in the solvent and hence good separation, while a lower distribution coefficient would require multistage extraction and large amounts of solvent. Classical solvent extraction using, for example, alcohols, ketones, esters, or aliphatic hydrocarbons is not effective in the DSP of dicarboxylic acids because of low distribution coefficient (<1), resulting in low yield and purity. One way to increase the distribution coefficient is adding a reactive compound, e.g. an amine, through a process called reactive extraction (Yang et al., 2013; Kertes and King, 1986). The chosen reactive compound should ideally have low water solubility, high distribution coefficient for the target dicarboxylic acid, and low distribution coefficient for other organic acids or impurities (Thongchul, 2013). Generally, a reactive extraction system comprises the steps of (A) reactive extraction followed by (B) back extraction as shown in Fig. 9.2, in this particular case using a displacer. The entire process also includes prior cell separation as well as subsequent concentration, crystallization, and drying.

In the reactive extraction stage, the cell-free fermentation broth with the target acid is mixed with the organic phase containing the reactive compound. Long-chain aliphatic amines have been used successfully as reactive compounds for dicarboxylic acids (Hong and Hong, 2004; Kurzrock and Weuster-Botz, 2010; Kurzrock, 2010; Lee et al., 2008, Tung and King, 1994). At the interphase of aqueous and solvent phase, the amine forms a complex with the acid. The acid-amine complex is solvated in the organic phase. Some amines benefit from the presence of a diluent, which reduces the corrosiveness and high viscosity of the amines, while also

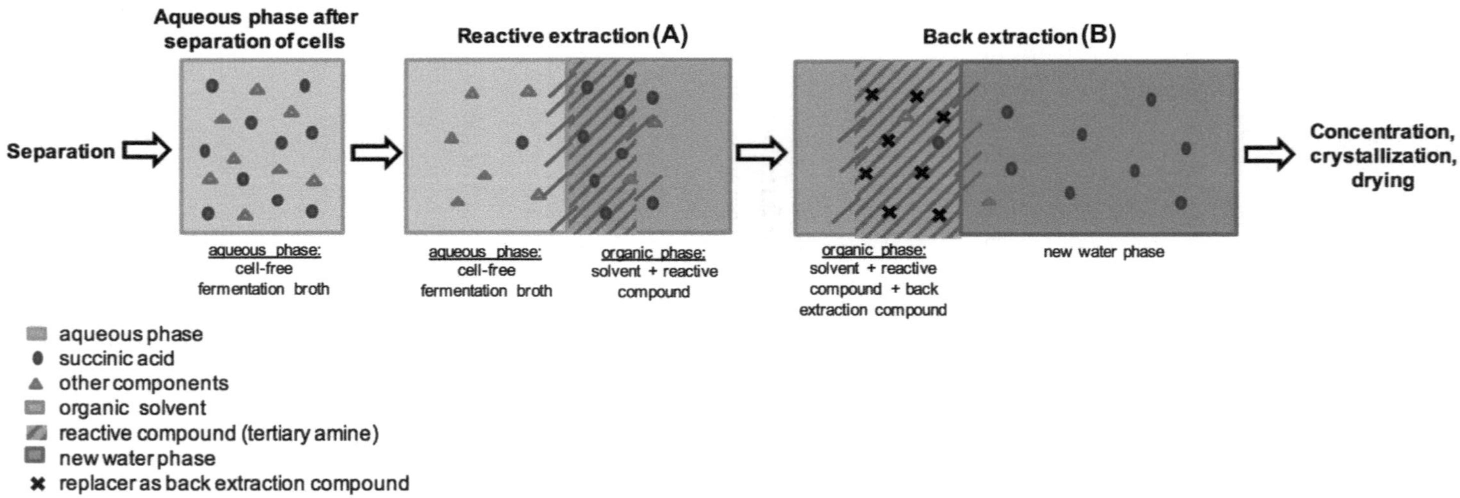

Figure 9.2
Schematic principle and processing steps during (A) reactive extraction and (B) back extraction system with displacer as back extraction compound.

having an effect on the extractive capabilities (Kurzrock and Weuster-Botz, 2011; Huh et al., 2006; Tamada and King, 1990). Active solvents (such as 1-butanol, 1-hexanol, and 1-octanol) are recommended to improve the solvation of hydrophilic organic acids by hydrophobic long-chain aliphatic amines (Yang et al., 2013; Thongchul, 2013) by stabilizing the acid-amine complex, while inert solvents (such as alkanes) act solely as solvating media (Kurzrock and Weuster-Botz, 2011). Generally, the efficiency of extraction with aliphatic amines depends largely on the pH, temperature, by-products in the fermentation broth, the concentration of target acid, the properties of the solvent and amine, the loading ratio, and distribution coefficient. In a similar fashion, other extractive elements can be employed, such as organophosphines. These substances, which are strong Lewis bases, were also tested for the recovery of succinic acid, albeit no application in recovery from fermentation broths has been reported. After the reactive extraction stage, it is necessary to recover the dicarboxylic acid, which is solvated in the organic phase as a nonvolatile acid-amine complex, and to recycle the costly amines and the organic phase. Back extraction is often suggested as the best alternative. It is based on a change in distribution coefficient between the organic phase and a new water phase, depending on operating variables such as a change of solvent, swing in temperature, and/or disintegration of the acid-amine complex by shift in pH or addition of a displacer, typically a stronger acid or base (e.g. Yabannavar and Wang, 1991; Lee et al., 2008; Kurzrock, 2010; Sadaka and Garcia, 1998). Typically, a combination of these methods yields better results. After back extraction, the solvent is separated from the initial fermentation broth, which can be recirculated into the fermentation process. The advantage is the backflow of nutrients and remaining dicarboxylic acid, which was not completely extracted. However, the recirculate might still contain residues of solvent, which should be biocompatible in order to prevent inhibition of the microorganisms. Overall, tertiary amines display better biocompatibility compared to primary or secondary amines. Nonpolar solvents tend to have better biocompatibility than polar solvents (Kurzrock, 2010). The toxicity of a solvent could also be avoided by applying aqueous two-phase systems (Yang et al., 2007; Planas et al., 1996).

Electrodialysis

Electrodialysis is a separation process where ionized compounds are separated from nonionized compounds in aqueous solutions based on transport through ion exchange membranes in an electric field. It has already been used in industry, for example, in the purification of lactic acid from whey, saltwater purification, and glycerine production. An early application on the purification of citric acid from fermentation broth has been described by Guenther (1963). The conventional monopolar electrodialysis stack consists of an alternating series of anion- and cation-selective membranes separated by flow distribution gaskets, and is often used for concentration and purification through separation of ions and uncharged molecules. The membranes are bound on one end by an anolyte compartment and an anode, while on the other end by a catholyte compartment and a cathode. The direct current moves the cations

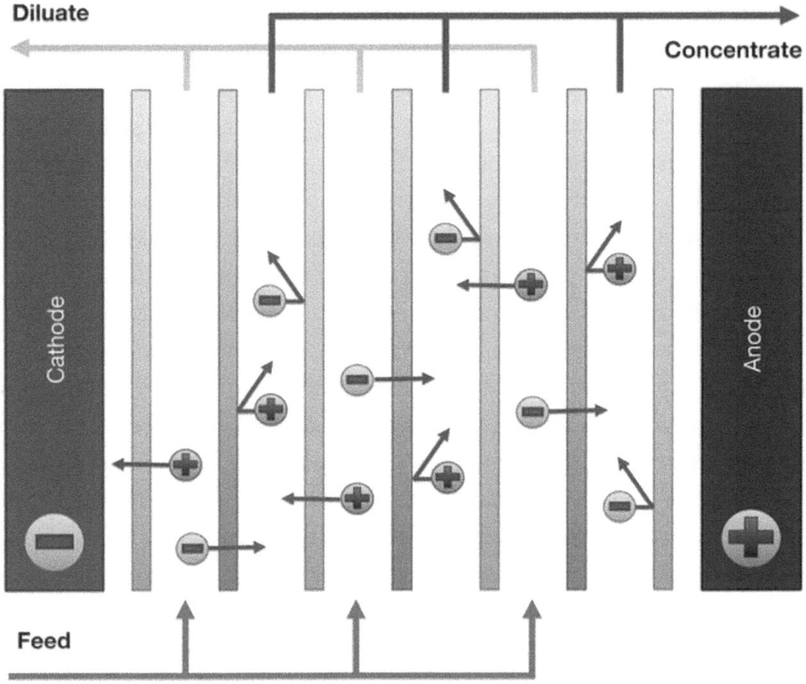

Figure 9.3
Schematic principle of electrodialysis.

through the cation exchange membranes and the anions through the anion exchange membranes. The net result is that salt in the feed compartment is depleted, and the adjacent concentration compartments are enriched with the salt. The negatively charged dicarboxylic acid crosses the anion exchange membrane into the concentrate stream, according to Fig. 9.3. The water splitting bipolar electrodialysis process allows the recovery of the dicarboxylic acid, since it efficiently converts salts into the corresponding alkali and acid, while water is split into H^+ and OH^-. It has the advantage of converting the carboxylate into the corresponding free carboxylic acid without requiring acidifying chemicals, and the recovered base can be reused to control the pH during fermentation. However, the high energy consumption, the cost of the membranes, and the low selectivity for succinic acid are disadvantages. Indeed, the presence of acetate and formate in the fermentation medium hinder the process efficiency due to the inability to selectively separate these acids from succinate (Song et al., 2007). A more specific problem is the presence of bivalent ions, like calcium and magnesium, often used as neutralization ions during the fermentation, which cause fouling of the electrodialysis membranes. More specifically to water splitting electrodialysis, the low pH of the acid/concentrate stream may cause the crystallization of the dicarboxylic acid on the membrane itself (Urbanus et al., 2010).

Other processes

Adsorption techniques include a high number of processes useful for the purification, particularly in polishing steps at the end of DSP of dicarboxylic acids using different solid phase materials, including activated carbon, alumina, silica, and zeolite molecular sieves (Efe et al., 2011; Inci et al., 2012; Jun et al., 2007). Similarly, ion exchange is often applied at the end of entire DSP to remove remaining cations and anions. Simulated moving bed chromatography may be considered to allow for continuous *in situ* separation during fermentation. The retained dicarboxylic acid can usually be eluted or desorbed with methanol, ammonia, or sulfuric acid (Yang et al., 2007). In general, high-capacity, complete and low-cost regenerability, as well as specificity for the target product are desired properties for sorption and ion exchange materials. The pH of the fermentation broth may need to be adjusted before this DSP process, and additional chemicals are needed to regenerate the solid phase (Yang et al., 2007; Thongchul, 2013).

Further, the distillation of fermentation broths can be used to recover carboxylic acids with a high degree of purity through the separation of volatile and nonvolatile compounds. The distillation processes for the recovery of dicarboxylic acids are based on processing a clarified fermentation broth, free from biomass and insoluble elements, optionally concentrated at high temperatures in distillation apparatuses in order to extract ammonia from the ammonium carboxylates and remove contaminants. The final product is obtained after crystallization by cooling. Operational parameters that influence the process include the pressure, the pH of the medium, the allowed water content in the distillation vessel, the presence of ammonia in the overhead of the distillation and in the liquid phase of the remaining mother liquor, as well as the presence of ammonium carboxylates in the liquid phase. Distillation has the clear disadvantage concerning the energetic costs, and relatively low yields are reported so far. Usually crystallization is used as final purification step, but some cases of direct crystallization are also described in literature (Li et al., 2010; Lin et al., 2010; Luque et al., 2009).

9.4.4 Designing an Effective Downstream Process

The most important unit operations for the DSP of dicarboxylic acids are discussed in the previous section. Even though they are technically feasible, their industrial realization depends on the ability to scale-up, their robustness, the overall separation yield, investment and operating costs, as well as environmental factors. Further, each process needs to be tailored to the properties of the dicarboxylic acid of interest, to the remaining elements capable of influencing the DSP such as carbon source, or producing microorganisms, and the requirements for yield and purity.

In general, recovery and purification of dicarboxylic acids requires a combination of different unit operations. The process should be kept simple with as few steps as possible and be implementable in conventional processing equipment in order to reduce capital and operational expenditure, while at the same time still complying with the required product specification. Importantly, the DSP should not be seen alone but as an integral part of the entire production chain.

9.5 Challenges for Sustainable Microbial Production of Dicarboxylic Acids Platform Chemicals

A competitive biotechnological production route of dicarboxylic acids needs to consider the following factors:

- Low-cost but efficient feedstock: Feedstock can represent a major source of costs for bulk chemicals. Therefore, a cost-effective alternative source for carbohydrates needs to be identified and established in the production process. Waste by-products from e.g. chemical sector (such as glycerol from biodiesel production) and waste biomass from e.g. agricultural or forestry sectors (such as straw, leftovers after fruit and vegetable processing) could be promising (Cheng et al., 2012b).
- Low organic and inorganic impurities: Several recovery methods are hampered in efficiency in the presence of organic and inorganic impurities, mostly coming from the unclean feedstock. Further, pretreatment of the raw material may need to be implemented, as some of the compounds entrained with the raw material can affect the growth of the producing microorganism.
- Optimized fermentation with yeasts: Production of dicarboxylic acids using yeast has several advantages, of which the most relevant is the high yeast tolerance to low pH. This reduces the need for chemical input for pH adjustment and consequently less by-product formation. Moreover, the dicarboxylic acid can predominantly be recovered in the form of free acid if pH is below the pKa (<4.21 for succinic acid).
- Increased productivity and yields while at the same time lower production of by-products: A higher concentration of target product results in less extensive DSP. The selective separation of different dicarboxylic acids is difficult if not impossible; thus the reduction of by-product formation during fermentation is relevant. Further, by avoiding the formation of by-products, the carbon source is not significantly diverted toward the production of unwanted products, and thus, the raw material usage is reduced, with both economic and environmental benefits. For this to be possible, a combination of the isolation or design of high-performing strains with the implementation of an optimized fermentation strategy will be required.
- Integrated production of other high value-added products: Simultaneous production of additional substances of commercial interest within the same fermentation is a suggested

path toward an efficient biorefinery (Cheng et al., 2012a). Parallel production of propanediol and succinate, isoamyl acetate and succinate, or polyhydroxybutyrate and succinate were suggested using *E. coli* (Dittrich et al., 2009; Kang et al., 2010). In the case of industrial production using yeast, another value-added product might be bioethanol that could be produced in the same infrastructure.

- Reduced by-product production and biocompatibility of the process: Production of unprofitable by-products during DSP such as calcium sulfate (gypsum) should be avoided. Moreover, the recycling and reuse of effluents needs to be integrated in the process to save costs and reduce the environmental impact of the overall process.
- *In situ* product recovery: The *in situ* product recovery results in several advantages for the production of dicarboxylic acids. The effect of product inhibition is reduced, along with the requirement for the pH-controlling element (e.g. less chemicals). At the same time, productivity may be increased, and the overall number of processing steps may be reduced.

9.6 Conclusion

In order to produce biobased dicarboxylic acids capable of competing with the petrochemical process in a cost-effective and sustainable manner, a strategy for an integrated biorefinery system for the entire value chain, including raw material sourcing and treatment, selection or design of an effective microbial strain, design of a robust and scalable fermentation and DSP process, and whenever possible, the integration of these various steps, needs to be considered.

References

Abbott, D.A., Zelle, R.M., Pronk, J.T., van Maris, A.J.A., 2009. Metabolic engineering of *Saccharomyces cerevisiae* for production of carboxylic acids: current status and challenges. FEMS Yeast Research 9, 1123–1136.

Agren, R., Otero, J.M., Nielsen, J., 2013. Genome-scale modeling enables metabolic engineering of *Saccharomyces cerevisiae* for succinic acid production. Journal of Industrial Microbiology and Biotechnology 40, 735–747.

Arikawa, Y., Kobayashi, M., Kodaira, R., Shimosaka, M., Muratsubaki, H., Enomoto, K., Okazaki, M., 1999. Isolation of sake yeast strains possessing various levels of succinate- and/or malate-producing abilities by gene disruption or mutation. Journal of Bioscience and Bioengineering 87, 333–339.

Attfield, P.V., 1997. Stress tolerance: the key to effective strains of industrial baker's yeast. Nature Biotechnology 15, 1351–1357.

Beauprez, J.J., de Mey, M., Soetaert, W.K., 2010. Microbial succinic acid production: natural versus metabolic engineered producers. Process Biochemistry 45, 1103–1114.

Berglund, K.A., Yedur, S., Dunuwila, D.D., 1999. Succinic Acid Production and Purification. US Patent 5958744.

Boy, C., Lappe, P., 2012. Dicarboxylic acids, aliphatic. Ullmann's Encyclopaedia of Industrial Chemistry, 11. Wiley-VCH Verlag, Weinheim, pp. 287–304.

BREW: The Brew Project, 2006. Medium and Long-term Opportunities and Risks of the Biotechnological Production of Bulk Chemicals from Renewable Resources – The Potential of White Biotechnology.

Cheng, K.K., Zhao, X.B., Zeng, J., Wu, R.C., Xu, Y.Z., Liu, D.H., Zhang, J.A., 2012a. Downstream processing of biotechnological produced succinic acid. Applied Microbiology and Biotechnology 95, 841–850.

Cheng, K.K., Zhao, X.B., Zeng, J., Zhang, J.-A., 2012b. Biotechnological production of succinic acid: current state and perspectives. Biofuels, Bioproducts and Biorefining 6, 302–318.

Cheng, K.K., Wang, G.Y., Zeng, J., Zhang, J.A., 2013. Improved succinate production by metabolic engineering. Biomedical Research International 2013, 1–12.

Coustou, V., Besteiro, S., Rivi, L., 2005. A mitochondrial NADH-dependent fumarate reductase involved in the production of succinate excreted by procyclic *Trypanosoma brucei*. Journal of Biological Chemistry 280, 16559–16570.

Datta, R., 1992. Process for the Production of Succinic Acid by Anaerobic Fermentation. US Patent 5143833.

Datta, R., Glassner, D.A., Jain, M.K., Vicky Roy, J.R., 1992. Fermentation and Purification Process for Succinic Acid. US Patent 5168055.

De Haan, A.B., van Breugel, J., van der Weide, P.L.J., Jansen, P.P., Vidal Lancis, J.M., CerdàBaró, A., 2013. Recovery of Carboxylic Acid from Their Magnesium Salts by Precipitation Using Hydrochloric Acid, Useful for Fermentation Broth Work-up. WO 2013025105.

De Klerk, J.L., 2010. Succinic Acid Production by Wine Yeasts (Thesis M.Sc. Agriculture). University of Stellenbosch, Viticulture and Oenology.

Dean, J.A., 1999. Langes's Handbook of Chemistry, fifteenth ed. McGraw-Hill, New York, pp. 8–68.

Dittrich, C.R., Bennett, G.N., San, K.Y., 2009. Metabolic engineering of the anaerobic central metabolic pathway in *Escherichia coli* for the simultaneous anaerobic production of isoamyl acetate and succinic acid. Biotechnology Progress 25, 1304–1309.

Drewke, C., Thielen, J., Ciriacy, M., 1990. Ethanol formation in adh0 mutants reveals the existence of a novel acetaldehyde-reducing activity in *Saccharomyces cerevisiae*. Journal of Bacteriology 172, 3909–3917.

Duetz, W.A., Rüedi, L., Hermann, R., O'Connor, K., Büchs, J., Witholt, B., 2000. Methods for intense aeration, growth, storage, and replication of bacterial strains in microtiter plates. Applied and Environmental Microbiology 66, 2626–2641.

Dunn, B., Levine, R.P., Sherlock, G., 2005. Microarray karyotyping of commercial wine yeast strain reveals shared, as well as unique genomic signatures. BMC Genomics 6, 53–63.

Efe, C., Pieterse, M., Van der Wielen, L.A.M., Straathof, A.J.J., 2011. Separation of succinic acid from its salts on a high-silica zeolite bed. Chemical Engineering Process 50, 1143–1151.

Flikweert, M.T., Johannes, P., 1997. Metabolic responses of pyruvate decarboxylase-negative *Saccharomyces cerevisiae* to glucose excess. Applied and Environmental Microbiology 63, 3399–3404.

Flikweert, M., Zanden, L., Janseen, W., Steensma, H., Van Dijken, J., Pronk, J., 1996. Pyruvate decarboxylase: an indispensable enzyme for growth of *Saccharomyces cerevisiae* on glucose. Yeast 12, 247–257.

Gallmetzer, M., Meraner, J., Burgstaller, W., 2002. Succinate synthesis and excretion by *Penicillium simplicissimum* under aerobic and anaerobic conditions. FEMS Microbiology Letters 210, 221–225.

Guenther, S., 1963. Process of Producing Citric Acid. US Patent 3086928.

GVR, March 2015. Fumaric Acid Market Analysis by Application (Food & Beverages, Rosin Paper Sizes, UPR, Alkyd Resins) and Segment Forecasts to 2020. Grand View Research, San Francisco, USA.

Hawksworth, D.L., 2001. The magnitude of fungal diversity: the 1.5 million species estimate revisited. Mycological Research 105, 1422–1432.

Hong, Y., Hong, W., 2004. Influence of chain length of tertiary amines on extractability and chemical interactions in reactive extraction of succinic acid. Korean Journal of Chemical Engineering 21, 488–493.

Huh, Y.S., Jun, Y.-S., Hong, Y.K., Song, H., Lee, S.Y., Hong, W.H., 2006. Effective purification of succinic acid from fermentation broth produced by Mannheimia succiniciproducens. Process Biochemistry 41, 1461–1465.

Inci, I., Asci, Y.S., Uslu, H., 2012. LSER modeling of extraction of succinic acid by tridodecylamine dissolved in 2-octanone and 1-octanol. Journal of Industrial and Engineering Chemistry 18, 152–159.

Ito, Y., Hirasawa, T., Shimizu, H., 2014. Metabolic engineering of *Saccharomyces cerevisiae* to improve succinic acid production based on metabolic profiling. Bioscience, Biotechnolology and Biochemistry 78, 151–159.

Jun, Y.S., Huh, Y.S., Park, H.S., Thomas, A., Jeon, S.J., Lee, E.Z., Won, H.J., Hong, W.H., Lee, S.Y., Hong, Y.K., 2007. Adsorption of pyruvic and succinic acid by amine-functionalized SBA-15 for the purification of succinic acid from fermentation broth. Journal of Physical Chemistry C 111, 13076–13086.

Kamzolova, S.V., Yusupova, A.I., Dedyukhina, E.G., Chistyakova, T.I., et al., 2009. Succinic acid synthesis by ethanol grown yeast. Food Technology and Biotechnology 47, 144–152.

Kamzolova, S.V., Vinokurova, N.G., Yusupova, A.I., Morgunov, I.G., 2012. Succinic acid production from n-alkanes. Engineering in Life Sciences 5, 1–7.

Kamzolova, S.V., Vinokurova, N.G., Dedyukhina, E.G., Samoilenko, V.A., Lunina, J.N., Mironov, A.A., Allayarov, R.K., Morgunov, I.G., 2014. The peculiarities of succinic acid production from rapeseed oil by *Yarrowia lipolytica* yeast. Applied Microbiology and Biotechnology 98, 4149–4157.

Kang, Z., Gao, C.J., Wang, Q., Liu, H.M., Qi, Q.S., 2010. A novel strategy for succinate and polyhydroxybutyrate co-production in *Escherichia coli*. Bioresource Technology 101, 7675–7678.

Kertes, A.S., King, C.J., 1986. Extraction chemistry of fermentation product carboxylic acids. Biotechnology and Bioengineering 28, 269–282.

Kurzrock, T.M., 2010. Integrierte Reaktivextraktionzur Gewinnung von biotechnologisch hergestellter Bernsteinsäure (Dissertation). Technischen Universität München.

Kurzrock, T., Weuster-Botz, D., 2010. Recovery of succinic acid from fermentation broth. Biotechnology Letters 32, 332–339.

Kurzrock, T., Weuster-Botz, D., 2011. New reactive extraction systems for separation of bio-succinic acid. Bioprocess and Biosystems Engineering 34, 779–787.

Lee, E.Z., Huh, Y.S., Jun, Y.S., Won, H.J., Hong, Y.K., Hong, W.H., 2008. Effect of operating variables on back-extraction characteristics of succinic acid from organic phase. Biotechnology and Bioprocess Engineering 13, 342–346.

Li, Q., Wang, D., Wu, Y., Li, W.L., Zhang, Y.J., Xing, J.M., Su, Z.G., 2010. One step recovery of succinic acid from fermentation broths by crystallization. Separation and Purification Technology 72, 294–300.

Lin, S.K.C., Du, C.Y., Blaga, A.C., Camarut, M., Webb, C., Stevens, C.V., Soetaert, W., 2010. Novel resin-based vacuum distillation–crystallization method for recovery of succinic acid crystals from fermentation broths. Green Chemistry 12, 666–671.

Ling, E.T.M., Dibble, J.T., Houston, M.R., 1978. Accumulation of 1-trans-2,3-epoxysuccinic acid and succinic acid by *Paecilomyces varioti*. Applied and Environmental Microbiology 35, 1213–1215.

Liti, G., Carter, D.M., Moses, A.M., Warringer, J., Parts, L., James, S.A., Davey, R.P., Roberts, I.N., et al., 2009. Population genomics of domestic and wild yeasts. Nature 458, 337–341.

Luque, R., Lin, C.S.K., Du, C., Macquarrie, D.J., Koutinas, A., Wang, R., Webb, C., Clark, J.H., 2009. Chemical transformations of succinic acid recovered from fermentation broths by a novel direct vacuum distillation–crystallisation method. Green Chemistry 11, 193–200.

Mandeva, R.D., Ermakova, I.T., Lozinov, A.B., 1981. Metabolite excretion by yeasts of the genus Candida in media lacking sources N, P, S, or Mg and having different carbon sources. Mikrobiologiia (Moscow) 50, 62–68.

van Maris, A.J.A., Geertman, J.A., Vermeulen, A., Groothuizen, M.K., Winkler, A.A., Piper, M.D.W., van Dijken, J.P., Pronk, J.T., 2004. Directed evolution of pyruvate decarboxylase-negative *Saccharomyces cerevisiae*, yielding a C2- independent, glucose-tolerant, and pyruvate-hyperproducing yeast. Applied and Environmental Microbiology 70, 159–166.

Negi, B., Sharma, P., Kashyap, S., Seth, S., Dey, G., 2013. Screening of yeast strains for vinification of fruits from cold desert regions of North West India. International Food Research Journal 20, 975–979.

Novo, M., Bigey, F., Beyne, E., Galeote, V., Gavory, F., Mallet, S., Cambon, B., Legras, J.L., Wincker, P., Casaregola, S., Dequin, S., 2009. Eukaryote-to-eukaryote gene transfer events revealed by the genome sequence of the wine yeast *Saccharomyces cerevisiae* EC1118. Proceedings of the National Academy of Sciences United States of America 106, 16333–16338.

OECD, 2009. The Bioeconomy to 2030: Designing a Policy Agenda. OECD Publishing, Paris.

Otero, J.M., Cimini, D., Paril, K.R., Poulsen, S.G., Olsson, L., Nielsen, J., 2013. Industrial systems biology of *Saccharomyces cerevisiae* enables novel succinic acid cell factory. PLoS One 8, e54144.

Ottens, M., Wesselingh, J.A., van der Wielen, L.A.M., 2006. Downstream processing. In: Ratledge, C., Kristiansen, B. (Eds.), Basic Biotechnology, third ed. Cambridge University Press, Cambridge, pp. 219–249.

Papagianni, M., 2007. Advances in citric acid fermentation by *Aspergillus niger*: biochemical aspects, membrane transport and modeling. Biotechnology Advances 25, 244–263.

Planas, J., Radstrom, P., Tjerneld, F., Hahn-Hägerdal, B., 1996. Enhanced production of lactic acid through the use of a novel aqueous two-phase system as an extractive fermentation system. Applied Microbiology and Biotechnology 45, 737–743.

Querol, A., Belloch, C., Ferández-Espinar, M.T., Barrio, E., 2003. Molecular evolution in yeast of biotechnological interest. International Microbiology 6, 201–205.

Raab, A.M., Gebhardt, G., Bolotina, N., et al., 2010. Metabolic engineering of Saccharomyces cerevisiae for the biotechnological production of succinic acid. Metabolic Engineering 12, 518–525.

Raab, A.M., Lang, C., 2011. Oxidative versus reductive succinic acid production in the yeast *Saccharomyces cerevisiae*. Bioengineered Bugs 2, 120–123.

Rezaei, M.N., Aslankoohi, E., Verstrepen, K.J., Courtin, C.M., 2015. Contribution of the tricarboxylic acid (TCA) cycle and the glyoxylate shunt in *Saccharomyces cerevisiae* to succinic acid production during dough fermentation. International Journal of Food Microbiology 204, 24–32.

Ritter, S.K., June 2011. Small-Business Award: BioAmber, Minneapolis. Chemical and Engineering News, twenty-seventh ed. (online). Available from: http://pubs.acs.org/cen/news/89/i26/8926notw1c.html.

Sadaka, M.G., Garcia, A.A., 1998. The effect of temperature on forward/back extraction using an amine extractant for the recovery of cyclic hydroxy carboxylic acids. Separation Science and Technology 33, 1667–1680.

Song, H., Huh, Y.S., Lee, S.Y., Hong, W.H., Hong, Y.K., 2007. Recovery of succinic acid produced by fermentation of a metabolically engineered *Mannheimia succiniciproducens* strain. Journal of Biotechnology 132, 445–452.

Tamada, J.A., King, C.J., 1990. Extraction of carboxylic acids with amine extractants. 2. Chemical interactions and interpretation of data. Industrial & Engineering Chemical Research 29, 1327–1333.

TMR, September 2013. Succinic Acid Market for 1,4-BDO, Resin, Coatings, Dyes & Inks, Pharmaceutical, Polyurethane, Food, Plasticizers, Cosmetics, Solvents & Lubricants and De-icing Solutions Applications – Global Industry Analysis, Size, Share, Growth, Trends and Forecast, 2012–2018. Transparency Market Research, Albany, USA.

TMR, July 2015. Itaconic Acid Market – Global Industry Analysis, Size, Share, Growth, Trends and Forecast 2015–2023. Transparency Market Research, Albany, USA.

Tung, L.A., King, C.J., 1994. Sorption and extraction of lactic and succinic acids at pH>pKa1. I. Factors governing equilibria. Industrial & Engineering Chemical Research 33, 3217–3223.

van den Tweel, W., October 2010. Sustainable Succinic Acid. DSM.

Thongchul, N., 2013. Production of lactic acid and polylactic acid for industrial applications. In: Yang, S.-T., El-Enshasy, H.S., Thongchul, N. (Eds.), Bioprocessing Technologies in Biorefinery for Sustainable Production of Fuels, Chemicals, and Polymers, first ed. John Wiley, Hoboken, pp. 293–316.

Urbanus, J., Bisselink, R.J.M., Nijkamp, K., ter Horst, J.H., Verdoes, D., Roelands, C.P.M., 2010. Integrated product removal of slightly water-soluble carboxylates from fermentation by electrochemically induced crystallization. Journal of Membrane Science 363, 36–47.

Vemuri, G.N., Eiteman, M., Altman, E., 2002a. Effects of growth mode and pyruvate carboxylase on succinic acid production by metabolically engineered strains of *Escherichia coli*. Applied and Environmental Microbiology 68, 1715–1727.

Vemuri, G.N., Eiteman, M.A., Altman, E., 2002b. Succinate production in dual-phase *Escherichia coli* fermentations depends on the time of transition from aerobic to anaerobic conditions. Journal of Industrial Microbiology and Biotechnology 28, 325–332.

Wakai, Y., Shimazaki, T., Hara, S., 1980. Formation of succinate during fermentation of sake mash and grape must. Hakkokogaku Kaishi-Journal of the Society of Fermentation Technology 58, 363–368.

Xiao, H., Shao, Z., Jiang, Y., Dole, S., Zhao, H., 2014. Exploiting *Issatchenkia orientalis* SD108 for succinic acid production. Microbial Cell Factories 13, 12.

Yabannavar, V.M., Wang, D.I.C., 1991. Extractive fermentation for lactic acid production. Biotechnology and Bioengineering 37, 1095–1100.

Yan, D., Wang, C., Zhou, J., Liu, Y., Yang, M., Xing, J., 2014. Construction of reductive pathway in *Saccharomyces cerevisiae* for effective succinic acid fermentation at low pH value. Bioresource Technology 156, 232–239.

Yang, S.-T., Huang, H., Tay, A., Qin, W., de Guzman, L., San Nicolas, E.C., 2007. Extractive fermentation for the production of carboxylic acids. In: Yang, S.-T. (Ed.), Bioprocessing for Value-Added Products from Renewable Resources, first ed. Elsevier, Amsterdam, pp. 421–446.

Yang, S.-T., Yu, M., Chang, W.-L., Tang, I.-C., 2013. Anaerobic fermentations for the production of acetic and butyric acids. In: Yang, S.-T., El-Enshasy, H.S., Thongchul, N. (Eds.), Bioprocessing Technologies in Biorefinery for Sustainable Production of Fuels, Chemicals, and Polymers, first ed. John Wiley, Hoboken, pp. 351–374.

Yedur, S., Berglund, K., Dunuwila, D.D., 2001. Succinic Acid Production and Purification. US Patent 6265190.

Yuzbashev, T.V., Yuzbasheva, E.Y., Sobolevskaya, T.I., Laptev, I.A., Vybornaya, T.V., Larina, A.S., Matsui, K., Fukui, K., Sineoky, S.P., 2010. Production of succinic acid at low pH by a recombinant strain of the aerobic yeast *Yarrowia lipolytica*. Biotechnology and Bioengineering 107, 673–682.

CHAPTER 10

Production, Stabilization, and Uses of Enzymes From Fruit and Vegetable Byproducts

L.A. Prado Barragán, J.J. Buenrostro-Figueroa
Universidad Autonoma Metropolitana Unidad Iztapalapa, Mexico, DF, Mexico

C.N. Aguilar González
Universidad Autonoma de Coahuila, Saltillo, Coahuila, México

I. Marañon
Tecnalia, Parque Tecnológico de Álava, Miñano, Álava, Spain

10.1 Introduction

Enzymes are biological macromolecules acting as biocatalysts to bring about a specific biochemical reaction accelerating the biological/biochemical reactions. Microbial enzymes play a crucial role as catalysts in everyday life, leading to their use in various industries and applications. The end use market for industrial enzymes is extremely widespread, with numerous industrial commercial applications.

Microbial enzymes are of great importance in the development of industrial bioprocesses. Current applications are focused on many different markets including pulp and paper, leather, detergents and textiles, pharmaceuticals, chemical, food and beverages, biofuels, animal feed, and personal care, among others. Many industrial processes, including chemical synthesis for production of chemicals and pharmaceuticals, have several disadvantages: low catalytic efficiency; lack of enantiomeric specificity for chiral synthesis; need for high temperature, low pH, and high pressure. Also, the use of organic solvents leads to organic waste and pollutants. Enzymes are more useful for these applications as they work under mild reaction conditions (e.g. temperature, pH, atmospheric conditions), do not need protection of substrate functional groups, have a long half-life, a high stereoselectivity yielding stereo- and regio-chemically defined reaction products at an acceleration of a million fold, and, in addition, they work on unnatural substrates. Microbial enzymes are convenient production systems, due to economic feasibility, high yields, consistency, ease of product modification and optimization, regular supply due to absence of seasonal fluctuations, rapid growth of

Biotransformation of Agricultural Waste and By-Products. http://dx.doi.org/10.1016/B978-0-12-803622-8.00010-0

microbes on inexpensive media, stability, and optimum catalytic activity. Enzymes with desired activity under industrial conditions have been obtained. These enzymes have been achieved in several ways, isolating new microorganisms, often thermotolerant, by searching in metagenomic libraries for new gene sequences, or by mutagenesis, screening for optimizing process conditions, and by protein engineering.

10.2 Markets for Microbial Enzymes

Today, there is a need for new and more versatile enzymes in order to develop more novel, sustainable, and economically competitive production processes. Microbial diversity and modern molecular techniques, such as metagenomics and genomics, are being used to discover new microbial enzymes with optimum catalytic properties. Enzymes are employed in a diverse array of applications in industries and scientific research, ranging from the degradation of various natural substances in the starch processing, detergent, and textile industries. The pulp and paper industry, textile industry, laundry detergents, dairy industry, baking industry, juice industry, starch processing, brewing industry, animal feeds industry, organic synthesis industry, cosmetics industry, waste treatment, and azo dye remediation and cleanup are the principle sectors that deploy enzymes in industrial processes. The use of enzymes frequently results in many benefits that cannot be obtained with traditional chemical treatment. These often include higher product quality and lower manufacturing cost, less waste, and reduced energy consumption. Key factors driving the market growth include new enzyme technologies endeavoring to enhance cost efficiencies and productivity, and growing interest among consumers in substituting petrochemical products with other organic compounds such as enzymes. Other market growth factors include surging demand from textile manufacturers, animal feed producers, detergent manufacturers, pharmaceutical companies, bioethanol producers, and cosmetics vendors.

10.3 Enzymes in Technical Applications

Technical enzymes are typically used as bulk enzymes in detergents, textile, pulp and paper industries, organic synthesis, and biofuels industry. Technical enzymes are valued at just over $1 billion in 2010 by several research associations. It is estimated that the technical enzymes market will increase at a 6.6% compound annual growth rate to reach $1.5 billion in 2015, with the highest sales in the leather market and bioethanol market. Commercially available enzymes used in these areas are amylases, proteases, lipases, laccases, cellulases, xylanases, and catalases.

Over 500 industrial products are being made using enzymes. The demand for industrial enzymes is on a continuous rise, driven by a growing need for sustainable solutions. Microbes have served and continue to serve as one of the largest and useful sources of many enzymes. Around 20 enzymes are produced on truly industrial scale. With the improved understanding of the enzyme production biochemistry, fermentation processes, and recovery methods, an increasing number of industrial enzymes can be foreseeable. The world enzyme demand is satisfied by about 12 major producers and 400 minor suppliers. The global industrial enzymes market is very competitive, with Novozymes being the largest player in the industry, followed by DSM, and DuPont (that acquired a majority stake in Danisco and its Genencor division), among others. The companies mainly compete on the basis of product quality, performance, use of intellectual property rights, and the ability to innovate, among other such factors. North America and Europe are the largest consumers of industrial enzymes, although the Asia-Pacific regions, China, Japan, and India are undergoing a rapid increase in enzyme demand.

The use of lipases (and of most enzymes) in industrial processes is still limited by their low stability under operational conditions and low activity or specificity on particular or nonnatural substrates. Two approaches are considered as the way to increase or maintain enzyme stability: (1) genetic manipulation of existing strains or (2) search for new microorganisms from extreme environments (extremophiles), i.e. extreme temperature, pH, or salt concentration, or organisms that cannot be cultured in the laboratory.

The market for enzymes used in cosmetic and toiletries is set to grow by 5% per year up to 2015, driven by the technological progress (achieving or maintaining enzyme stability and activity) and consumer awareness of the seemingly potent power of enzymes. Although interests in the use of enzymes in cosmetics have been continuous increased, past and present areas of related applications seem to be few. One successful example is the use of superoxide dismutase (SOD) to capture free radicals and prevent damage to skin caused by environmental pollution, bacteria, and other harmful factors. It is proposed to use a combination of SOD and peroxidase as free radical scavengers to reduce UV-induced erythema in sunscreen cream. Peroxidase is also used to prevent cosmetic formulations from bacterial attack, for it can consume the oxygen present in the contents. Another example is proteases used in skin creams to clean and smoothen the skin by peeling off dead or damaged skin. Other enzymes with cosmetic application are: laccase in hair dye, lipase to prevent treat skin rash or diaper rash, and endoglycosidase and papain in toothpaste and mouthwash to whiten teeth and remove plaque and odor-causing deposits on teeth and gum tissue; enzymes are also used in contact lens cleaners to remove protein films.

The hydrolytic lipases are commercially very important, and their addition to detergents is mainly in laundries and household dishwashers. Enzymes reduce the environmental damage of detergent products, as they save energy by enabling a lower wash temperature to be used, use of chemicals in detergents is reduced, mostly biodegradable, leaving no harmful residues has no negative impact on sewage treatment processes; and does not possess any kind of risk to aquatic life.

10.4 Encapsulation of Enzymes and Use in the Industry

Enzymes are catalytic proteins and, like the most proteins, they are highly liable to different physiological parameters such as pH or heat, but also to some chemicals like denaturation agents. On account of this sensitivity, enzymes are commonly encapsulated or immobilized to be used as food supplements, pharmaceutical products, hygiene products, enzyme-linked immunosorbent assays (ELISA), biocatalysis or biosensors, household detergent, etc. The first use of immobilized enzymes for food application was in the 1970s, and since then, more and more applications of immobilized or encapsulated enzymes have appeared (Barbosa et al., 2015; Krajewska, 2004; Olson and Richardson, 1974; Sassolas et al., 2012).

Immobilization involves different techniques specifically designed to greatly restrict the freedom of movement of the enzymes using a physical support. Physical methods (adsorption, entrapment, or membrane confinement) or chemical methods (covalent bonds or cross-linking) are different possible options to immobilize the most of enzymes for industrial applications (Fig. 10.1). The support material or carrier could be a natural carrier (alginate, gelatin, dextran, albumin, etc.), a synthetic organic polymer (polyvinyls, epoxide, acrylate, etc.), or an inorganic carrier (glass, silica gel, bentonite, etc.). The carrier selection is influenced by the characteristics of the use or application medium for the selected enzymes: pH, temperature, organic solvent presence, shear, pressure, etc. Even the support material would be influenced by the method of immobilization desired, and the combination of both would have effect on the enzyme activity and half-life.

The principle advantages of the enzyme immobilization are convenience, economical, and stability aspects:

- Enhancing of enzyme properties: stability, modulation of specificity and activity (Mateo et al., 2007)
- The ease to recovery and reuse (Chang et al., 2011)
- The improvement of the reaction economy (it could accelerate the chemical reaction)
- Easy separation and the possibility of continuous operations
- Improvement of the reaction control and the possibility of multienzyme reactions

However, in 2015, the most challenging aspects of enzyme encapsulation continues to be on the development of the matrix materials and encapsulation methods for enzyme integration in the host matrix and retaining full activity and increase stability. In the coming years, it will be very important to use methods and new smart materials that easily anticipate encapsulated enzyme behavior (Fonseca and Meesters, 2013).

10.4.1 Immobilization

Enzyme immobilization has been used to enhance enzyme productivity and development of novel techniques for increasing their shelf-life for industrial enzymatic reactions. These requirements are inevitable to facilitate large-scale and economic formulation.

Between the immobilization methods showed in Fig. 10.1, *adsorption* is the simplest and most economic alone, and involves mostly electrostatic forces creating reversible linkages between

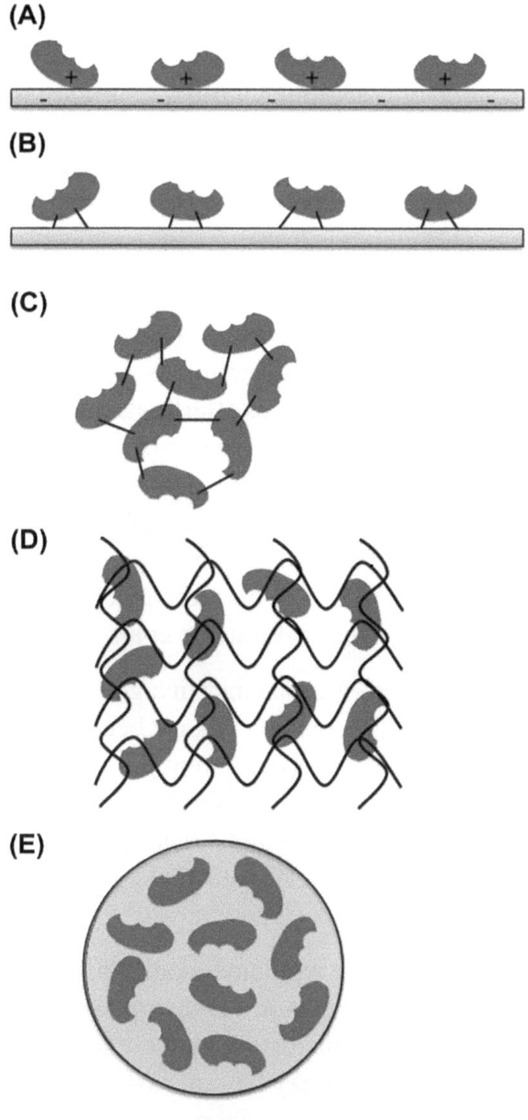

Figure 10.1
Main methods of enzyme immobilization: (A) electrostatic adsorption, (B) covalent binding, (C) cross-linking, (D) entrapment, and (E) encapsulation.

the support material and the enzymes. This sort of forces is very weak (like van der Waals forces or hydrogen bonding interactions) and usually strongly dependent on pH value. The leakage of enzymes from the support to the product is one of the principle disadvantages of the adsorption method, linked to the exposure of enzymes to microbial attack. Therefore, the yield is often low due to desorption and inactivation. The formation of *covalent bonds* between the enzymes and the active groups of the support material (like —OH, —COOH, or $—NH_2$)

helps to solve the leakage inconvenience, and the reusability of the enzyme immobilized could be improved, but the reaction procedure could have an effect on the enzyme activity. Regarding the material, the support hydrophobicity/hydrophilicity is one of the most important factors for maintaining enzyme activity, followed by surface charge, but also ambient conditions, like temperature, pH, and enzyme concentration, have influence on the enzyme activity (Gashtasbi et al., 2014; Talbert and Goddard, 2012).

Cross-linking is a support-free immobilization option: the enzymes are bonded to each other to form a large 3D structure. A physical method is largely used in the biotechnology industry by flocculation in order to achieve high cell densities. However, the resulting product has poor stability and mechanical properties, so the flocculation technique is more useful to recover and concentrate the enzymes from dilute systems (Berlin, 2013). Chemical cross-linking method involves the use of bi- or multifunctional reagents to form covalent bonds between the enzymes. These cross-linking reagents contain reactive ends to specific functional groups, such as primary amines or sulfhydryls, and some examples could be glutaraldehyde, formaldehyde, or toluene diisocyanate. The chemical method provides better mechanical properties and stability. However, the toxicity of such reagents, the inactivation of active sites, and the cost are the main limiting factors (Fernandez-Lafuente et al., 1995; Govardhan, 1999).

Immobilization by *entrapment* involves the enzymes in a gel structure. Their movement is restricted by this 3D gel structure, but there is no linkage between the enzymes and the gel, so the enzymes are free to move inside. There are different possible methods to achieve the enzyme entrapment, such as ionotropic gelation of polyionic macromolecules (e.g. alginate, kappa-carrageenan, or chitosan) (Bayramoğlu and Arıca, 2002), temperature-induced gelation (e.g. gelatin, agarose, or kappa-carrageenan) (Pernetti et al., 2003; Vujčić et al., 2011), organic polymerization (Elagli et al., 2014), and precipitation.

The permeability of the matrices could be regulated modifying the material porosity to allow the transport of low molecular weight compounds (substrate and products) without leaking of the entrapped enzymes. So, this is an interesting methodology due to that there is not chemical modification, and the entrapped enzymes are easy to handle and reused. However, the accessibility may be reduced due to free radical polymerization of gel, and it is necessary to consider the diffusional aspects to assure the suitable diffusion of substrate and products.

10.4.2 Encapsulation

This immobilization technology is similar to entrapment in that the enzymes are free in the solution but restricted in space; the encapsulation allows the confinement of enzymes without totally loss of its freedom, but restricts unfolding movements. The formation of beads varying 1–1000 μm in diameter is the difference in front of the large structures formed in the entrapment approach (films, membranes, and other 3D gel structures). Moreover, the enzyme encapsulation methods use very mild conditions that hardly affect enzyme activity, and the capsule mimics the natural mode of occurrence of enzymes inside of the cells, offering a protection from the changes in the reaction environment. The principal disadvantage is similar to the entrapment immobilization method: the diffusion control of enzymes, substrates, and products.

Encapsulation can be accomplished through many technologies, and they have demonstrated their utility in different applications such as drug delivery (Das et al., 2011; Saez-Martinez et al., 2012), food ingredients or nutraceutical compounds stabilization (McClements et al., 2009; Munin and Edwars-Lévy, 2011), living probiotics administration (Chávarri et al., 2010, 2012), and cell therapy (Acarregui et al., 2014; Rokstad et al., 2014). Among the different encapsulation technologies, the approach to encapsulation of enzymes is dependent on the specific application or challenge.

However, the enzyme encapsulation is used not only to immobilize the enzymes, but also to control his delivery at a specific point in the application of the product. In addition, each product has different requirements for the enzyme stability to achieve a specific delivery; the product processing and the consumption cycle are critical to design and develop the suitable encapsulated enzyme approach, depending on the product requirements and on the trigger of delivery proposed (heat, solubility, pH, etc.).

Stability of solid formulation is often much better than in a liquid formulation, since the water activity (a_w) is low. The enzyme encapsulation helps to maintain the intrinsic activity of enzymes included in an aqueous formulation; a thermogel microbead is a suitable environment for enzymes included in cosmetic, food or feed products, and detergents. Other important aspects considered to encapsulate the enzymes are the allergenicity of enzymes: the largest application of enzymes today is the detergence with 1000 tons per year, and the regulation requires to avoid any trace of enzymes in air during manipulations (Ahmadian and Ghadiri, 2007). The solution could be the enzyme immobilization in mechanically resistant microcapsules and to avoid any dust during manipulation of enzymes.

The cost of the product and the investment required to produce encapsulated enzymes are the principle disadvantages. However, the improvement of stability allows controlling the dosage on the final products and improving their effectiveness.

Here, some technologies used to enzyme encapsulation are described.

10.4.2.1 Spray Drying

In this method, protein solution or emulsion is sprayed into the air by atomization, usually at elevated temperatures, to evaporate the solvent, and the enzymes are collected, encapsulated in the carrier material, and mixed in the feed solution or emulsion (Fig. 10.2). The properties of final microspheres depend on the nature of the feeding flow as well as the operating parameters, such as flow rate and inlet temperature. The dry solid formulation improves the

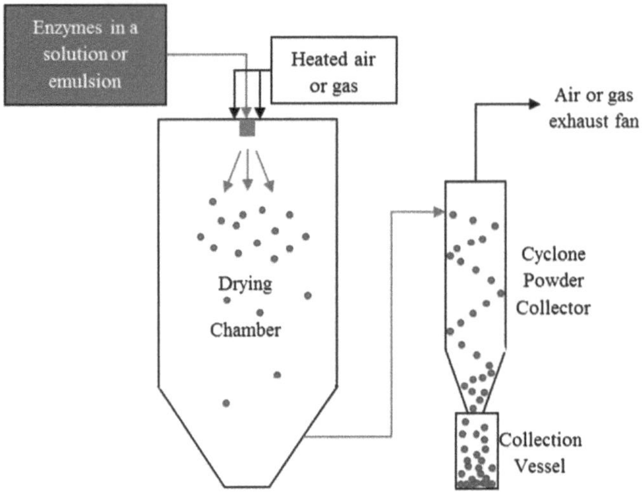

Figure 10.2
Schematic diagram of a spray drying encapsulation process.

enzyme shelf life due its low a_w value. Depending on the enzyme, even its application, the literature proposes many different biopolymers as carriers (Belghith et al., 2001; Estevinho et al., 2014; Millqvist-Fureby et al., 1999).

10.4.2.2 Granulation

There are two possibilities: high and low shear granulation.

Low shear granulation is a technique where a solid powder is mixed with a liquid sprayed between some mixer blades, where there is short contact time, and the mixing is not intense. Compared with a spray drying process, the enzyme is not in the liquid feed; the solids contain the enzyme. The result is an open wet agglomerated granule, with a relative low bulk density (depending on the amount of liquid added per weight of powder proceeding) and good instant properties (1–5 s). The wet product is usually dried in a fluid bed dryer.

In high shear granulation, much denser particles are formed (Hoornaert et al., 1998). The advantage is to obtain much stronger particles with less dust formed, which makes the handling much safer. The disadvantage is not as quick dissolution as the powders obtained by low shear granulation (a few minutes).

The enzyme is dispersed through the whole particle. Some common binders included in the enzymes formulations are polyvinylpyrrolidone (PVP) or sugar. In addition, to prevent allergic reactions and achieve safer handling, the granules could be coated in the same high shear granulator or, most of the time, in a fluid bed coater.

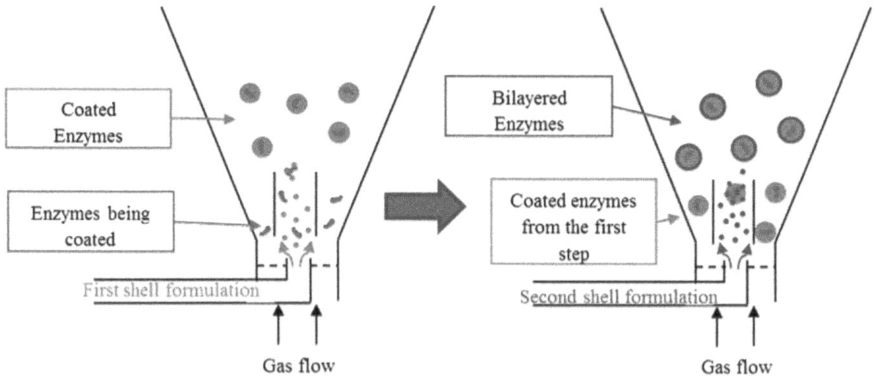

Figure 10.3
Schematic diagram of a bottom spray fluid bed coating process (Wurster process) to obtain enzymes with a double layer coating, changing the shell formulation atomized at the appropriated time.

10.4.2.3 Fluid Bed Coating

To reach the release of the enzyme at a specific stage of the application (e.g. soaking or washing for detergent products), different layers could be applied on the initial particles obtained by other encapsulation technologies using a fluid bed coater (Fig. 10.3). Some layers could offer mechanical protection, some other protections against acidity, or oxidation, etc. (Dale, 2001; Falholt, 1990).

10.4.2.4 Fluid Bed Agglomeration

The top spray fluid bed is characterized by the placement of nozzle above the fluidizing bed solid particles (Fig. 10.4). The countercurrent collisions benefit the agglomeration of the enzymes. The heavier the agglomerated particles are, they have less fluidization, so the agglomeration is selective for the smaller particles, and a uniform ending size is possible. The result is the enzyme agglomerated in a "blackberry" structure (open structure) with very good instant properties.

10.4.2.5 Combination of Spray Drying and Fluid Bed Agglomeration

The combination of spray drying and fluid bed drying technologies is also called multistage drying (from GEANiro, www.niro.com) or integrated spray drying. Through the process, it is possible to produce coarse, agglomerated, free-flowing dustless powder. Feed material is sprayed into a fluidized drying chamber, where it is dried to single particles, which pass to the fluid bed. The smallest particles are swept along by the current and recirculated to the bottom of the drying chamber and to the fluid bed agglomerator (Fig. 10.5). The new single particles sprayed collide with these recycled small ones to form agglomerates, allowed by the stickiness of the drying particles.

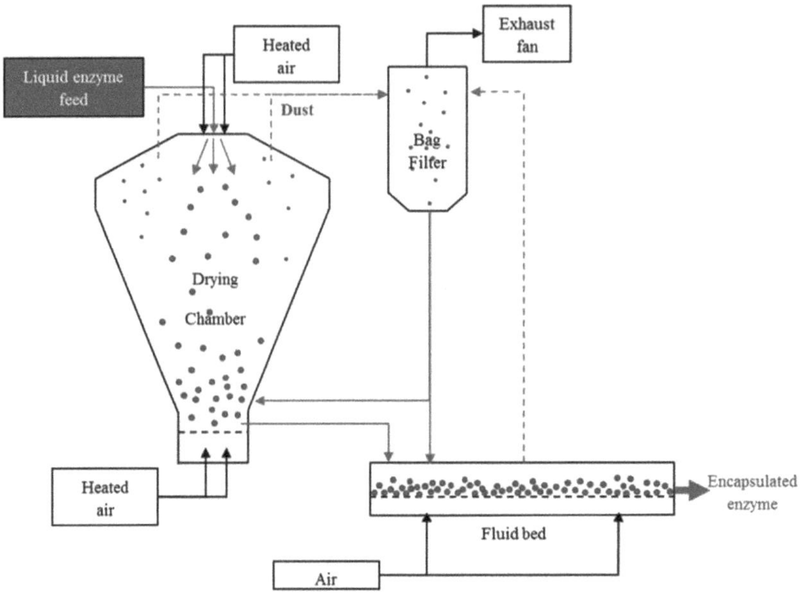

Figure 10.4
Schema of a process combining spray drying and a fluid bed.

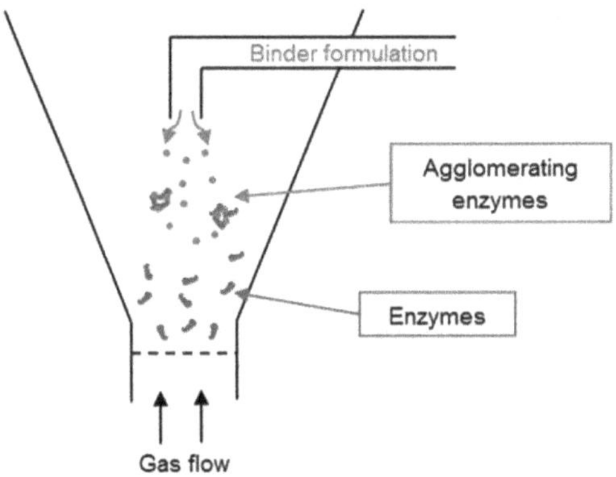

Figure 10.5
Schematic diagram of a top spray fluid bed agglomerating process.

10.4.2.6 Gel Entrapment

The encapsulation of enzymes could be accomplished using the materials and methods of gelification indicated in the immobilization by entrapment description. The beads are obtained dropwise by the polymer solution containing the enzyme into a hardening bath.

The methods to encapsulate enzymes in microspheres include two principle steps. First, the polymer solution containing the enzyme is dispersed in small drops by dripping methods, for

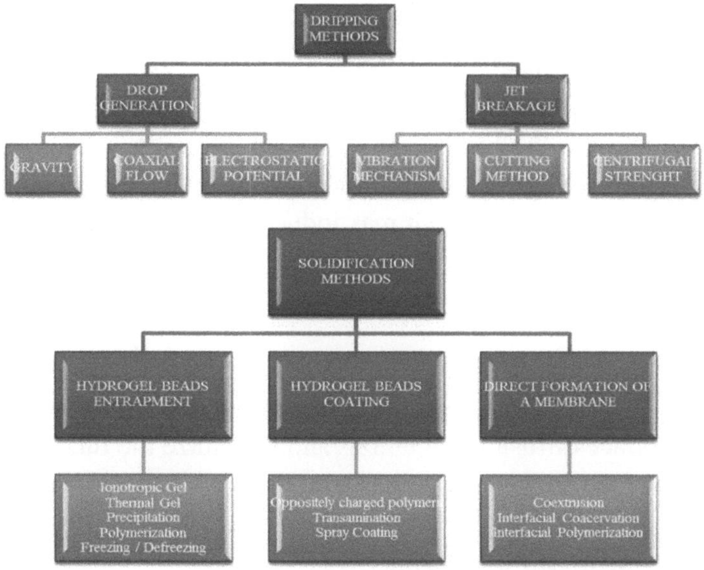

Figure 10.6
Classification of methods to make and solidify drops.

example. Second, these drops will solidify by gelation or formation of a membrane in their surface. There are different technologies available for this purpose, and the selection of the best one is related to different aspects, such as desired size, acceptable dispersion size, production scale, and the maximum shear that the enzyme cells can support.

When a liquid is pumped to go through a nozzle, first this is extruded as individual drops. Increasing enough the flow rate, the drop is transformed in a continuous jet, and this continuous jet has to be broken into small droplets. Regardless of the selected technique, the liquid obtained drops have to be solidified by gelation or external membrane formation (Fig. 10.6) (Chávarri et al., 2012).

Alginate is the most referenced polymer for bioactives encapsulation, including enzymes (Anjani et al., 2007; Rahim et al., 2013; Taqieddin and Amiji, 2004; Watanabe et al., 2001). Alginate is a polysaccharide made of two sugars: D-mannuronate and L-guluronate. It is a very interesting polymer for encapsulation due its biocompatibility and solidification options: gelation with divalent cations (e.g. calcium ions) and covalently cross-linked (e.g. polyethylene glycol (PEG)). To obtain a membrane in the bead surface in a single step, the enzyme mixed with one polymer solution is dropped into a bath containing an oppositely charged polymer, forming a polyelectrolyte complex. That is the complex coacervation process. It requires the mixing of two colloids at a pH at which both polymers are oppositely charged, such as chitosan (+) and alginate (−) (Sankalia et al., 2007), and also gelatin (+) and gum arabic (−). Other polymers as poly-L-lysine, PEG, and different combinations between them and even with other biopolymers have been evaluated (TRANSBIO project, 2012).

10.4.2.7 Liposomes

Liposomes are used to encapsulate a large number of molecules, even proteins such as enzymes (Laloy et al., 1998; Qu et al., 2010). The encapsulation efficiency of enzymes depends on the preparation method and conditions used. Some disadvantages are the difficulty to entrap large enzymes in small liposomes and the possibility of interactions between the enzyme and the lipid vesicle because it may induce aggregation. Even if the liposomes are relatively stable in liquid systems, the enzymes may leak out during the product shelf-life.

10.4.2.8 Nanogels or Nanoparticles

Recent interest in nanotechnology has also provided a wealth of diverse nanoparticles and nanoscaffolds that could potentially support enzyme encapsulation and immobilization. These nanoscale structures reduce diffusion limitations and maximize the functional surface area, which allows increasing the enzyme loading (Steinhilber et al., 2013).

10.5 Uses of Immobilized Enzymes

During the initial years of the development of the field of immobilized enzymology, researchers used to find only the advantage of the immobilized enzymes in comparison to their soluble/free counterparts. From the begining of the XXI century, immobilized enzyme technology has advanced into ever-expanding and multidisciplinary fields to analyze clinical, industrial, and environmental samples (Khan and Alzohairy, 2010).

The academic literature and the patents described many methods for enzyme immobilization, and a substantial number of applications are cited. However, relatively few methods and a limited number of applications have been successful in the marketplace. Di Cosimo et al. offer a critical review to understand the apparent discrepancy between the literature and the industrial applicability. The cost of the immobilized enzyme has to be justified by the process advantages related to the immobilization. However, the rapid evolution of immobilized enzyme technology and current industry trends, such as biocatalysis, pharmaceutical and food applications, and biosensors, favor the development of this field (DiCosimo et al., 2013).

Some industrial applications are here presented:

- Chemical applications:
  - Biofuel obtaining by breakdown of starch (cellulases for producing biomass ethanol)
  - Pulping of wood to paper production
- Food industry:
  - Clarification of beer and juices in the food industry
  - Diagnostic test for medical and biotechnological applications
  - Sweetener production (glucose conversion into fructose by isomerase)
  - Cheese ripening

  - Production of lactose-free milk (encapsulated lactase to optimize the enzyme reuse)
  - Bakery applications (dough conditioners such as xylanase, amylase, etc.)
- Hygiene products:
  - Detergent for laundry containing proteases and lipases
  - Oral care
- Medical and pharmaceutical applications:
  - Diagnostic assays and biosensors (e.g. ELISA, autoanalyzers, etc.)
  - Antibiotic production
  - Controlled delivery of enzymes, stable and active in the required site
  - Bioreactors for extracorporeal enzyme therapy

10.5.1 Laundry and Dish Applications

In 1913, Otto Röhm discovered the efficacy of pancreatic trypsin, extracted from pig pancreas for the removal of proteinaceous stains from clothes. The first enzyme-containing detergent introduced to the household market was Burnus, from Röhm und Hass, GmbH (Germany; nowadays Evonik Degussa). Due to the increasing consumer requirements, the laundry and household detergents need to remove more and more efficiently any type of stain from a broad range of complex fiber surfaces: milk, egg, blood, chocolate, starch, etc. The second generation of enzymes is being developed for further improved performance under difficult washing conditions: low temperature, single wash effects, new requirements of improved compatibility with new detergent technology, and for broader applications.

The encapsulation technology appeared in the detergent industry in the early 1970s to reduce the enzyme exposure of detergent industry employees, in response to its health effects (skin irritation and occupational asthma). This allowed the safe introduction and use of enzymes in many products and production processes. From this moment, the enzyme encapsulation technology has been really useful to improve the detergent products: increasing tolerance to temperature and alkali, and developing liquid enzyme preparations.

Detergents contain many components that can modify or denature enzymes, such as other enzymes, surfactants, metal chelating agents such as ethylenediaminetetraacetic acid (EDTA), bleach, and high alkalinity. Oxidation is one of the mechanisms that can more affect the enzyme activity and performance. Encapsulation has been used to achieve a physical separation of enzymes from each other and from other components of the detergent matrix, but the encapsulated form has to ensure its disintegration and the rapid release active enzyme during a wash cycle.

The matrix formulation is more complex than the choice of an encapsulating polymer. The encapsulated enzyme development requires the definition of type and appropriated amount for the encapsulating material and other excipients, such as stabilizers, disintegrants, and others. An example about the complexity of a detergent formulation development is the granulation of subtilisin (bacterially fermented protease) in alginate microspheres realized by researchers from the Queen's University (Kingston, Canada) and DuPont (Palo Alto, USA). The microencapsulated enzyme showed much higher temperature and lower pH resistance than the free enzyme, but it is necessary also to improve the enzyme release once in liquid suspension. Disintegrants as microcrystalline cellulose (MCC), polyvinyl alcohol, sucrose, and starch were evaluated, obtaining the best results with MCC 10% for 15 min dissolution time (Chan et al., 2006).

However, in the early 2010s, the most industrial enzymes are fermentation-derived products, and the enzyme liquid has a characteristic yellow or brown color and a distinctive odor. What's more, most detergent new products are liquid formulations, and the use of water as trigger of enzyme delivery is not useful. Companies developing new products for this market are working with more sophisticated solutions (such as nanotechnology) to solve the new problems. Unfortunately, the structural basis of enzyme encapsulation is only starting to be understood, and there is a long trial-and-error process to optimize it. Due the large market related to this application and its economical relevance, encapsulation methods and materials have been investigated and developments protected by patent (Boutique et al., 2011).

10.6 Conclusion and Perspectives

Different microbes as well as Genetically Modified Organisms (GMO) with recombinant genes in bioreactors are extensively used for the production of enzymes. Many microorganisms and their enzymes with unique functions have been discovered by means of extensive screening, and now they are commonly used in different industrial and medical fields. Amylase, protease, and lipase are enzymes that have a wide-spectrum role in this sector. They are used in almost in every industry, whether it may be detergent, dairy, food, or medicine. Recent advances in genomics, metagenomics, proteomics, efficient expression systems, and emerging recombinant DNA techniques have facilitated the discovery of new microbial enzymes from nature (through genome and metagenome) or by creating (or evolving) enzymes with improved catalytic properties. The ongoing progress and interest in enzymes provides further success in areas of industrial biocatalysis. The next years should see a lot of exciting developments in the area of biotransformations. There are many factors influencing the growing interest in biocatalysis. Advances in investigations will make use of combinations of engineered and de novo designed enzymes, coupled with chemistry, to generate chemicals and materials from cheaper (and renewable) resources. Immobilization and encapsulation will add stability and prolonged activity, avoiding enzyme inactivation.

References

Acarregui, A., Herrán, E., Igartua, M., Blanco, F.J., Pedraz, J.L., Orive, G., Hernandez, R.M., 2014. Multifunctional hydrogel-based scaffold for improving the functionality of encapsulated therapeutic cells and reducing inflammatory response. Acta Biomaterialia 10, 10.

Ahmadian, H., Ghadiri, M., 2007. Analysis of enzyme dust formation in detergent manufacturing plants. Advanced Powder Technology 18, 1.

Anjani, K., Kailasapathy, K., Philips, M., 2007. Microencapsulation of enzymes for potential application in acceleration of cheese ripening. International Dairy Journal 17.

Barbosa, O., Ortiz, C., Berenguer-Murcia, Á., Torres, R., Rodrigues, R.C., Fernandez-Lafuente, R., 2015. Strategies for the one-step immobilization–purification of enzymes as industrial biocatalysts. Biotechnology Advances 33, 5.

Bayramoğlu, G., Arıca, M.Y., 2002. Procion Green H-4G immobilized on a new IPN hydrogel membrane composed of poly(2-hydroxyethylmethacrylate)/chitosan: preparation and its application to the adsorption of lysozyme. Colloids and Surfaces A: Physicochemical and Engineering Aspects 202, 1.

Belghith, H., Ellouz Chaabouni, S., Gargouri, A., 2001. Stabilization of *Penicillium occitanis* cellulases by spray drying in presence of Maltodextrin. Enzyme and Microbial Technology 28, 2–3.

Berlin, A., 2013. Flocculants for Enzyme Recovery and Recycling. Lignol Innovations Ltd.

Boutique, J.P., Vandenberghe, F., Bouilliche, M., Laurent, J.C.T.R.B.S., 2011. Compact Fluid Laundry Detergent Composition, US8940677. The Procter & Gamble Company.

Chan, A.W.J., Mazeaud, I., Becker, T., Neufeld, R.J., 2006. Granulation of subtilisin by internal gelation of alginate microspheres for application in detergent formulation. Enzyme and Microbial Technology 38, 1–2.

Chang, K.-L., Thitikorn-amorn, J., Chen, S.-H., Hsieh, J.-F., Ratanakhanokchai, K., Huang, P.-J., Lin, T.-C., Chen, S.-T., 2011. Improving the remaining activity of lignocellulolytic enzymes by membrane entrapment. Bioresource Technology 102, 2.

Chávarri, M., Marañón, I., Ares, R., Ibáñez, F.C., Marzo, F., Villarán, M.C., 2010. Microencapsulation of a probiotic and prebiotic in alginate-chitosan capsules improves survival in simulated gastro-intestinal conditions. International Journal of Food Microbiology 142, 1–2.

Chávarri, M., Marañón, I., Villarán, M.C., 2012. Encapsulation technology to protect probiotic bacteria. In: Rigobelo, E.C. (Ed.), Probiotics, pp. 501–540.

Dale, D.A., 2001. Comprising Inert Seed or Carrier Particle, Low-density Filler Layer Having Enzyme Contained Therein Coated on Seed or Particle, Outer Coating; Liquid Detergents. US6310027 B1. Genencor International, Inc.

Das, S., Chaudhury, A., Ng, K.-Y., 2011. Preparation and evaluation of zinc–pectin–chitosan composite particles for drug delivery to the colon: role of chitosan in modifying in vitro and in vivo drug release. International Journal of Pharmaceutics 406, 1–2.

DiCosimo, R., McAuliffe, J., Poulose, A.J., Bohlmann, G., 2013. Industrial use of immobilized enzymes. Chemical Society Reviews 42, 15.

Elagli, A., Belhacene, K., Vivien, C., Dhulster, P., Froidevaux, R., Supiot, P., 2014. Facile immobilization of enzyme by entrapment using a plasma-deposited organosilicon thin film. Journal of Molecular Catalysis B: Enzymatic 110, 0.

Estevinho, B.N., Damas, A.M., Martins, P., Rocha, F., 2014. Microencapsulation of β-galactosidase with different biopolymers by a spray-drying process. Food Research International 64, 0.

Falholt, P., 1990. Enteric Coated Detergent Enzymes. Novo Industri A/S.

Fernandez-Lafuente, R., Rosell, C.M., Rodriguez, V., Guisan, J.M., 1995. Strategies for enzyme stabilization by intramolecular crosslinking with bifunctional reagents. Enzyme and Microbial Technology 17, 6.

Fonseca, L., Meesters, G.M.H., 2013. Enzyme Encapsulation. Bioencapsulation Innovations December 2013.

Gashtasbi, F., Ahmadian, G., Noghabi, K.A., 2014. New insights into the effectiveness of alpha-amylase enzyme presentation on the *Bacillus subtilis* spore surface by adsorption and covalent immobilization. Enzyme and Microbial Technology 0, 64–65.

Govardhan, C.P., 1999. Crosslinking of enzymes for improved stability and performance. Current Opinion in Biotechnology 10, 4.

Hoornaert, F., Wauters, P.A.L., Meesters, G.M.H., Pratsinis, S.E., Scarlett, B., 1998. Agglomeration behaviour of powders in a Lödige mixer granulator. Powder Technology 96, 2.

Khan, A.A., Alzohairy, M.A., 2010. Recent advances and applications of immobilized enzyme technologies: a review. Research Journal of Biological Sciences 5, 8.

Krajewska, B., 2004. Application of chitin- and chitosan-based materials for enzyme immobilizations: a review. Enzyme and Microbial Technology 35, 2–3.

Laloy, E., Vuillemard, J.-C., Dufour, P., Simard, R., 1998. Release of enzymes from liposomes during cheese ripening. Journal of Controlled Release 54, 2.

Mateo, C., Palomo, J.M., Fernandez-Lorente, G., Guisan, J.M., Fernandez-Lafuente, R., 2007. Improvement of enzyme activity, stability and selectivity via immobilization techniques. Enzyme and Microbial Technology 40, 6.

McClements, D.J., Decker, E.A., Park, Y., Weiss, J., 2009. Structural design principles for delivery of bioactive components in nutraceuticals and functional foods. Critical Reviews in Food Science and Nutrition 49, 6.

Millqvist-Fureby, A., Malmsten, M., Bergenståhl, B., 1999. Spray-drying of trypsin – surface characterisation and activity preservation. International Journal of Pharmaceutics 188, 2.

Munin, A., Edwars-Lévy, F., 2011. Encapsulation of natural polyphenolic compounds: a review. Pharmaceutics 3, 4.

Olson, N.F., Richardson, T., 1974. Symposium: immobilized enzymes in food systems. Journal of Food Science 39, 4.

Pernetti, M., Poncelet, D., Thouand, G., Annesini, M.C., Merli, C., 2003. Characterization of agarose as immobilization matrix model for a microbial biosensor. Hemijska industrija 57, 12.

Qu, B., Guo, L., Chu, X., Wu, D.-H., Shen, G.-L., Yu, R.-Q., 2010. An electrochemical immunosensor based on enzyme-encapsulated liposomes and biocatalytic metal deposition. Analytica Chimica Acta 663, 2.

Rahim, S.N.A., Sulaiman, A., Hamzah, F., Hamid, K.H.K., Rodhi, M.N.M., Musa, M., Edama, N.A., 2013. Enzymes encapsulation within calcium alginate-clay beads: characterization and application for Cassava slurry saccharification. Procedia Engineering 68, 0.

Röhm, O., 1913. A Method for Cleaning Domestic and Commercial Objects and Materials of All Kinds DE 283 923. Röhm & Hass.

Rokstad, A.M.A., Lacík, I., de Vos, P., Strand, B.L., 2014. Advances in biocompatibility and physico-chemical characterization of microspheres for cell encapsulation. Advanced Drug Delivery Reviews 0, 67–68.

Saez-Martinez, V., Argarate, N., Morin, F., Garagorri, N., 2012. Comparative study of dexamethasone and vancomycin release behavior from stimuli-sensitive microgel aqueous dispersions. Journal of Drug Delivery Science and Technology 22, 4.

Sankalia, M.G., Mashru, R.C., Sankalia, J.M., Sutariya, V.B., 2007. Reversed chitosan–alginate polyelectrolyte complex for stability improvement of alpha-amylase: optimization and physicochemical characterization. European Journal of Pharmaceutics and Biopharmaceutics 65, 2.

Sassolas, A., Blum, L.J., Leca-Bouvier, B.D., 2012. Immobilization strategies to develop enzymatic biosensors. Biotechnology Advances 30, 3.

Steinhilber, D., Witting, M., Zhang, X., Staegemann, M., Paulus, F., Friess, W., Küchler, S., Haag, R., 2013. Surfactant free preparation of biodegradable dendritic polyglycerol nanogels by inverse nanoprecipitation for encapsulation and release of pharmaceutical biomacromolecules. Journal of Controlled Release 169, 3.

Talbert, J.N., Goddard, J.M., 2012. Enzymes on material surfaces. Colloids and Surfaces B: Biointerfaces 93, 0.

Taqieddin, E., Amiji, M., 2004. Enzyme immobilization in novel alginate–chitosan core-shell microcapsules. Biomaterials 25, 10.

TRANSBIO project (FP7-KBBE-2011-5), 2012. Biotransformation of by-products from fruit and vegetable processing industry into valuable bioproducts. http://www.transbio.eu

Vujčić, Z., Miloradović, Z., Milovanović, A., Božić, N., 2011. Cell wall invertase immobilisation within gelatin gel. Food Chemistry 126, 1.

Watanabe, H., Matsuyama, T., Yamamoto, H., 2001. Preparation of immobilized enzyme gel particles using an electrostatic atomization technique. Biochemical Engineering Journal 8, 2.

CHAPTER 11

Sustainability Considerations on the Valorization of Organic Waste

L. Six, F. Velghe, S. Verstichel
Organic Waste Systems, Ghent, Belgium

S. De Meester
Ghent University, Ghent, Belgium

Abbreviations

E-LCA Environmental life cycle assessment
LCA Life cycle assessment
LCC Life cycle costing
PHB Polyhydroxybutyrate
PLA Polylactic acid
S-LCA Social life cycle assessment

11.1 Introduction

The global solid waste production rate has reached over 11 billion tons per year and is expected to more than double by 2025. The generation of this huge mass of waste is not distributed evenly among the world population: on average, people with high income levels produce more than three times as much waste as those with lower incomes. Income also influences the waste composition: lower-income countries show lower fractions of wasted paper and cardboard and higher fractions of organic waste. The position in the supply chain at which most of the organic waste is generated differs: low-income countries produce most waste early on in the supply chain due to poor storage facilities, transportation, and packaging, while high-income countries generate more organic waste at the end of the chain due to esthetic defects, poor environmental conditions during display at stores, wrong interpretation of best-before date, and leftovers (Bandara et al., 2007; UNEP, 2011; Gustavsson et al., 2011; World Bank, 2014).

Organic waste does not only differ by origin in the supply chain; it is also a very diverse group of substances. It contains both industrial (e.g. agricultural) and postconsumer waste, including fruit and vegetable waste, animal or human manure, meat products, plants and flowers, yard waste, etc. The streams all contain hydrocarbons, can contain nutrients, vitamins and spore elements, and range from very dry (such as wood shavings) to very wet (such as pig manure).

Biotransformation of Agricultural Waste and By-Products. http://dx.doi.org/10.1016/B978-0-12-803622-8.00011-2

Inappropriate management of organic waste can lead to several environmental problems, depending on organic waste type and management tactic: application of too much nutrient-rich organic material to agricultural fields for fertilization leads to eutrophication of surface waters and soil. Five percent of global greenhouse gas emissions originate from organic waste decay, which can lead to emissions of ammonia, odors, pathogens, and so on (Sims and Wolf, 1994; El-Fadel et al., 1997; Corell, 1998; Westerman and Bicudo, 2005; Shah, 2008; UNEP, 2011). Selecting the most appropriate and effective waste processing solutions can thus have important environmental benefits, both on the global and the local scale.

11.2 Valorization Pathways for Organic Waste and the Waste Hierarchy

Although efforts have been directed at lowering the amount of products or by-products wasted, the agricultural and food industries continue to create large waste streams which are difficult to avoid. These include crop residues leftover after harvesting the edible plant parts or rejected food products. A large fraction of this food waste is still landfilled despite valorization or diversion options such as composting and anaerobic digestion (FAO, 2013). Moreover, Europe stipulates a mandatory separate collection for biowaste by 2020 (European Parliament, 2015), resulting in organic streams that must be treated appropriately.

Biomass valorization methods can be very simple, such as using the stream as an animal feed, or very complex, as can be the case when extracting chemicals from the feedstock. They can be divided into three broad groups:

- Direct use: unaltered or slightly altered, e.g. for feed applications.
- Material recovery: biochemical or chemical extraction and/or conversion of the biomass into other useful products such as platform chemicals, solvents, or fertilizer.
- Energy recovery: burning biomass or biogas originating from anaerobic digestion in order to recover part of its energy contents.

Following the waste hierarchy (Fig. 11.1), reuse is the preferred waste management choice, preceded only by reduction and prevention (Hansen et al., 2002). Direct use of the biomass as an animal feed is also the most simple valorization route and one that can be applied on-farm. Many by-products, however, cannot be reused directly or entirely because they are not edible, of too low quality or of too large quantity. Alternative valorization methods are needed in order to avoid wasting these potentially valuable streams.

Energy recovery is another relatively simple biomass valorization strategy. If the biomass is reasonably dry, it can be burnt in order to recover the heat for direct use onsite or for district heating, or the heat can be used to produce steam that is fed to a generator for electricity production. If the biomass is rather wet, it might be preferable to convert it into biogas by anaerobic digestion, avoiding the energy wasted by evaporation of water. The biogas is

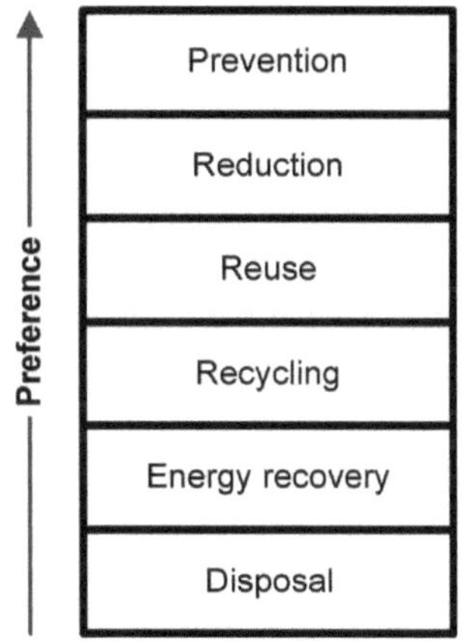

Figure 11.1
The waste hierarchy (based on Young and Stanner, 2008).

typically burnt for heat recovery and electricity production. The remaining liquid or solid fraction can be used as a fertilizer. But energy recovery is positioned at the bottom of the waste hierarchy, only preceding disposal (landfilling) and is thus not a preferred waste treatment tactic.

Material recovery (recycling) sits in-between reuse and energy recovery in the waste hierarchy and should thus generally be preferred to the latter. If humanity is to divert itself from fossil hydrocarbon resources, which it inevitably will have to, considering shrinking fossil reserves, biomass-based products are a necessity. Potential products include chemical building blocks with a wide range of applications or fertilizers that can be recycled within the agricultural chain. The necessary extraction and conversion processes can, however, be complex, of very low efficiency or simply unavailable for specific biomass streams. So in practice, energy recovery is often a better choice in terms of environmental impact. Novel and more efficient processes should be developed and optimized in order to alleviate these problems and create environmentally competitive solutions that cannot only create maximum value from agricultural by-products, but also serve purposes that are now met using nonrenewable products.

Although the waste hierarchy is a good general guideline for selecting the most environmentally friendly waste management scenario, going into more detail to assess the sustainability of different solutions may lead to other, more preferable outcomes. Take, for example, the case of an old washing machine: recycling and replacing it with a new, much more energy efficient machine is preferable from an environmental point of view, even though reuse scores higher in the waste hierarchy (Moberg et al., 2005; Schmidt et al., 2007; RDC, 2008; JRC, 2015). The following sections go into more detail and discuss sustainability assessment and its application to agricultural organic waste streams.

11.3 Sustainability

11.3.1 The Concept of Sustainability

Sustainable development is defined as "meeting the needs of the present without compromising the ability of future generations to meet their own needs" (Brundtland et al., 1987) or, in other words, to guarantee human well-being, both today and in the future. Achieving this state of sustainability may appear trivial, but it is anything but. Problems arise related to social injustice, to damage to our natural environment, and to economic instability. These three *pillars of sustainability* often referred to as people, planet, and profit, inevitably interact with one another (Benoît and Mazijn, 2009). True sustainability cannot be achieved without optimizing all three simultaneously (Khan and Islam, 2007).

In order to illustrate this interaction between sustainability pillars, we shall discuss a simplified example: the sustainability of a corporation. The main goal of this corporation is simple: make a profit. To reach this goal year after year, however, the corporation has more complex needs. In order to achieve a profit, at least the long-term expenses have to be lower than the income of the corporation. Its expenses include the required resources for its business, both natural and human resources. The company needs energy and materials as well as people to create and market its product. Energy and materials originate from the natural environment. When these natural resources are depleted and not regenerated at an adequate pace, the company is no longer capable of achieving its goal. When people working for the corporation are unable to enjoy a healthy lifestyle, both physically and mentally, their inputs to the company will degrade quickly, leaving the company with a lack of human resources needed for continuation of its activities. Furthermore, in order to create a profit, the organization also needs customers able to pay for its products. When its customer base is no longer able to gather the needed funds due to poverty, the corporation is unable to maintain its revenue stream. It is clear that any corporation that neglects any of the three sustainability pillars will fail in the long run unless external intervention provides the necessary framework. In principle, the same can be concluded for any human activity.

Natural resource depletion

Natural resource depletion is a specific sustainability issue with growing importance due to exploding human populations and strongly increasing consumption rates (Anderson, 2012; McLellan et al., 2014).

Natural resources can be divided into two categories, depending on their natural growth rate:

- *Renewable resources* are those that have a natural production rate. For example, wood is a renewable resource since it grows on its own without human intervention. This does not imply that the use of wood is, by definition, sustainable over the long term. When the harvest rate exceeds the production rate, renewable resources are harvested unsustainably, leading to the depletion of the resource. Other renewable resources, such as solar energy, have a continuous production rate, independent of consumption and cannot be depleted by overconsumption.

- *Nonrenewable resources* are those without a natural production rate or a natural production rate that is so low that it is irrelevant on a human timescale. Metal ores and fossil fuels are examples of nonrenewable resources. The available amount is limited, so long-term use of these is, by definition, unsustainable unless the resources can be recycled in a very effective manner. Unlike metals, nonrenewable energy sources such as fossil fuels are not recoverable, and thus, their use is unsustainable in the long run as reserves shrink.

11.3.2 Sustainability Assessment

In order to effectively improve the sustainability of production processes, products, or entire societies, measuring tools are needed to calculate how sustainable they are and what their strengths and weaknesses look like. Without adequately precise measuring tools, efforts made to improve sustainability may completely miss their goal and even worsen the situation, despite best intentions. Sustainability assessments should alleviate these problems by providing objective results that can aid with making the most sustainable decisions (Klöpffer and Grahl, 2014).

Many different sustainability measurement tools exist, depending on the application and analyzed sustainability pillar (Ness et al., 2007). In order to assess the true sustainability, it is necessary to analyze all three pillars and the entire life cycle of the subject (Klöpffer, 2008; Candice, 2014). The *life cycle* of a product includes all stages that it undergoes during its lifetime (Fig. 11.2): the production phase (starting from raw resource extraction), the use phase, and the end-of-life phase. Depending on the goal of the assessment, it may also be limited to (a part of) one of these stages, but the entire cycle should always be kept in mind for effective sustainability improvement (Maxwell and van der Vorst, 2002).

Typically, the three sustainability pillars are assessed individually because it is very difficult, if not impossible, to objectively assess sustainability as a whole (Rotmans, 2006). Their assessment methods are briefly discussed below. The bulk of this chapter only focuses on the environmental aspects of sustainability, as, at the time of writing, more sustainability

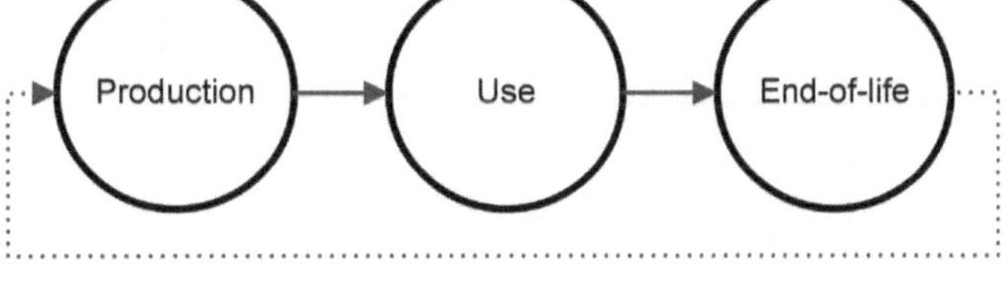

Figure 11.2
Life cycle phases of a product (*dotted line* = recycling).

assessments are needed in order to provide an apprehensive overview of the economic and social life cycle impacts of different biowaste valorization options.

Environmental sustainability assessment

Environmental sustainability is measured by life cycle assessment (LCA), also called environmental LCA (E-LCA) in order to avoid confusion with social LCA (S-LCA) (Paragahawewa et al., 2009). The method is used to calculate environmental impacts, determining the damage caused by a product throughout its production, use, and end-of-life phases. The framework for performing an LCA is standardized in the ISO 14000 series of standards (especially ISO 14040:2006 and ISO 14044:2006), but the actual impact calculation depends on the selected method.

Following ISO guidelines (ISO 14040:2006), an LCA is performed in four interacting phases (Fig. 11.3): (1) the goal and scope definition in which the aim of the LCA is defined, as well as the system borders within which impacts are taken into account; (2) the life cycle inventory (LCI) analysis, during which data is collected describing material and energy in- and output streams of the involved processes; (3) the life cycle impact assessment phase (LCIA) in which the inventory data is translated into environmental impacts by using impact assessment methods; and (4) the interpretation phase during which results are discussed and conclusions are drawn.

The LCA practitioner is free to select and use an impact assessment method of choice for the LCIA, even though this may lead to different results for similarly named impact categories (Dreyer et al., 2003; Renou et al., 2008). The selected method should thus always be taken into account when comparing results, and ideally conclusions are only drawn when comparing results using the same impact assessment method.

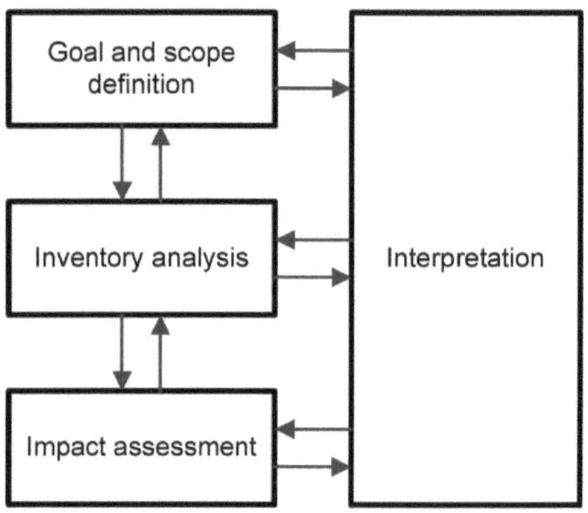

Figure 11.3
Life cycle assessment phases and their interactions. *Based on ISO (2006a) ISO 14040:2006. Environmental Management – Life Cycle Assessment – Principles and Framework.*

Several environmental impact assessment methods exist: International Life Cycle Data System (ILCD), ReCiPe, Eco-Indicator 99, Cumulative Exergy Demand, etc. They can be divided into two distinct groups:

- *Midpoint methods* are used to calculate very specific impacts in a relatively precise manner by weighted sum of the exchanges with the natural environment quantified in the LCI. Midpoint impact categories include climate change, freshwater use, natural land use, freshwater eutrophication, ozone layer depletion, etc. The ILCD method for example is used to calculate 16 different impact categories (Table 11.1).

Table 11.1: Impact Assessment Methods Used in the International Life Cycle Data System Midpoint Method (JRC, 2011).

Impact Category	Recommended Default Life Cycle Impact Assessment Method	Indicator	Unit	Level[a]
Climate change	Baseline model of 100 years of the Intergovernmental Panel on Climate Change (IPCC)	Radiative forcing as global warming potential (GWP100)	kg CO_2 eq.	I
Ozone depletion	Steady-state ozone depletion potentials 1999 as in World Meteorological Organization (WMO) assessment	Ozone depletion potential	kg CFC-11 eq.	I
Human toxicity, cancer effects	USEtox model (Rosenbaum et al., 2008)	Comparative toxic unit for humans	CTU_h	II/III
Human toxicity, noncancer effects	USEtox model (Rosenbaum et al., 2008)	Comparative toxic unit for humans	CTU_h	II/III
Particulate matter/ respiratory inorganics	RiskPoll model (Rabl and Spadaro, 2004; Greco et al., 2007)	Intake fraction for fine particles (kg PM 2.5-eq./kg)	kg PM 2.5 eq.	I
Ionizing radiation, human health	Human health effect model as developed by Dreicer et al. (1995) (Frischknecht et al., 2000)	Human exposure efficiency relative to U^{235}	kBq U^{235} eq.	II
Ionizing radiation, ecosystems	No methods recommended	Comparative toxic unit for ecosystems	CTU_e	Interim
Photochemical ozone formation	LOTOS-EUROS (Van Zelm et al., 2008) as applied in ReCiPe	Tropospheric ozone concentration increase	kg NMVOC eq.	II
Acidification	Accumulated Exceedance (Seppälä et al., 2006; Posch et al., 2008)	Accumulated exceedance	molc H^+ eq.	II
Eutrophication, terrestrial	Accumulated Exceedance (Seppälä et al., 2006; Posch et al., 2008)	Accumulated exceedance	molc N eq.	II
Eutrophication, aquatic	EUTREND model (Struijs et al., 2009)	Fraction of nutrients reaching freshwater end compartment (P)/marine end compartment (N)	kg P eq./kg N eq.	II

Continued

Table 11.1: Impact Assessment Methods Used in the International Life Cycle Data System Midpoint Method (JRC, 2011).—cont'd

Impact Category	Recommended Default Life Cycle Impact Assessment Method	Indicator	Unit	Level[a]
Ecotoxicity (freshwater)	USEtox model (Rosenbaum et al., 2008)	Comparative toxic unit for ecosystems	CTU_e	II/III
Ecotoxicity (terrestrial and marine)	No methods recommended			
Land use	Model based on soil organic matter (SOM; Milà I Canals et al., 2007)	Soil organic matter	kg C deficit	III
Resource depletion, water	Model for water consumption as in Swiss ecoscarcity (Frischknecht et al., 2008)	Water use related to local scarcity of water	m^3 water eq.	III
Resource depletion, mineral, fossil, and renewable[b]	CML method (Guinée et al., 2002)	Scarcity	kg Sb eq.	II

[a]Level I: recommended and satisfactory; level II: recommended but in need of some improvements; level III: recommended, but to be applied with caution.
[b]Depletion of renewable resources is included in the analysis, but none of the analyzed methods are mature for recommendation.

- *Endpoint methods* are the result of combining several midpoint results by means of a weighted sum in order to determine more general overall impact scores. Due to uncertainties in this process, the endpoint results are less precise than the midpoint scores, but they do offer an easy to interpret overview that can be used to identify environmental hot spots and can simplify decision making (Bare et al., 2000). As an example, the relationships between the inventory data and midpoint and endpoint scores of the ReCiPe method are shown in Fig. 11.4.

Economic sustainability assessment

Economic sustainability is assessed by life cycle costing (LCC). LCC was initially developed in order to calculate the total price of, for example, long-living products, which also have significant usage or disposal costs besides the initial production price (Sherif and Kolarik, 1981). Similar to LCA, this method adds up all costs associated with the production of a product, its use phase, and end-of-life phase (Baker, 1978; Asiedu and Gu, 1998). This allows the identification of the total life cycle cost, expressed in monetary value, which is useful for product selection procedures (Woodward, 1998). The method also allows identification of economic hot spots throughout the life of a product, revealing which processes or life stages contribute most to the total cost (Kirk and Dell'isola 1995; Janz et al., 2007).

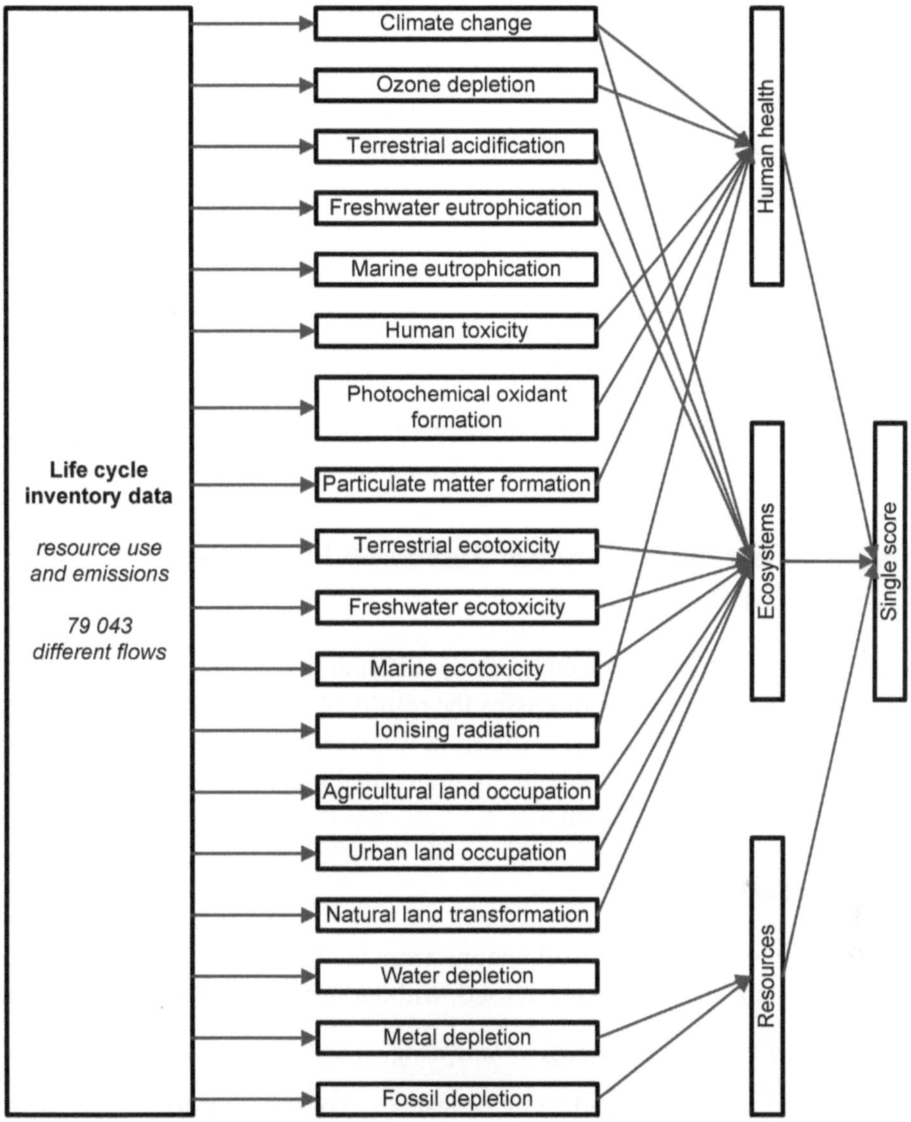

Figure 11.4

Relationship between life cycle inventory, midpoint indicators, endpoint indicators, and single endpoint score in ReCiPe 2008 (from left to right). *Based on Pre-sustainability (2014) Relationship Between LCI Parameters, Midpoint Indicator and Endpoint Indicator in ReCiPe 2008.* http://www.lcia-recipe.net/.

The results of an LCC can be of great value for producers, since they need to compete on both production price and product quality in a more and more competitive global market. Both can be assessed by LCC, providing valuable information and insight that can already be applied during the design phase of the product (Asiedu and Gu, 1998).

Social sustainability assessment

Social sustainability is calculated by S-LCA. This method is also based on the LCA methodology, taking into account all life cycle stages, from production to consumption and end-of-life (Dreyer et al., 2005). Its aim is to improve social and socioeconomic conditions (Benoît and Mazijn, 2009).

S-LCA is more complex than E-LCA in the sense that E-LCA deals exclusively with quantified data defining in- and outputs, while S-LCA also takes into account a lot of qualitative data and relationships between different factors. Social impacts are a function of politics, economy, ethics, psychology, legal issues, culture, etc. (Benoît and Mazijn, 2009). This makes data gathering and formulation of conclusions much more complicated than is the case with E-LCA and has led to more diverse methodologies and differences of opinion within the S-LCA community, leading to difficulties in comparing study results (Jørgensen et al., 2007).

11.4 Case Studies: Environmental Sustainability of Organic Waste Valorization Pathways

Organic waste streams can be valorized by many different means, ranging from reuse as feed to disposal in landfills. This section discusses the results of a selection of environmental sustainability assessments performed for different valorization pathways in order to provide an overview of existing technologies, their environmental impacts, and challenges related to selecting the most sustainable solution.

11.4.1 Reuse as Feed

Organic waste streams, such as those originating from the food industry, can contain valuable nutrients and energy sources that can be reused, without further modification, as animal feeds.

Food waste conversion to feed, composting, or landfilling

Kim and Kim (2010) have performed an LCA evaluation of different food waste treatment scenarios. They considered food waste reuse for dry feed and wet feed manufacturing, composting and landfilling in the Jungnang District, South Korea, taking into account curbside collection and transportation to the treatment plants. The dry feed production process involved cracking of the input stream, removal of plastics (vinyl) and metals, dehydration, selection, grinding, drying, and cooling, resulting in a production rate of 35.8 tons of dry feed per 260 tons of food waste. The wet feed production also started with cracking, followed by two sorting stages, sterilization, partial dewatering, and mixing, resulting in a production rate of 8.3 tons wet feed per 20.0 tons of food waste. At the composting facility, the food waste is mixed with other waste streams and sawdust, which reduces the moisture content, lowers the food waste concentration, and limits odors. The resulting compost is used as a fertilizer in agriculture. At the landfilling plant, the food waste biodegrades, and 45% of the landfill gas is

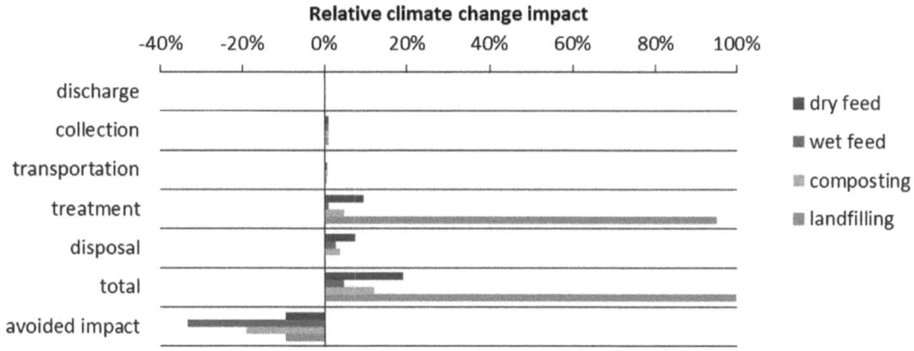

Figure 11.5
Climate change impact of food waste valorization routes: dry feed production, wet feed production, composting, and landfilling.

captured and either flared (88% of landfill gas) or burned to produce electricity (12% of captured landfill gas). The comparison is limited to climate change impact.

The study results (Fig. 11.5) show that wet feed production performs better than composting due to low treatment impact and large avoided impact. Dry feed production is outperformed by composting due to high treatment and disposal impacts combined with relatively little avoided impacts due to equivalent product displacement. Landfilling scores much worse than the other options because of the large impact of the landfilling process and relatively low electricity generation benefits. Note that the avoidance of producing equivalent products (feeds, fertilizer, and electricity) plays an important role to the total impact of each scenario but depends on the impact of the equivalent products. If the latter were improved, e.g. due to reference electricity being generated from renewable resources such as wind instead of fossil sources, the avoided impact could be reduced significantly. Furthermore, the authors stress that the disposal of the remaining waste can play a much larger role, depending on the precise treatment scenario, especially for the wet feed production and composting processes.

11.4.2 Recycling by Conversion to Higher Value Chemicals

Large waste streams of relatively high purity are generated in the fruit and vegetable producing industries. At least at first sight, these complex organic by-products offer major opportunities, both from an environmental and an economic perspective, to produce high value products such as polyhydroxybutyrate (PHB), enzymes, succinic acid, polylactic acid (PLA), fibers, ethanol, gibberellic acid, cellulose acetate, carboxymethylcellulose, benzyl cellulose, pectin, polyphenols, citric acid, carotenoids, and so on. Despite their seemingly high potential, more research into this field is required in order to develop, evaluate, and improve these pathways converting waste streams to higher value chemicals. Their long-term application will strongly depend on their economic and environmental sustainability.

Polyhydroxybutyrate, enzymes, and succinic acid from agricultural waste

The European research project TRANSBIO is developing production pathways for PHB, enzyme, and succinic acid production from agricultural waste products. The project is not finalized yet, but the preliminary results indicate that it is indeed possible to produce these higher value products from agricultural waste streams. More research and development are, however, needed to further improve production efficiencies in order to match the environmental performance of more mature processes using more conventional feed streams.

11.4.3 Recycling by Composting

Composting is a controlled microbiological process in which organic materials are degraded in the presence of oxygen by different microorganisms like bacteria, actinomycetes, yeasts, and fungi. The organic matter is partly converted to CO_2, and the remaining biomass (compost) is biologically stabilized and used as soil improver. Composting results in a volume reduction of the waste of about 50%. The composting process is a well-known and well-documented process to convert organic waste to a stable product, and is practiced both at small-scale and low-tech levels (e.g. home composting), as at large-scale and high-tech levels (e.g. industrial composting of organic household waste). More information about composting can be found in chapter "Organic Recycling of Agricultural Waste Today: Composting and Anaerobic Digestion."

Composting technologies compared: tunnel composting and confined windrow composting

Cadena et al. (2009) have compared the environmental impacts of two composting plants in Catalonia applying different technologies: one applies tunnel composting, and the other works with confined windrows. Both plants process the organic fraction of municipal solid waste to which wood chips or pruning wastes are added, respectively. The environmental impact is assessed in terms of global warming potential (over 100 years), acidification, human toxicity, and ozone layer depletion.

The results (Fig. 11.6) show that both systems have their own strengths and weaknesses: they perform rather similarly in terms of global warming and human toxicity, but confined windrows composting scores better in terms of acidification and eutrophication, and much worse in terms of photochemical oxidation and ozone layer depletion.

Municipal solid waste: 65% composting and 35% landfilling, or 100% landfilling

Abduli et al. (2011) have analyzed two scenarios for mixed solid waste treatment in Tehran: landfilling the entire stream or separating the organic waste from the stream for composting and landfilling the remainder. The landfill has a gas collection system, which captures 50% of the gaseous emissions and burns them for electricity production. Eighty percent of the landfill leachate is captured and treated in a municipal sewage treatment plant. In the composting

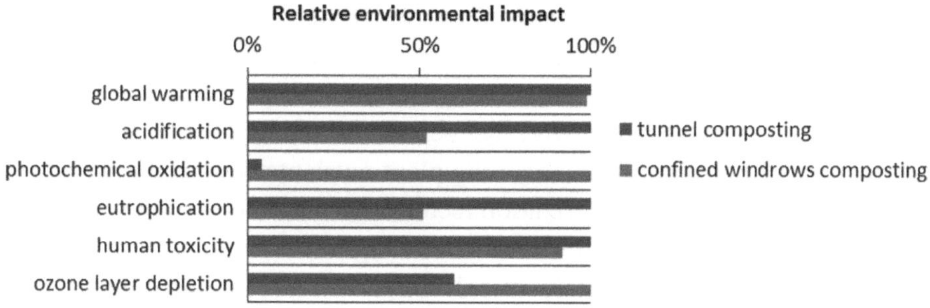

Figure 11.6
Environmental impact of different composting technologies.

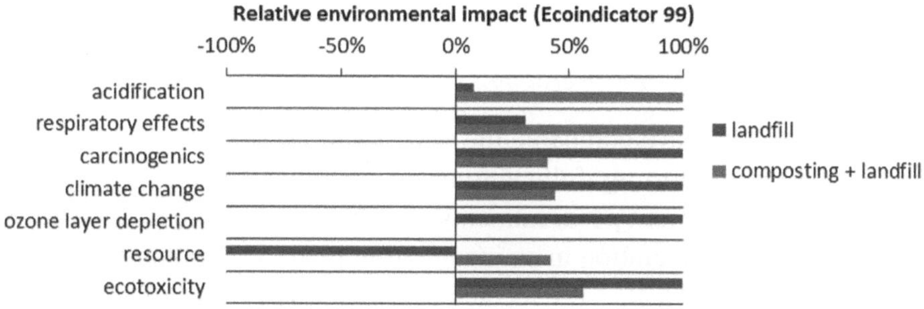

Figure 11.7
Landfilling everything or separating for 65% composting and 35% landfilling.

scenario, 65% of the waste is composted, and 35% is landfilled. The composting plant has no emission control system, emitting pollutants into the air without any restrictions. The environmental impact is assessed using the Eco-indicator 99 method.

The comparison shows mixed results (Fig. 11.7). The 100% landfill scenario performs better in terms of acidification, respiratory effects, and resources, while the combination of composting and landfilling is better in terms of carcinogenic effects, climate change, ozone layer depletion, and ecotoxicity. The advantage of landfilling in terms of resources is remarkable, as the compost produced in the alternative system recycles nutrients and displaces mineral fertilizer. Apparently, the energy recovery by burning the biogas captured in the landfill has a much larger positive effect in this aspect.

11.4.4 Energy Recovery and Recycling by Anaerobic Digestion With Composting

Anaerobic digestion is a microbiological process in which organic material is degraded by microorganisms in the absence of oxygen. The input material is largely converted to biogas, a mixture of CH_4 and CO_2, and digester sludge (digestate). The biogas can either be purified and injected into the natural gas grid or burned onsite for electricity and heat production. Depending on the input material and desired application, the digestate can be used, such as for organic fertilizer, or can be converted to compost, fertilizer pellets, nutrient rich liquor, etc., using different types of posttreatment. Independent of the treatment, the digestate can thus be applied as an organic fertilizer, recycling plant nutrients back to the soil. Anaerobic digestion is thus an organic waste valorization technology that leads to both energy recovery and nutrient recycling. It can be applied on a wide range of organic input streams—from very wet to relatively dry waste streams—by selecting the most optimal digester design. Nevertheless, due to the absence of fungi, wood and other products rich in lignin are not or are hardly degradable in anaerobic digestion. Because of the higher complexity compared to composting, anaerobic digestion is usually performed at (semi)industrial scale. More information about the anaerobic digestion (AD) process can be found in chapter "Organic Recycling of Agricultural Waste Today: Composting and Anaerobic Digestion."

Anaerobic digestion, composting, or incineration

Edelmann and Baier (2005) have compared the environmental impacts of several composting systems (open windrow composting; fully automated, enclosed tunnel composting), anaerobic digestion combined with different posttreatments (aerobic posttreatment, open and enclosed composting systems) and incineration in an incineration plant with exhaust gas scrubbing but without energy recovery. All plants were located in Switzerland.

Using the Eco-indicator 99 impact assessment method to calculate the climate change impact over 100 years, the study concluded that anaerobic digestion performs better than fully enclosed composting and incineration but worse than open windrow composting (Fig. 11.8). The latter is caused by a larger amount of methane emissions from the anaerobic digestion with composting scenario than from the direct composting scenarios. Using the Eco-indicator 95+ impact assessment method expressing the results in endpoints (Fig. 11.9), the difference is more pronounced with a clear lead for anaerobic digestion with aerobic posttreatment, but surprisingly, fully enclosed composting performs even worse than incineration in this

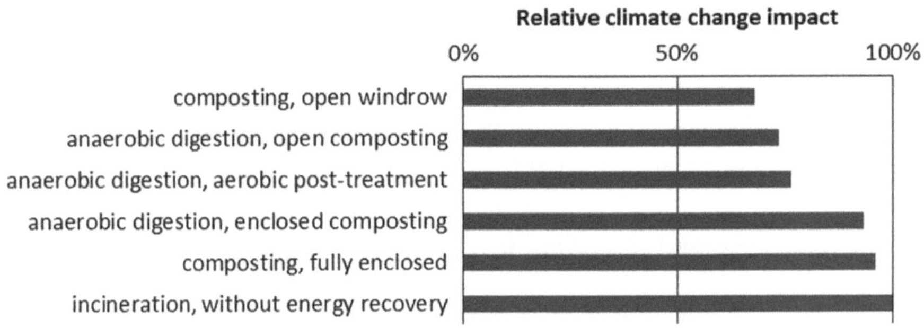

Figure 11.8
Climate change impact of composting, anaerobic digestion, and incineration technologies (over 100 years).

comparison. The study concludes that this is caused by net energy use, which plays a dominant role in the impact assessment, explaining the low impact of anaerobic digestion as it produces energy from the waste, while fully enclosed composting performs badly because it is an energy intensive process. Incineration performs badly due to the lack of energy recovery and the lack of fertilizer production, which is achieved by anaerobic digestion and composting. It should, however, be noted that the processing of, for example, strong ammonia emitting waste streams would warrant the use of fully enclosed systems with biofilters in order to avoid emissions to air, and that the precise composition of the waste stream can also impact the biogas yield of the digester plant, and thus, also its energy production rate.

Khoo et al. (2010) have compared different food waste end-of-life scenarios in Singapore, combining incineration with energy recovery, anaerobic digestion with composting, and small-scale composting in different ratios. Several municipal solid waste incinerators are analyzed with food waste input fractions ranging from 13% to 36%. The benefits associated with replacement of mineral fertilizer by compost are taken into account. Residual waste streams are landfilled.

The normalized results (Fig. 11.10) show that diverting (part of) the food waste to the anaerobic digestion and small-scale composting plants improves the overall environmental performance of the municipal waste treatment scheme.

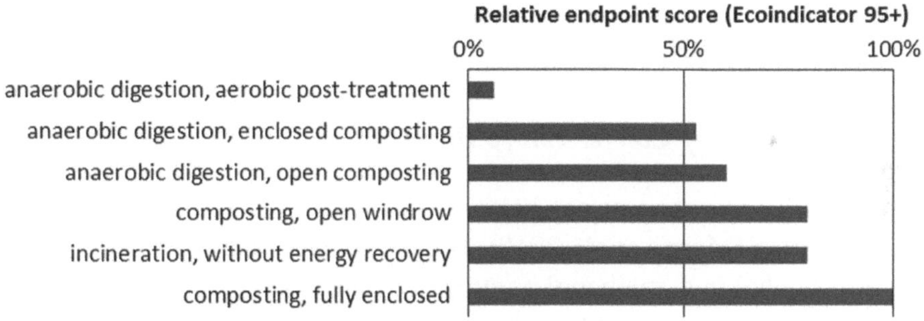

Figure 11.9
Endpoint impact of composting, anaerobic digestion, and incineration technologies (Eco-indicator 95+).

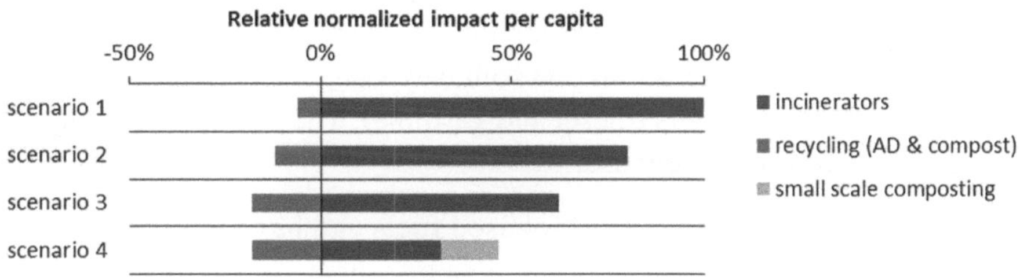

Figure 11.10
Environmental impact of municipal waste treatment with increasing diversion of food waste from incineration to anaerobic digestion with composting and small-scale composting.

Anaerobic digestion for electricity production, fertilizer production and the effect of different feeds

De Meester et al. (2013) have analyzed the environmental sustainability of anaerobic digestion from different angles. They compared electricity production from organic waste and energy crops by anaerobic digestion to average electricity generation in the German grid, they studied the impact of applying digestate as a fertilizer for maize silage production, and they analyzed different feed streams in an agricultural context. The results are calculated using the cumulative exergy extraction from the natural environment (CEENE) and ReCiPe impact assessment methods.

In terms of electricity production, the analysis shows that power generation by anaerobic digestion of domestic organic waste and from maize silage performs markedly better than grid average production in most impact categories.

Comparing maize silage production using different fertilizers, the results are more mixed. Taking into account uncertainty of the results, using digestate for fertilization, organic or mineral fertilizer yields relatively similar results in most impact categories. Digestate and organic fertilizer outperform mineral fertilizer in terms of fossil fuels and minerals. Organic fertilizer outperforms the others in terms of water resources, freshwater and marine eutrophication.

When comparing different input mixtures to produce electricity in an anaerobic digestion plant, it is concluded that this does lead to significant differences: generating power from a corn stover and manure mixture performs better than from a sugar beet and grass silage mixture, which in turn outperforms generation from silage maize in almost every impact category.

11.4.5 Energy Recovery by Conversion to Biodiesel

Fat and oil containing waste streams can be converted to biodiesel in order to recover the energy contained within (Morais et al., 2010). Biodiesel is produced by transesterification of triglycerides with alcohol (e.g. methanol or ethanol). This reaction requires both heat and catalysts in order to achieve acceptable reaction rates.

Biodiesel from food waste or other renewable and nonrenewable fuels

Xunmin et al. (2009) have analyzed and compared the climate change impact of biodiesel produced from food waste (used cooking oil), biodiesel, and ethanol produced from several agricultural products (biodiesel from Jatropha fruit, soybean, and corn; ethanol from cassava and sorghum) and conventional fuels (diesel and gasoline). The impact was analyzed over the entire life cycle of the fuels, including use.

The study results (Fig. 11.11) show that biobased fuels can have lower climate change impacts than conventional fuels, but they can also be worse, depending on feedstock. Jatropha fruit biodiesel, cassava ethanol, and used cooking oil biodiesel showed better than conventional results. Soybean and corn biodiesel and sorghum ethanol, however, had larger greenhouse gas impacts than conventional fuels.

11.4.6 Energy Recovery by Incineration

Organic waste can also be incinerated in order to produce heat or electricity. If this valorization pathway is selected, it is important to take into account the moisture content of the stream, as the vaporization of water is very energy intensive, lowering the amount of heat available for energy recovery.

Heat and electricity from wood chips or from other renewable and nonrenewable sources

Cherubini et al. (2009) have compared the greenhouse gas emissions associated with both heat and electricity production from the burning of biomass (wood chips or pellets), fossil, and renewable resources. Wood chips can be produced as a by-product in the woodworking industry or from crops grown specifically for the purpose of energy generation. The study, however, does not appear to differentiate by precise origin of the biomass.

In terms of electricity production, the results (Fig. 11.12) show that biomass does not perform as well as wind, geothermal, or hydroelectricity production, but can compete with nuclear,

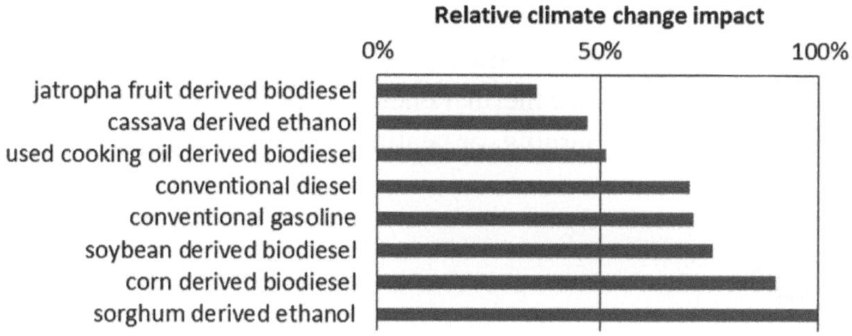

Figure 11.11
Life cycle climate change impact of fuels derived from food waste, agricultural products, and fossil resources (per MJ energy).

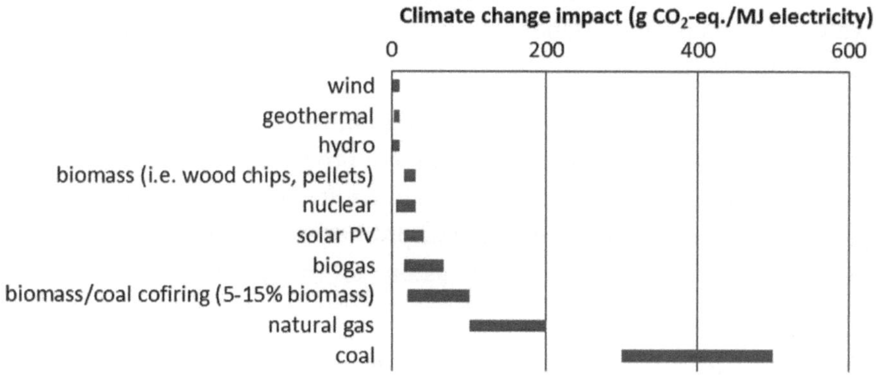

Figure 11.12
Climate change impact ranges of electricity generation by renewable and nonrenewable sources.

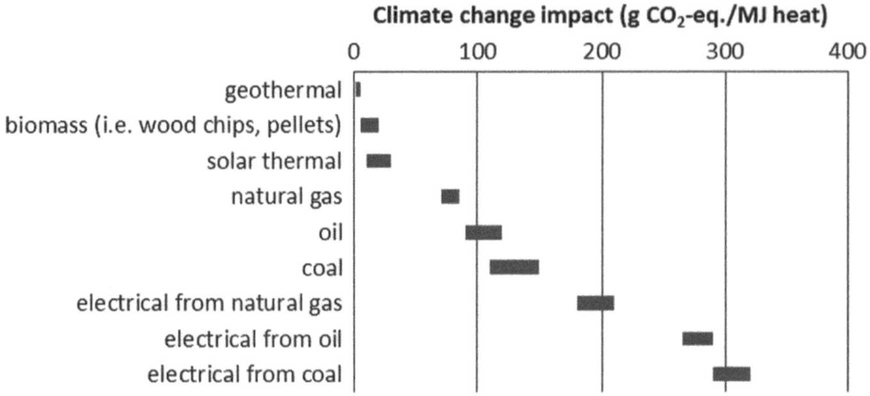

Figure 11.13
Climate change impact ranges of heat production by renewable and nonrenewable sources.

photovoltaics, and biogas production, while performing clearly better than generation from natural gas or coal.

Biomass is more competitive for heat production (Fig 11.13) in which it is only outclassed by geothermal heat and on par with solar thermal energy. It performs much better than direct heat from natural gas, oil, or coal burning, or heating with electricity generated from natural gas, oil, or coal.

11.5 Conclusions

It can be concluded that the waste hierarchy is not sufficient for decision making in the field of organic waste valorization. Several LCA studies clearly show that similar waste valorization techniques can lead to strongly varying results, depending on specific details of the considered processes. Solutions which are less preferred in the waste hierarchy, such as disposal by landfill, can be more environmentally friendly than more preferred solutions, such as recycling by composting, depending on which environmental impacts are considered to be more important.

Furthermore, the referenced studies show that certain biowaste-based products are capable of competing with and even outperforming conventional processes. This was shown to be the case for specific biodiesel production pathways, electricity, and heat production. It should be noted though that the impact of bio(waste)-based production pathways can strongly differ by feedstock and process specifics, and that conclusions for one process should not be generalized.

Finally, there appears to be a shortage of studies describing the environmental impact associated with production pathways of high value compounds from organic waste streams. This is somewhat disturbing, as these pathways are often developed with the aim of solving environmental problems. As this literature study has shown, LCA studies are needed to verify green claims in this field, as simply following the waste hierarchy can lead to false conclusions.

References

Abduli, M.A., Naghib, A., Yonesi, M., Akbari, A., 2011. Life cycle assessment (LCA) of solid waste management strategies in Tehran: landfill and composting plus landfill. Environmental Monitoring and Assessment 178, 487–498.

Anderson, R., 2012. Resource Depletion: Opportunity or Looming Catastrophe? BBC News.

Asiedu, Y., Gu, P., 1998. Product life cycle cost analysis: state of the art review. The International Journal of Production Research 36 (4), 883–908.

Baker, W.R., 1978. Life-cycle costing. Plant Engineering 32, 11.

Bandara, N.J.G.J., Hettiaratchi, J.P.A., Wirasinghe, S.C., Piapiiya, S., 2007. Relation of waste generation and composition to socio-economic factors: a case study. Environmental Monitoring and Assessment 135, 31–39.

Bare, J.C., Hofstetter, P., Pennington, D.W., de Haes, H.A.U., 2000. Midpoints versus endpoints: the sacrifices and benefits. International Journal of Life Cycle Assessment 5 (6), 319–326.

Brundtland, G.H., Ahalid, M., et al., 1987. Our Common Future: Report of the World Commission on Environment and Development.

Benoît, C., Mazijn, B., et al., 2009. Guidelines for Social Life Cycle Assessment of Products. United Nations Environment Programme, Belgium Federal Public Planning Service Sustainable Development.

Cadena, E., Colón, J., Artola, A., Sánchez, A., Font, X., 2009. Environmental impact of two aerobic composting technologies using life cycle assessment. International Journal of Life Cycle Assessment 14, 401–410.

Candice, S., 2014. Sustainability Assessment Methodologies. OECD Sustainable Development Advisor.

Cherubini, F., Bird, N.D., Cowie, A., Jungmeier, G., Schlamadinger, B., Woess-Gallasch, S., 2009. Energy- and greenhouse gas-based LCA of biofuel and bioenergy systems: key issues, ranges and recommendations. Resources, Conservation and Recycling 53, 434–447.

Corell, D.L., 1998. The role of phosphorus in the eutrophication of receiving waters: a review. Journal of Environmental Quality 27 (2), 261–266.

De Meester, S., Dewulf, J., Roes, L., Patel, M., Hellweg, S., 2013. Prospective Sustainability Assessment of Technologies: Development of Basic Engineering Modules for Prospective Estimations of the Material Flows and Energy Requirements. Prosuite.

Dreicer, M., Tort, V., Manen, P., 1995. ExternE, Externalities of Energy, Vol. 5. Nuclear, Centre d'étude sur l'Evaluation de la Protection dans le domaine Nucléaire (CEPN), edited by the European Commission DGXII, Science, Research and Development JOULE, Luxembourg.

Dreyer, L.C., Niemann, A.L., Hauschild, M.Z., 2003. Comparison of three different LCA methods: EDIP97, CML2001 and Eco-indicator 99. International Journal of Life Cycle Assessment 8 (4), 191–200.

Dreyer, L., Hauschild, M., Schierbeck, J., 2005. A framework for social life cycle impact assessment. International Journal of Life Cycle Assessment 11 (2), 88–97.

Edelmann, W., Baier, U., 2005. Environmental aspects of the anaerobic digestion of the organic fraction of municipal solid wastes and of solid agricultural wastes. Water Science & Technology.

El-Fadel, M., Findikakis, A.N., Leckie, J.O., 1997. Environmental impacts of solid waste landfilling. Journal of Environmental Management 50 (1), 1–25.

European Parliament, 2015. European Parliament Resolution of 9 July 2015 on Resource Efficiency: Moving towards a Circular Economy (2014/2208(INI)).

FAO, 2013. Food Wastage Footprint – Impacts on Natural Resources. Summary Report.

Frischknecht, R., Braunschweig, A., Hofstetter, P., Suter, P., 2000. Modelling human health effects of radioactive releases in Life Cycle Impact Assessment. Environmental Impact Assessment Review 20 (2), 159–189.

Frischknecht, R., Steiner, R., Jungbluth, N., 2008. Methode der ökologischen Knappheit –Ökofaktoren 2006, ö.b.u. und Bundesamt für Umwelt, Bern.

Greco, S.L., Wilson, A.M., Spengler, J.D., Levy, J.I., 2007. Spatial patterns of mobile source particulate matter emissions-to-exposure relationships across the United States. Atmospheric Environment (41), 1011–1025.

Guinée J. B. (Ed.), Gorrée, M., Heijungs, R., Huppes, G., Kleijn, R., de Koning, A., Van Oers, L., Wegener Sleeswijk, A., Suh, S., Udo de Haes, H.A., De Bruijn, J.A., Van Duin, R., Huijbregts, M.A.J., 2002. Handbook on Life Cycle Assessment: Operational Guide to the ISO Standards. Series: Eco-efficiency in industry and science. Kluwer Academic Publishers. Dordrecht (Hardbound, ISBN 1-4020-0228-9; Paperback, ISBN 1-4020-0557-1).

Gustavsson, J., Cederberg, C., Sonesson, U., 2011. Global food losses and food waste. In: International Safe Food Congress, Düsseldorf 16 May 2011.

Hansen, W., Christopher, M., Verbuecheln, M., 2002. EU waste policy and challenges for regional and local authorities. In: Background Paper for the Seminar on Household Waste Management "capacity Building on European Community's Environmental Policy".

ISO, 2006a. ISO 14040:2006. Environmental Management – Life Cycle Assessment – Principles and Framework.

ISO, 2006b. ISO 14044:2006. Environmental Management – Life Cycle Assessment – Requirements and Guidelines.

Janz, D., Sihn, W., Warnecke, H.-J., 2007. Product redesign using value-oriented life cycle costing. CIRP Annals – Manufacturing Technology 54 (1), 9–12.

Jørgensen, A., Le Bocq, A., Nazarkina, L., Hauschild, M., 2007. Methodologies for social life cycle assessment. International Safe Food Congress, Düsseldorf 16 May 2011 International Journal of Life Cycle Assessment 13 (2), 96–103.

JRC, 2011. Recommendations for Life Cycle Impact Assessment in the European Context – Based on Existing Environmental Impact Assessment Models and Factors.

JRC, 2015. Life Cycle Thinking and Assessment for Waste Management. Newsletter.

Khan, M.I., Islam, M.R., 2007. True Sustainability in Technological Development and Natural Resource Management. Nova Science Publishers, Inc.

Khoo, H.H., Lim, T.Z., Tan, R.B.H., 2010. Food waste conversion options in Singapore: environmental impacts based on an LCA perspective. Science of the Total Environment 408, 1367–1373.

Kim, M.-H., Kim, J.-W., 2010. Comparison through a LCA evaluation analysis of food waste disposal options from the perspective of global warming and resource recovery. Science of the Total Environment 408, 3998–4006.

Kirk, S.J., Dell'isola, A.J., 1995. Life Cycle Costing for Design Professionals, second revised edition. ISBN: 0-07-034804-9.

Klöpffer, W., 2008. Life cycle sustainability assessment of products. International Journal of Life Cycle Assessment. March 2008 13 (2), 89–95.

Klöpffer, W., Grahl, B., 2014. Life Cycle Assessment (LCA): A Guide to Best Practice. Wiley-VCH VerlagGmbH & Co. KGaA.

Maxwell, D., van der Vorst, R., 2002. Developing sustainable products and services. Journal of Cleaner Production 11 (8), 883–895.

McLellan, R., Iyengar, L., Jeffries, B., Oerlemans, N., Grooten, M., Guerraoui, M., Sunters, P., 2014. Living Planet Report 2014. WWF.

Milà i Canals, L., Bauer, C., Depestele, J., Dubreuil, A., Freiermuth Knuchel, R., Gaillard, G., Michelsen, O., Müller-Wenk, R., Rydgren, B., 2007. Key elements in a framework for land use impact assessment within LCA. Int J LCA 12, 5–15.

Moberg, Å., Finnveden, G., Johansson, J., Lind, P., 2005. Life cycle assessment of energy from solid waste – part 2: landfilling compared to other treatment methods. Journal of Cleaner Production 13 (3), 231–240.

Morais, S., Mata, T.M., Martins, A.A., Pinto, G.A., Costa, C.A.V., 2010. Simulation and life cycle assessment of process design alternatives for biodiesel production from waste vegetable oils. Journal of Cleaner Production 18, 1251–1259.

Ness, B., Urben-Pürsalu, E., Anderberg, S., Olsson, L., 2007. Categorising tools for sustainability assessment. Ecological Economics 60, 498–508.

Paragahawewa, U., Blacket, P., Small, B., 2009. Social Life Cycle Analysis (S-LCA): Some Methodological Issues and Potential Application to Cheese Production in New Zealand. AgResearch.
Posch, M., Seppälä, J., Hettelingh, J.P., Johansson, M., Margni, M., Jolliet, O., 2008. The role of atmospheric dispersion models and ecosystem sensitivity in the determination of characterisation factors for acidifying and eutrophying emissions in LCIA. International Journal of Life Cycle Assessment 13, 477–486.
Pre-sustainability, 2014. Relationship between LCI Parameters, Midpoint Indicator and Endpoint Indicator in ReCiPe 2008. http://www.lcia-recipe.net/.
Rabl, A., Spadaro, J.V., 2004. The RiskPoll software, version 1.051 (dated August 2004). www.arirabl.com.
RDC, 2008. Evaluation des Bénéfices Environnementaux, Économiques et Sociaux de Différents Scenarios de Reutilisation des Déchets par les Entreprises D'économie Sociale. Rapport final. Pour L'office Wallon des Déchets.
Renou, S., Thomas, J.S., Aoustin, E., Pons, M.N., 2008. Influence of impact assessment methods in wastewater treatment LCA. Journal of Cleaner Production 16 (10), 1098–1105.
Rosenbaum, R.K., Bachmann, T.M., Gold, L.S., Huijbregts, M.A.J., Jolliet, O., Juraske, R., Köhler, A., Larsen, H.F., MacLeod, M., Margni, M., McKone, T.E., Payet, J., Schuhmacher, M., van de Meent, D., Hauschild, M.Z., 2008. USEtox - The UNEP - SETAC toxicity model: recommended characterisation factors for human toxicity and freshwater ecotoxicity in Life Cycle Impact Assessment. International Journal of Life Cycle Assessment 13 (7), 532–546.
Rotmans, J., 2006. Tools for integrated sustainability assessment: a two-track approach. The Integrated Assessment Journal 6 (4), 35–37.
Schmidt, J.H., Holm, P., Merrild, A., Christensen, P., 2007. Life cycle assessment of the waste hierarchy – a Danish case study on waste paper. Waste Management 27 (11), 1519–1530.
Seppälä, J., Posch, M., Johansson, M., Hettelingh, J.P., 2006. Country-dependent characterisation factors for acidification and terrestrial eutrophication based on accumulated exceedance as an impact category indicator. International Journal of Life Cycle Assessment 11 (6), 403–416.
Sims, J.T., Wolf, D.C., 1994. Poultry waste management: agricultural and environmental issues. Advances in Agronomy 52.
Shah, R., 2008. Waste Statistics. Workshop on Environment Statistics. United Nations Statistics Division.
Sherif, Y.S., Kolarik, W.J., 1981. Life cycle costing: concept and practice. Omega International Journal of Management Science 9 (3), 287–296.
Struijs, J., Beusen, A., van Jaarsveld, H., Huijbregts, M.A.J., 2009. Aquatic eutrophication, Chapter 6 in: Goedkoop M., Heijungs R., Huijbregts M. A. J., De Schryver A., Struijs J., Van Zelm R. (2009). ReCiPe 2008 A life cycle impact assessment method which comprises harmonised category indicators at the midpoint and the endpoint level. Report I: Characterisation factors, first edition.
UNEP, 2011. Waste – Investing in Energy and Resource Efficiency. Towards a Green Economy. (Advance copy online release).
Van Zelm, R., Huijbregts, M.A.J., Den Hollander, H.A., Van Jaarsveld, H.A., Sauter, F.J., Struijs, J., Van Wijnen, H.J., Van de Meent, D., 2008. European characterization factors for human health damage of PM10 and ozone in life cycle impact assessment. Atmospheric Environment 42, 441–453.
Westerman, P.W., Bicudo, J.R., 2005. Management considerations for organic waste use in agriculture. Bioresource Technology 96, 215–221.
Woodward, D.G., 1998. Life cycle costing – theory, information acquisition and application. International Journal of Project Management 15 (6), 335–344.
World Bank, 2014. Waste Generation. Urban Development Series Knowledge Papers.
Xunmin, O., Xiliang, Z., Shiyan, C., Qingfang, G., 2009. Energy consumption and GHG emissions of six biofuel pathways by LCA in (the) People's Republic of China. Applied Energy 86, S197–S208.
Young, S., Stanner, E., 2008. Waste Hierarchy. Image From the Wikimedia Commons published Under the Creative Commons Attribution-Share Alike 3.0 Unported License.

CHAPTER 12

GMO Acceptance in the World and Issues for the Overcoming of Restrictions: Cisgenesis, RNA Transfer, Rootstock to Shoot Delivery, Novel Methods of Transformation

P. Poltronieri
CNR-ISPA, Lecce, Italy

12.1 Introduction

Alternative and accelerated breeding methods are well-accepted methods of production of nongenetically modified (non-GM) plants. In this chapter, we compare methods based on mutagenesis and the plant products that are used in every country presently, with transgenic plants and the constraints that limit their approval in Europe and in other countries. New plant products based on cisgenesis, intragenesis, nuclease-based methods such as zinc-finger nucleases (ZFNs), transcription activator-like effector nucleases (TALENs), clustered regularly interspaced short palindromic repeats (CRISPR)-associated nuclease9, oligonucleotide-directed mutagenesis (ODM), seed production technology (SPT), reverse breeding (RB), and rootstock-to-scion transfer are presented and discussed. Finally, an overview on the needs of harmonization among different regulatory processes around the world is introduced.

12.1.1 Genetically Modified Plants and Their Regulation Process in Different Countries

Biotechnology products in the United States are regulated according to a system, the Coordinated Framework, established by the Office of Science and Technology Policy in 1986. Deriving its mandate from existing laws regulating food safety and agriculture, the Coordinated Framework assigns lead responsibility for biotechnology products to the appropriate regulatory agency and sets out principles for cooperative reviews in areas where responsibilities or authorities overlap. The regulation of agricultural biotechnology products is handled by three agencies:

The Department of Agriculture Animal and Plant Health Inspection Service (APHIS), in US, oversees the field testing of biotechnology-derived plants as "regulated articles" to ensure that

Biotransformation of Agricultural Waste and By-Products. http://dx.doi.org/10.1016/B978-0-12-803622-8.00012-4

the environment is protected. A petition for nonregulated status must be granted by APHIS prior to commercial growth and sale of any bioengineered crop.

The Environmental Protection Agency (EPA) is responsible for ensuring that pest-resistant biotech varieties are safe to grow and consume. It regulates environmental exposure to these crops to ensure there are no adverse effects to the environment or any beneficial, nontargeted insects and other organisms. The agency also regulates bioengineered microorganisms under the Toxic Substances Control Act.

The Center for Food Safety and Nutrition of the Food and Drug Administration (FDA) imposes on foods developed through biotechnology the regulatory requirements used to safeguard all foods in the marketplace. The FDA has both premarket and postmarket authority to regulate the safety and labeling of all foods and animal feed. Foods from biotechnology are judged on their individual safety and nutrition, not the methods used to produce them. Under federal law, the producer of a food has the legal obligation to ensure its safety to consumers, and the FDA may pull from the market any foods found to be unsafe. Since 1992, FDA has used a voluntary review process for biotechnology foods. Over 50 such products have been reviewed, and none has been found to pose a safety concern. To improve consumer confidence, proposed rules issued by FDA in 2001 have made premarket review of biotech foods mandatory. However, regulatory authorities in other countries require that GM plants be revised by different authorities and review processes. Table 12.1 provides a list of GM plants introduced in various countries around the world.

The regulation of health safety and environmental risks are generally much stricter in EU than in USA. The EU accepted Cartagena Protocol on Biosafety to the Convention on Biological Diversity, an international agreement that aims to ensure the safe handling, transport, and use of living modified organisms (LMOs) resulting from modern biotechnology that may have adverse effects on biological diversity, taking also into account risks to human health (http://bch.cbd.int/protocol). The precautionary principle and recently also so-called social and economic aspects, which were included to the decision process of the risks assessment, are the main reasons that delay genetically modified organisms' (GMOs') approval in Europe, while in other advanced countries, these constraints are less tight. It is the responsibility of individual EU member states to formulate guidelines for the growing of GM crops. National guidelines already exist in several countries. These rules have been designed to allow the coexistence of GM or conventional agriculture in those countries with a more open attitude to the introduction of GM plants. There is an EU register of GM plants approved for authorized use (http://ec.europa.eu/food/dyna/gm_register/index_en.cfm): Foods and food ingredients containing, consisting of, or produced from; Foods containing, consisting of, or produced from; Products other than food and feed containing or consisting of) until authorization expiration date. As for the 2015, different plant varieties are approved for agricultural use (http://www.gmo-compass.org/eng/agri_biotechnology/gmo_planting/) in the EU, but only one transgenic crop, *Bacillus thuringiensis* (Bt) maize, is commercially planted in EU

Table 12.1: List of Genetically Modified Crops Cultivated Around the World.

Alfalfa (*Medicago sativa*)
Apple (*Malus × domestica*)
Argentine canola (*Brassica napus*)
Bean (*Phaseolus vulgaris*)
Carnation (*Dianthus caryophyllus*)
Chicory (*Cichorium intybus*)
Cotton (*Gossypium hirsutum* L.)
Creeping bentgrass (*Agrostis stolonifera*)
Eggplant (*Solanum melongena*)
Eucalyptus (*Eucalyptus* sp.)
Flax (*Linum usitatissimum* L.)
Maize (*Zea mays* L.)
Melon (*Cucumis melo*)
Papaya (*Carica papaya*)
Petunia (*Petunia hybrida*)
Plum (*Prunus domestica*)
Polish canola (*Brassica rapa*)
Poplar (*Populus* sp.)
Potato (*Solanum tuberosum* L.)
Rice (*Oryza sativa* L.)
Rose (*Rosa hybrida*)
Soybean (*Glycine max* L.)
Squash (*Cucurbita pepo*)
Sugar beet (*Beta vulgaris*)
Sugarcane (*Saccharum* sp.)
Sweet pepper (*Capsicum annuum*)
Tobacco (*Nicotiana tabacum* L.)
Tomato (*Lycopersicon esculentum*)
Wheat (*Triticum aestivum*)

presently. Bt maize was originally planted in six countries (Spain, Czech Republic, Slovakia, Portugal, Romania, and Poland) from 27 EU member states at a very low area of agricultural land (about 0.3 mil ha). Today, MON810, Monsanto's maize, is cultivated in Spain and Portugal. In Germany, five Bt maize seeds have obtained approval, and all of them are derived from MON810.

EU approved early in 2015 the import of 10 new types of GM maize, soybeans, cotton, and oilseed rape, produced by Monsanto, BASF, and Bayer Crop Science, as either human food or animal feed, and two GM carnations for sale as cut flowers. The GM crops obtained a positive scientific assessment by the European Food Safety Authority (EFSA). The EU Commission renewed the licenses for seven GM organisms (Waltz, 2015c). Transgenic soy obtained approval for import as a feed component of domestic animals (source: http://www.biotradestatus.com/). In Table 12.2 are listed all the GM plants approved for import in the EU.

Table 12.2: Genetically Modified Crops Approved in European Union.

Canola—*Brassica napus*		
Event Name	**Code**	**Trade Name**
GT73 (RT73)	MON-73-7	Roundup Ready™ Canola
HCN28 (T45)	ACS-BNØØ8-2	InVigor™ Canola
HCN92 (Topas 19/2)	ACS-BNØØ7-1	Liberty Link™ Innovator™
MON88302	MON-883Ø2-9	TruFlex™ Roundup Ready™ Canola
MS1 (B91-4)	ACS-BNØØ4-7	InVigor™ Canola
MS1 × RF1 (PGS1)	ACS-BNØØ4-7 × ACS-BNØØ1-4	InVigor™ Canola
MS1 × RF2 (PGS2)	ACS-BNØØ4-7 × ACS-BNØØ2-5	InVigor™ Canola
MS8	ACS-BNØØ5-8	InVigor™ Canola
MS8 × RF3	ACS-BNØØ5-8 × ACS-BNØØ3-6	InVigor™ Canola
RF1 (B93-101)	ACS-BNØØ1-4	InVigor™ Canola
RF2 (B94-2)	ACS-BNØØ2-5	InVigor™ Canola
RF3	ACS-BNØØ3-6	InVigor™ Canola
Maize—*Zea mays* L.		
Event Name	**Code**	**Trade Name**
TC1507 × 59122 × MON88017	Code: DAS-Ø15Ø7-1 × DAS-59122-7 × MON-88Ø17-3	Not available
TC1507	DAS-Ø15Ø7-1	Herculex™ I, Herculex™ CB
TC1507 × 59122	DAS-Ø15Ø7-1 × DAS-59122-7	Herculex XTRA™
T25	Code: ACS-ZMØØ3-2	Maize Liberty Link™
NK603	MON-ØØ6Ø3-6	Roundup Ready™ 2 Maize
NK603 × MON810	MON-ØØ6Ø3-6 × MON-ØØ81Ø-6	YieldGard™ CB + RR
NK603 × T25	MON-ØØ6Ø3-6 × ACS-ZMØØ3-2	Roundup Ready™ Liberty Link™ Maize
TC1507 × 59122 × NK603	DAS-Ø15Ø7-1 × DAS-59122-7 × MON-ØØ6Ø3-6	Herculex XTRA™ RR
TC1507 × NK603	DAS-Ø15Ø7-1 × MON-ØØ6Ø3-6	Herculex™ I RR
MON89034 × TC1507 × NK603	MON-89Ø34-3 × DAS-Ø15Ø7-1 × MON-ØØ6Ø3-6	Power Core™
MON89034 × TC1507 × MON88017 × 59122	MON-89Ø34-3 × DAS-Ø15Ø7-1 × MON-88Ø17-3 × DAS-59122-7	Genuity® SmartStax™
MON89034 × NK603	MON-89Ø34-3 × MON-ØØ6Ø3-6	Genuity® VT Double Pro™
MON89034 × TC1507	MON-89Ø34-3 × DAS-Ø15Ø7-1	Not available
MON89034 × TC1507 × 59122	MON-89Ø34-3 × DAS-Ø15Ø7-1 × DAS-59122-7	Not available
MON89034 × TC1507 × MON88017	MON-89Ø34-3 × DAS-Ø15Ø7-1 × MON-88Ø17-3	Not available
MON89034 × MON88017	MON-89Ø34-3 × MON-88Ø17-3	Genuity® VT Triple Pro™
MON89034 × 59122	MON-89Ø34-3 × DAS-59122-7	Not available

Table 12.2: Genetically Modified Crops Approved in European Union.—cont'd

Maize—*Zea mays* L.		
Event Name	**Code**	**Trade Name**
MON89034 × 59122 × MON88017	MON-89Ø34-3 × DAS-59122-7 × MON-88Ø17-3	Not available
MON89034	MON-89Ø34-3	YieldGard™ VT Pro™
MON88017	MON-88Ø17-3	YieldGard™ VT™ Rootworm™ RR2
MON87460	MON-8746Ø-4	Genuity® DroughtGard™
MON863 × NK603	MON-ØØ863-5 × MON-ØØ6Ø3-6	YieldGard™ RW + RR
MON863 × MON810 × NK603	MON-ØØ6Ø3-6 × MON-ØØ81Ø-6 × MON-ØØ863-5	YieldGard™ plus with RR
MON863 × MON810	MON-ØØ863-5 × MON-ØØ81Ø-6	YieldGard™ plus
MON863	MON-ØØ863-5	YieldGard™ Rootworm RW, MaxGard™
MON810 × MON88017	MON-ØØ81Ø-6 × MON-88Ø17-3	YieldGard™ VT Triple
MON810	MON-ØØ81Ø-6 Y	YieldGard™, MaizeGard™
MIR604 × GA21	SYN-IR6Ø4-5 × MON-ØØØ21-9	Agrisure™ GT/RW
MIR604	SYN-IR6Ø4-5	Agrisure™ RW
MIR162	SYN-IR162-4	Agrisure™ Viptera
GA21 × MON810	MON-ØØØ21-9 × MON-ØØ81Ø-6	YieldGard™ maize
GA21	MON-ØØØ21-9	Roundup Ready™ Maize, Agrisure™GT
Bt176 (176)	SYN-EV176-9	NaturGard KnockOut™, Maximizer™
BT11 × MIR604 × GA21	SYN-BTØ11-1 × SYN-IR6Ø4-5 × MON-ØØØ21-9	Agrisure™ 3000 GT
Bt11 × MIR604	SYN-BTØ11-1 × SYN-IR6Ø4-5	Agrisure™ CB/LL/RW
Bt11 × GA21	SYN-BTØ11-1 × MON-ØØØ21-9	Agrisure™ GT/CB/LL
Bt11 (X4334CBR, X4734CBR)	SYN-BTØ11-1	Agrisure™ CB/LL
59122 × NK603	DAS-59122-7 × MON-ØØ6Ø3-6	Herculex™ RW Roundup Ready™ 2
59122 × MON88017	DAS-59122-7 × MON-88Ø17-3	Not available
59,122	DAS-59122-7	Herculex™ RW
Potato—*Solanum tuberosum* L.:		
Name	**Code**	**Event**
EH92-527-1	BPS-25271-9	Amflora™
Soybean—*Glycine max* L.:		
Name	**Code**	**Event**
MON89788	MON-89788-1	Genuity® Roundup Ready 2 Yield™
MON87769	MON87769-7	Not available
MON87708 × MON89788	MON-877Ø8-9 × MON-89788-1	Not available

Continued

Table 12.2: Genetically Modified Crops Approved in European Union.—cont'd

Soybean—*Glycine max* L.:		
Name	**Code**	**Event**
MON87708	MON-877Ø8-9	Genuity® Roundup Ready™ 2 Xtend™
MON87705	MON-877Ø5-6	Vistive Gold™
MON87701 × MON89788	MON-877Ø1-2 × MON-89788-1	Intacta™ Roundup Ready™ 2 Pro
MON87701	MON-877Ø1-2	Not available
GTS 40-3-2 (40-3-2)	MON-Ø4Ø32-6	Roundup Ready™ soybean
DP356043	DP-356Ø43-5	Optimum GAT™
DP305423	DP-3Ø5423-1	Treus™, Plenish™
CV127	BPS-CV127-9	Cultivance
A5547-127	ACS-GMØØ6-4	Liberty Link™ soybean
A2704-12	ACS-GMØØ5-3	Liberty Link™ soybean
Sugar Beet—*Beta vulgaris*		
Name	**Code**	**Event**
H7-1	KM-ØØØH71-4	Roundup Ready™ sugar beet

Sources: EU Register of Authorized GMOs; GMO Compass. http://www.isaaa.org/gmapprovaldatabase/approvedeventsin/default.asp?CountryID=EU&Country=European%20Union.

GM plants containment. Another issue related to the release of GM plants is to avoid the gene transfer by impollination of other varieties. Severe containment procedures for planting the GM plants have been set up, or specific traits are introduced to control male sterility (MS) have been put in action in order to avoid crosses with non-GM plants and their wild relative species (cultivated rice and red rice, for instance). Terminator technology refers to research of seeds/plants that produce sterile seeds. While research of this technology has been conducted in conjunction with the US Department of Agriculture (USDA), no agricultural biotechnology company currently uses this technology to prevent gene flow between biotechnology and traditional crops.

Several traits and genes have been modified to obtain a desired phenotype: herbicide tolerance (HT) as weed control method; MS; agronomic (AG) trait genetic improvement for biotic stress (nematode resistance) and abiotic stress (drought, salt, heat, cold); product quality (PQ) traits include genetic improvement for quality, higher content of vitamins, essential amino acids, yield benefit, better processability with extended shelf-life, lower content in antinutritional factors and toxic factors such as acrylamide, modification in composition of starches and oils, and modification of cell walls for better biomass conversion in biofuels.

In the range of field trials, both for intended commercialization and research purposes, many projects were performed or are still continuing (e.g. potatoes with modified sugar content and inhibition of synthesis of acrylamide, changed composition of starch polysaccharides or resistance against infection, virus-resistant plum trees, flax with changed linseed oil properties, trees for bioremediation, resistance against diseases, growth acceleration, or changed technological properties of wood).

Plants can suffer from infections caused by fungi, bacteria, viruses, nematodes, and other pathogens. Various high-tech approaches have been proposed to protect plants from harmful afflictions, introducing genes conferring insect resistance (IR), combined HT and insect resistance (HT-IR), virus resistance (VR) for resistance against viruses, and resistance against bacterial diseases. To date, most interest has been focused on virus-resistant transgenic plants, but using biotechnology to confer resistance to fungi, bacteria, or nematodes has also been gaining attention.

Fungus resistant plants. Genetic engineering enables new ways of managing fungal infections. Several approaches have been taken: introducing genes from other plants or bacteria; encoding genes like *chitinase* or *glucanase*, producing enzymes that break down chitin or glucan, respectively, which are essential components of fungal cell walls; introducing plant genes to enhance innate plant defense mechanisms (e.g. activating phytoalexins, proteinase inhibitors, or proteins toxic to fungi); and involving the hypersensitive reaction. Plants varieties that are naturally resistant to specific types of fungal diseases are often programmed to have individual cells quickly die at the site of fungal infection. This response, known as the hypersensitive reaction, effectively stops an infection locally. Genetic engineering helps plant cells to know when a fungus is present, or may broaden and make adaptive a race-specific response such as an effector triggered immunity (ETI) response or may reinforce a pathogen associated molecular pattern (PAMP) until a strong innate immunity response is triggered.

Viral diseases. Viruses cause many economically important plant diseases. For example, the *beet necrotic yellow vein virus* (BNYVV) causes sugar beets to have smaller, hairier roots, reducing yields by up to 50%. The spread of most viruses is very difficult to control. Once infection sets in, no chemical treatment methods are available. Losses are usually very high and require longer rotation intervals and modified cropping systems. This translates into considerable losses. Viruses are often transmitted from plant to plant by insects. Insecticides are sometimes used to control viral infections, but success is very limited. The most effective ways of managing viruses are cultural controls (e.g. removing diseased plants) and using resistant cultivars. Although conventional methods of breeding have been able to provide some tolerant cultivars, they are not available for most crops.

Virus-resistant GM plants can be used to make virus-resistant crops. The most common way of achieving this objective was by transferring into the plant genome, the gene encoding the viral coat protein. The plant can then produce this viral protein before the virus infects the plant. This activates the phenomenon of co-suppression, shutting down the protein's expression. When the virus tries to infect the plant, the production of its essential coat protein is blocked. GM virus-resistant plants (e.g. papaya and squash) have coat protein mediated resistance. It may also be possible to confer resistance by taking a resistance gene naturally found in a wild plant species and then transferring it to the plant crop.

Insect resistance. Delta endotoxin is a natural insecticide, produced from *cry* genes in Bt, applied in organic farming to control to insect pests. The gene has been introduced in several GM plants to confer resistance to their pest insects. More than 100 different variations of Bt toxins have been identified in diverse strains of Bt. The different variations have different target insect specificity. The toxins classified under Cry1a group target Lepidoptera (butterflies), while toxins in the Cry3 group are effective against beetles.

In several countries, GM maize plants possessing increased resistance to insects have been approved, such as sweet corn Bt11×MIR162×GA21 by Syngenta (allowed in Argentina, Australia, New Zealand, Brazil, Canada, Colombia, Japan, Korea, Mexico, Philippines, Russian Federation, South Africa, US, Uruguay, and Taiwan), combining an IR gene (MIR162) associated to delta endotoxin expression and dwarf phenotype by suppression of gibberellin synthesis.

IR eggplant (Bt brinjal). Eggplant is a popular crop in the subtropics and tropics, especially in India and Bangladesh, where it is grown on about 1.5 million acres. Cornell University contributed to develop a pest-resistant eggplant, the first GM food crop in South Asia since 2009. Before that date, farmers have grown GM cotton in India since 2002. Eggplant (Brinjal) varieties cultivated in Bangladesh have been transformed with Bt delta endotoxin, with activity against insects, largely used in organic farming. The fruit and shoot borer (FSB), *Leucocinoides orbonalis*, is an insect that can cause crop losses as high as 70%. FSB showed to be reduced by 98% in eggplants transformed with the Bt endotoxin. The tiny larvae account for up to 40% of eggplant crop losses each year in India, Bangladesh, and the Philippines, and other areas of South and Southeast Asia. The first trial started in 2013 on 150 ha, and now has been extended to a larger number of farmers. The cultivation of GM eggplant requires 80% less chemicals. The eggplant itself has been sold on the market at a higher price, since it contains much less chemicals. It is estimated that the Bt eggplant can reduce insecticide use by 30% while doubling the yield of marketable eggplants. All the safety tests for the Bt eggplant have been conducted in India, starting in greenhouses and then moved to large-scale field trials. The eggplant has been found to be nontoxic to fish, chickens, rabbits, goats, rats, and cattle, as well as nonallergenic. Ongoing tests are examining whether the plant will continue to resist FSB in the field and for how long, whether the Bt eggplant cross-pollinates with other eggplants in the field, the distance required by Bt plants to be from other eggplant fields, whether nontarget insect populations are affected in the long term, and how yields compare with those of other eggplant varieties.

Product quality traits. Polyphenol oxidase (PPO) is the enzyme that causes browning in fruits and vegetables that require blanching for enzyme inactivation. The Arctic GM apple is modified by Okanagan Specialty Fruits, Canada, with a PPO targeting oligonucleotide that silences the PPO genes by RNA interference (RNAi). Arctic GM apple has been designed for the precut fruit market and is marketed with a packaging including the logo by Arctic (Waltz, 2015a).

Nonbrowning potato, produced by JR Simplot, Boise, Idaho, has been approved by USDA in November 2014. Fragments of a single potato PPO were reintroduced into potato, activating the RNAi pathway. Differently from apple, the potato double-stranded RNA is formed by an inverted repeat transcribed in the tuber and processed into small interfering RNAs. The potato was also modified to have reduced content of acrylamide through the silencing of the *asparagine synthetase-1* gene (*Asn1*) (Waltz, 2015b).

Agronomic traits. The exploitation of gene modification and use of antisense technologies have already translated in new plant varieties adapted to abiotic stresses. One such approach is the Senesco proprietary technology, used to produce transgenic plants and trees. Senesco targets the regulation of a polyamine synthesis and inhibition of eIF5A, an eukaryotic translation initiation factor that acts at a critical point in the process of cell death: eIF5A is a partner of proteins involved in endoplasmic reticulum (ER)-to-Golgi vesicle transport of the ribosome-bound nascent proteins (cotranslational translocation). A block in eIF5A activation upregulates stress-induced chaperones. By modulating the expression of eIF5A, the company obtains an increase in plant size, seed yield, shelf-life of perishable products, and in growth rates, with tolerance to environmental stresses. Target companies are those dealing with production of banana, corn, soybean, forestry species (cold adapted Eucalyptus trees), alfalfa, turfgrass, ornamental bedding plants, and plants for ethanol production. Recently, Senesco has gone through a rebranding process to a new company, Sevion Therapeutics (http://www.seviontherapeutics.com/technology/).

12.1.2 Novel Plant Products

Most regulatory authorities have shown to prefer modified plants devoid of unnecessary DNA, especially vector backbone sequences (VBSs), containing bacterial resistance genes and origins of replication (OR), and methods for genetic modification are perceived safer and are sustained by favorable opinion in a large majority of individuals in Europe. Simplified safety assessments are considered more appropriate in cases where there is no transmission of novel gene products to the food and no altered characteristics as a result of the genetic modification. The techniques that do not present a significantly greater food safety concern than other forms of mutagenesis are considered equivalent to standard transgenesis methods.

In the past, GM plants have fallen under USDA's regulation because they were made using plant pests, such as *Agrobacterium tumefaciens*, one of the most common delivery systems for shuttling foreign genes into plant genomes. Cauliflower mosaic virus, also on the plant pest list, is a popular tool among crop developers for its promoter, 35S, which is used to constitutively activate the transcription of transgenes. Scotts Miracle-Gro, Marysville, Ohio, used no plant pests in creating GM Kentucky bluegrass. Instead of using *Agrobacterium*, the company used to deliver novel genes, a gene gun method that accelerates DNA on pellets made of gold into plant cells through the cell walls. The gene used to confer herbicide resistance, *5-enolpyruvylshikimate-3-phosphate synthase* (*EPSPS*), was from *Arabidopsis thaliana,* while other genetic elements were from corn and rice. In July 2011, the USDA said an herbicide-tolerant variety of lawn grass fell outside its regulatory authority. The case represented the first time a large company has successfully taken advantage of a critical weakness in the regulation of GM plants (Waltz, 2011).

Novel plant products (NPPs) include GMOs that do not use traditional methods of genetic transformation (plasmids, markers for selection, antibiotic resistance) and may not contain stable transgenes (e.g. cisgenics, zinc-finger technology). Transient or secondary genetic alterations caused by the novel transformation process may still be detected.

Category 1: Transient introduction of recombinant DNA

Techniques that introduce recombinant DNA molecules transiently to plants are ZFNs introduced into the cell with or without a repair template (ZFN1 and ZFN2), ODM (Breyer et al., 2007; Pauwels et al., 2014), and agroinfiltration. These processes resemble transgenesis—in vitro synthesized nucleic acids and DNA delivery methods—but the end products are similar to, and indistinguishable from, plants obtained through conventional plant breeding. Therefore, the novel plant products (NPPs) are, in most cases, undetectable (Lusser et al., 2011). Definitions, according to the EU working group on new techniques, are based on the fact that ODM uses oligonucleotides for targeted (site-specific) induction of point mutations; ZFN1 generates site-specific random mutations by nonhomologous end joining (NHEJ), while ZFN2 generates site-specific desired point mutations by DNA repair processes through homologous recombination.

Agroinfiltration based on *Agrobacterium* vector allows inject of several DNA molecules contemporarily into the plant cells.

Category 2: Stable introduction of recombinant DNA during an intermediate step in the development of NPPs

Techniques that use stable GM intermediates include: ZFN1 and ZFN2, RNA-dependent DNA methylation (RdDM), and RB. Intermediate plants are GM plants, but the end products are similar to and indistinguishable from plants obtained through conventional plant breeding. Therefore, the NPP is, in most cases, undetectable (Lusser et al., 2011).

Definitions according to the EU working group on new techniques are consequent to ZFN1 and ZFN2 as already defined; and on the fact that RdDM is a technique based on the effect of small RNA sequences to alter gene expression through methylation of specific DNA sequences without changing the nucleotide sequence itself (epigenetic change). RB is able to reconstitute parental lines starting with an elite F1 hybrid whose genetic material is unknown. RB combines several other techniques, such as RNAi, to suppress meiotic recombination, tissue culture to regenerate plants from cells, and the double haploidization technique to create double haploid plants, which are used as the respective parental lines to produce new elite F1 hybrids.

Category 3: Stable integration of recombinant DNA

Integration-based plant breeding techniques include cisgenesis, intragenesis, grafting, and ZFNs (ZFN3). The process of generating cisgenic plants resembles transgenesis (random DNA insertion), but the product is similar to plants obtained through conventional breeding. Detection might be challenging. ZFN3 technique targets delivery of transgenes (insertions) by homologous recombination: cisgenic scab-resistant apple (Vanblaere et al., 2014) and herbicide-resistant oilseed rape produced by targeted mutagenesis, and new plant products (NPPs) obtained by using these techniques have been developed.

Cibus Genetics LLC has patented a Rapid Trait Development System (RTDS) based on ODM technology that has led to herbicide-tolerant canola. The conversion of an imidazolinone-sensitive plant to an imidazolinone-tolerant plant was achieved by introducing a single nucleotide change (G to A at nucleotide 1958) in *acetolactate synthetase* (*ALS*) gene, which converts a serine to asparagine at amino acid position 653 of the enzyme.

The SSN-1 and SSN-2 techniques can introduce subtle modifications, such as small deletions and single-base substitutions of target genes. The final plants derived by SSN-1 and SSN-2 techniques are similar to natural variants or those produced by physical or chemical mutagenesis in conventional breeding. SSN-3 techniques allow scientists to insert foreign genes at predefined sites, and this site-specific gene addition should prevent the "position effects" associated with random insertion of genes into plant genomes. The delivery of SSN DNAs using *Agrobacterium* or other delivery methods requires that the SSN DNAs integrate at different loci from the target loci; hence these foreign SSN DNAs can be easily eliminated from the genome during the segregation and recombination accompanying sexual reproduction. The final plants generated by SSN-1 or SSN-2 should fall outside the existing definitions and regulation affecting GM crops (Lusser et al., 2012; Podevin et al., 2013). Thus, SSN techniques have great advantages over existing transgenic breeding techniques.

The results of a written survey of a number of plant biotech companies revealed that ZFN technology had been used in breeding of maize, oilseed rape, and tomato (Lusser et al., 2012). Also, Dow AgroSciences has received assurance from the USDA that their GM corn developed by the ZFN technique will not require regulatory oversight. Evaluation of risk assessment will still be required before commercialization to ensure food and environment safety (Chen and Gao, 2014).

Potential uses of site-directed nucleases (SDNs) for plant genome editing are very large. DNA shuffling can generate sequence variation, producing proteins with desirable properties (kinetics, substrate specificity, temperature or pH optimum, and ligand binding) including improved disease resistance (e.g. R-gene shuffling). SDNs can target the most effective changes required.

Furthermore, different techniques such as Targeting-Induced Local Lesions in Genomes (TILLING), EcoTILLING, and next-generation sequencing (NGS)/whole genome sequence analysis have been applied to identify candidate genes and sequences polymorphisms that can be used in SDN applications for targeted phenotype development.

There are many examples that show the potential applications of SDN (Podevin et al., 2013): (1) VR (targeting translation initiation factors); (2) HT (target: *ALS* gene); (3) lowering antinutritional compounds (erucic acid in Brassicas, targeting fatty acid elongases) and allergens (target: conglutin genes in peanut; *Mal d 1*, an apple pathogenesis response protein allergen) (Gambino and Gribaudo, 2012); (4) improved nutritional value via elevated carotenoids, modified carotenoid balance (target: zeaxanthin epoxidase); (5) modified starches and fats for food and nonfood uses (targeting starch synthases, branching enzymes, and fatty acid desaturases); (6) longer shelf-life/reduced wastage (targets include aminocyclopropane (ACC) oxidase and polygalacturonase); (7) improved quality by reducing enzymic (target PPOs) and nonenzymic browning (high-quality and low-acrylamide potato, target invertase genes); (8) yield benefits via modified RuBisCO genes, increasing catalytic activity and/or decreasing oxygenation activity and improved seed set (e.g. by targeting homeodomain leucine zipper genes); and (9) improved biomass conversion for biofuels (lower lignin, target caffeic acid O-methyltransferase gene).

Food Standards Australia New Zealand (FSANZ) organized a meeting by an expert scientific panel to provide advice on a number of new plant breeding techniques that have come to the attention of regulators (New Plant Breeding Techniques. Report of a Workshop hosted by Food Standards Australia New Zealand, 2012). As a result of the panel discussion, the techniques were grouped into three categories.

Category 1 comprises cisgenesis/intragenesis, targeted gene addition or replacement using ZFN technology, and GM rootstock grafting. It was the view of the panel that foods produced using these techniques should be regarded as GM food and undergo premarket safety assessment.

In the case of cisgenesis/intragenesis and targeted gene addition or replacement using ZFN technology, the derived food would be similar to that produced using standard transgenic techniques. Consideration of GM rootstock grafting was more complicated because food produced by a non-GM scion grafted onto a GM rootstock would not contain any introduced DNA. However, it may contain novel RNA and/or protein as a result of the genetic modification to the rootstock. Depending on the genetic modification, the food may also have altered composition or other characteristics. The panel did, however, note the following:

- In the case of cisgenesis and intragenesis, a simplified form of food safety assessment may be warranted because the transferred genes will be derived from the same or a closely related species, which is likely to be commonly used as food and have a history of safe use;
- In the case of GM rootstock grafting, the majority of foods will not contain any novel genetic material or have altered characteristics and therefore should only require a simplified food safety assessment.

Category 2 comprises techniques used for targeted mutagenesis, including ODM and ZFN technology. It was the view of the panel that changes introduced using such techniques would be typically small and definable, and have predictable outcomes. Such techniques would therefore be similar to traditional mutagenic techniques used in conventional plant breeding, and food derived from these plants should not be regarded as GM food.

Category 3 comprises techniques that involve the use of gene technology at an early stage, which is separate from the final plant breeding process. The techniques in this category include SPT and RB.

SPT was developed by Pioneer Hi-bred International for use in corn to improve the efficiency of hybrid seed production. It involves using a GM plant line to propagate a male-sterile plant line, which is then used as one of the parents to produce hybrid seed. The genetic modification is not inherited by the hybrid plant line.

RB is a novel plant breeding technique that involves suppressing meiotic recombination in order to recreate homozygous parental lines that, once hybridized, reconstitute the composition of an elite heterozygous plant without the need for backcrossing or selection.

Although not specifically discussed, the panel noted that accelerated breeding following induction of early flowering using gene technology could also be included in this category. In the case of SPT, the panel was of the view that food produced using this technique should not be regarded as GM food, as a genetic separation exists between the early GM ancestor (known as the GM maintainer line) and the non-GM parents of the final food-producing line, which does not contain the genetic modification. The panel considered, however, that it would be useful to have more information confirming the reliability of the sorting technique for indicating the presence or absence of the introduced genes, as well as general compositional analysis confirming the equivalence of an F1 hybrid produced via SPT with a standard F1 hybrid.

While there are clear parallels with SPT, the panel did not reach a conclusions about RB because insufficient technical detail was available on how transgene-free end products are produced, as well as the reliability of the process overall. They noted that there are not any particular hazards associated with the GM component of the technique. The panel also considered that it would be helpful to develop some criteria for distinguishing techniques such as SPT, accelerated breeding, RB from those where the final food-producing lines are clearly GM, and also for ensuring that a complete barrier/genetic separation exists between the early GM breeding lines and the non-GM food-producing lines.

The panel concluded the following:

1. ZFN3, which involves introducing a new gene into a specific site in the genome, is equivalent to standard transgenesis.
2. ZFN1 and ZFN2 are mutagenic techniques that are conceptually similar to ODM. Such techniques do not present a significantly greater food safety concern than other forms of mutagenesis. The changes introduced using ZFN1 and ZFN2 will be small and definable and the outcomes predictable. Food derived from plants modified using ZFN1 and ZFN2 would be similar to food produced using traditional mutagenic techniques, and should therefore not be regarded as GM food.

12.1.3 Gene Modification Using TALEN and Zinc-finger Nucleases, CRISPR/Cas9 Nuclease System

Engineered nucleases and technologies applied to nuclease-based gene targeting (NBGT) are posing specific and novel safety and regulatory considerations, shifting the concerns from the organisms that are not containing external genetic material, to novel foods that have not had a tradition of consumption (EC Directive 258/97), thus not being supported by a long and sustained use (Lusser and Davies, 2013; Pauwels et al., 2014).

The critical step in NBGT is the introduction of DNA double-strand breaks (DSBs) at given genomic sites. It was soon discovered that engineered nucleases could generate DSBs and consequently activate DNA repair to seal the breaks, along with any modifications such as mutations, insertions, replacements, and chromosomal rearrangements.

The production of single and double breaks in the genome induces the DNA repair machinery through homology-directed repair (HDR). Nuclease-mediated DNA strand breaks can increase the homologous recombination (HR), favoring it in respect to NHEJ that is a predominant pathway in most higher eukaryotes. A possible approach to obtaining targeted integrations is to provide a donor with 5′ overhangs, complementary to those created by nuclease cleavage (Pauwels et al., 2014).

The EFSA published a scientific opinion to address the safety assessment of NPP produced by SDNs (EFSA, 2012a).

Gene modification by ZFNs, TALENs (Scholze and Boch, 2011; Deng et al., 2012; Mak et al., 2013; Mussolino and Chatomen, 2012; Wei et al., 2013 a,b), and the CRISPR/CRISPR-associated

Cas9 nuclease (Gaj et al., 2013; Upadhyay et al., 2013; Chu et al., 2015) is based on methods that make use of DNA endonucleases associated to oligonucleotide guides. Genome editing technologies using ZFNs, TALENs, and the CRISPR/Cas9 system have been applied to plants to generate genome modifications at defined positions in the genomes, based on DNA sequences complementary to the desired DNA regions (Jiang et al., 2013; Belhaj et al., 2013; Nekrasov et al., 2013; Hsu et al., 2013; Xie and Yang, 2013, Chen and Gao, 2014; Endo and Toki, 2014; Xu et al., 2015). Several improvements have been obtained to avoid unspecific genome editing, through mutated cas nucleases, and through inhibition of NHEJ, one of the main causes of undesired DNA modifications; NHEJ inhibition was achieved through removal of histone-like H2A.Z protein (Alatwi and Downs, 2015), through inhibitors of NHEJ key molecules Ku70, Ku50, and DNA ligase IV (Maruyama et al., 2015), or co-expressing Adenovirus four protein E1B55K, which mediates DNA ligase ubiquitination (Chu et al., 2015). These pathways, being blocked in the DNA deleting activity, allowed only DNA repair through homologous recombination (HR) mechanism by perfect matches with oligonucleotide guide and its insertion into the sequence to be edited. These new technologies are robust, affordable, and easy to engineer, and have been recently applied to plant engineering.

When introduced to cells via cDNAs, ZFNs create a DSBs at the mutation site, allowing the cell to replace the mutated DNA with a desired DNA sequence, also provided to the cell on a plasmid. But ZFNs often cut areas of the genome in addition to the intended site, an undesirable effect that needs to be avoided in order to introduce only the desired gene editing sites. A paper by Chen et al. (2013) presented an innovation in the method for delivering ZFNs to cells: to produce a ZFN as a protein by attaching it with a disulfide linker to the transferrin ligand, showing that this approach should enhance endocytosis of the ZFN. The idea was that the reducing environment inside the cell would split the disulfide bond in the linker, releasing the ZFN from the transferrin ligand. Once the protein reaches the endosome, the ZFN protein moves into the cytosol, the effect being ascribed to the ZFN's high positive charge. From there, the ZFN is targeted to the nucleus, where it cleaves a specific sequence of DNA. The gene correction activity was almost as high as in cells that had been treated with zinc-finger cDNA, while their method of delivery was highly efficient, providing better temporal control and control of the ZFN quantity delivered while maintaining roughly the same cutting efficiency as a ZFN expressed from cDNA. The Chen team is looking at other receptor-mediated pathways that are more specific to the cells used, experimenting to see whether the method could be used for other nuclease systems such as TALENs (Tsuji et al., 2013), which may have some competitive advantages due to a greater cutting efficiency and less off-target activity.

Recent advances in the study of the prokaryotic adaptive immune system, involving type II clustered, regularly interspaced, short palindromic repeats (CRISPR), provide an alternative genome editing strategy. Type II CRISPR systems, widespread in bacteria, make use of an endonuclease, CRISPR-associated protein (Cas9) to defend against invading viral and plasmid DNAs. Cas9 can form a complex with a synthetic single-guide RNA (sgRNA), consisting of a fusion of CRISPR RNA (crRNA) and *trans*-activating crRNA. The sgRNA guides Cas9 to recognize and cleave target DNA. Customizable sgRNAs directing Cas9 have been shown able to induce sequence-specific genome modifications in rice and common wheat (Shan et al., 2013). They reported the application of CRISPR-Cas-mediated genome editing to wheat (*Triticum aestivum*), the most important food crop plant with a very large and complex genome. The mutations were targeted in the inositol oxygenase (inox) and phytoene desaturase (pds) genes using cell suspension cultures. The expression of chimeric guide RNAs (cgRNA) targeting single and multiple sites resulted in insertion/deletion (indel) mutations in all the tested samples. The expression of Cas9 or sgRNA alone did not cause any mutation. The expression of duplex cgRNA with Cas9 targeting two sites in the same gene resulted in deletion of DNA fragment between the targeted sequences. Multiplexing the cgRNA could target two genes at one time.

Inhibition of NHEJ and an increase of the efficiency of precise genome editing with a modified Cas9 D10A substitution, which regulates the frequency of target gene modifications, allowed a reduction of off-target modifications by several fold (Maruyama et al., 2015), while conditional expression by doxocyclin tetracycline-inducible cDNAs (Dow et al., 2015; Mahiny et al., 2015) and short hairpin RNAs within a recombinase-mediated cassette exchange approach limited the duration of Cas9 expression.

Investigators are presently using the CRISPR/Cas9 protocol for reverse genetics, although the utility of the gain-of-function approach is often limited by suboptimal levels of transcriptional activation at some endogenous loci. Strikingly, Konermann et al. (2015) reported structure-based engineering of the CRISPR/Cas9 complex with optimized synergistic activation mediators, which resulted in higher efficiency and robustness in transcriptional activation than any previous designs. The success of simultaneous multiple-gene activation further demonstrated their novel Cas9-mediated transcription activation system to be an approach for interrogating gene networks involved in any cellular process.

12.1.4 Cisgenesis and Intragenesis

The main aims in the efforts to create a new GMP are to accelerate the breeding process and to address consumer concerns about GMPs, which are partly based on a perceived lack of "naturalness," by creating plants that could also have been obtained by conventional breeding. Although naturalness is a controversial concept, plant products produced by conventional breeding are more familiar to consumers. One concern about transgenic crops relates to the mixing of genetic materials between species that cannot hybridize by natural means. To overcome the interspecies differences in gene content, two new transformation approaches, cisgenesis and intragenesis, were developed as alternatives to transgenesis. These approaches imply that plants must only be transformed with genetic material derived from the species itself or from closely related species capable of sexual hybridization.

Cisgenesis seeks to maintain the familiarity of NPPs in the public opinion by relying on the existing genetic variation in the breeders' gene pool.

Recently, the European Food Safety Authority (EFSA) published a scientific opinion to address the safety assessment of plants developed through cisgenesis and intragenesis (EFSA, 2012b).

Definitions according to the EU working group on new techniques:

Cisgenesis is genetic modification of a recipient organism with a gene (cisgene) from a crossable—sexually compatible—organism; intragenesis is genetic modification of a recipient organism that involves the insertion of a reorganized, full, or partial coding region of a gene combined frequently with a promoter and/or terminator from another gene of the same species or a crossable species.

Cisgenic NPPs seek to maintain the consumers familiarity by relying on the existing genetic variation in the breeders' gene pool. Overall, cisgenic products are perceived to be more natural, less problematic for the environment, and generally safer and more promising. Another major premise for the development of NPPs is the regulation of GMPs. Cisgenic NPPs could fall outside the definitions of GMPs in some jurisdictions and might not be subjected to regulatory oversight beyond that applied to other conventionally bred plants.

Pastoral Genomics in New Zealand has registered the trademark Cisgenics and uses this trademark for their future GM ryegrass (Bajaj et al., 2008, 2010). The DNA introduced in the modified plants is not completely in agreement with the cisgenesis definition, since the plant-derived P-DNA borders have been introduced using a vector backbone of bacterial origin, so that in some countries, a definition of intragenesis may be applied.

These NPPs include cisgenic potatoes with a higher content of amylopectin for industrial applications, or with improved resistance to pathogens such as *Phytophthora infestans*, as well as pest-resistant plants grown on a GM rootstock.

12.1.5 Genetically Modified Rootstock Grafting

Grafting a non-GM scion onto a GM rootstock results in a fruit that does not contain the insert.

Overview of the technique. Grafting is used extensively in many plant-breeding programs and is typically done to modify plant architecture or to counter biotic and abiotic stresses. The lower part of the plant, which contributes roots and support, is called the rootstock and the upper part of the plant, contributing stems, leaves, flowers, and fruit, is called the scion.

The rootstock plus the scion can be regarded as a composite plant, which essentially functions as a single organism with one genotype in the rootstock and another genotype in the scion.

A graft union, which consists of a small mass of callus tissue, physically joins the two parts. For many years, the graft union was believed to be separate tissue that would act as a selective filter, but it is now known there is a vascular connection between the rootstock and scion through which there is a one-way movement of water and soluble mineral nutrients from roots to leaves via the xylem and a two-way movement of photosynthates and various macromolecules (RNA and proteins but not DNA) via the phloem. On each side of the graft union, all the components of the phloem and xylem are the same and in the same concentration. The graft union is, therefore, not a selective barrier, nor is it a source of other solutes moving up or down the plant.

The main points of the debate on root-to-scion transfer are summarized below.

- The food produced by a non-GM scion grafted onto a GM rootstock does not contain any modified DNA. However, the evidence indicates there could potentially be novel gene products (such as RNA or proteins) moving from the rootstock into the scion and potentially also into food products (such as the fruit). In some cases, the genetic modification to the rootstock may be done to intentionally alter characteristics in the food product. These changes, transmitted via the rootstock, would not be heritable through the seed, which is produced in the non-GM scion. Some examples where a genetic modification to the rootstock could alter food characteristics include the use of RNAi to produce non-browning apples or seedless fruit. Most of the applications of GM rootstock grafting would not be of that type, however, and the presence of novel gene products in the scion, should it occur, would typically not alter the characteristics of the food.
- Although the rootstock and the scion exist together as a composite plant consisting of two different genotypes, the grafted plant essentially functions as a single organism with both the rootstock and the scion being connected by a single vascular system. As a grafted plant can essentially be regarded as a single organism, a plant with a GM rootstock should therefore be regarded as a GMO.
- Where there is transmission of GM material to the food-producing parts of the plant or the modification of the rootstock is intended to target the fruit or other edible products derived from the scion, there is a need for regulatory oversight to ensure any potential human health risks associated with consumption of the food are adequately assessed.
- The reviewing departments, according to regulation of GMOs, have to undertake an assessment of risks to human health and safety and the environment.
- The reviewing departments can consider a dedicated approach to the assessment of foods derived from GM rootstock plants that reflects differences in potential food safety concerns, depending on the genetic modification in the rootstock. While a safety assessment would be essential in situations where there is transmission of GM material to the food or where the trait was intended to modify the food, a simplified process could be adopted for the majority of cases where there are no detectable effects or differences in the food compared with that from a conventional plant.

- The reviewing departments could consider whether assessment and approval of a particular GM rootstock would, in certain cases, be sufficiently protective of public health and safety to justify excluding the non-GM scion from assessment. This approach would ensure food safety while eliminating the costs associated with unnecessary assessments of multiple non-GM scions. In these cases, the developer of the GM rootstock would be responsible for generating data and seeking regulatory approval.

Rootstock-to-scion transfer of transgene-derived small interfering RNAs (siRNAs) was shown to affect VR in nontransgenic sweet cherry (Zhao and Song, 2014). The study provided the foundation for using transgenic rootstocks to produce products of nontransgenic scions in fruit trees. siRNAs are silencing signals in plants. Virus-resistant transgenic rootstocks developed through siRNA-mediated gene silencing may enhance VR of nontransgenic scions via siRNAs transported from the transgenic rootstocks. However, convincing evidence of rootstock-to-scion movement of siRNAs of exogenous genes in woody plants is still lacking. To determine whether exogenous siRNAs can be transferred, nontransgenic sweet cherry (scions) was grafted on transgenic cherry rootstocks (TRs), which were transformed with an RNAi vector expressing short hairpin RNAs of the genomic RNA3 of *Prunus necrotic ringspot virus* (PNRSV-hpRNA). Small RNA sequencing was conducted on RNA from bud tissues of TRs and RNA from grafted (rootstock/scion) trees. Comparison of the siRNA profiles revealed that the PNRSV-hpRNA was efficient in producing siRNAs and eliminating PNRSV in the TRs. Furthermore, the study confirmed the long-distance transfer of PNRSV-hpRNA-derived siRNAs from the transgenic rootstock to the nontransgenic scion in woody plants, and showed that the transferred siRNAs enhanced PNRSV resistance of the scions grafted on the TRs.

In a report of a workshop hosted by FSANZ (New Plant Breeding Techniques. Report of a Workshop hosted by Food Standards Australia New Zealand, 2012), the FSANZ panel concluded with the guidelines to be followed for these types of NPPs.

12.1.6 Other Plant Improvement Techniques

Gene overexpression, gene disruption, gene silencing, RNAi, targeting specific mRNAs by inducible overexpression of microRNAs, and "miRNA sponges" sequestering specific microRNA species are further potential tools to control gene expression, and will take stage in the next years. The exploitation of RNA silencing and antisense technologies for controlling gene expression have been already translated in new plant phenotypes and tree populations with novel traits.

Gene silencing or gene overexpression technologies deploy at first a plasmid vector to deliver the gene and integrate it, but during the final breeding process, the plants carrying the inserted transgene are segregated out. Double haploid plants screened for the absence of RNAi constructs produce by breeding plants devoid of GM-related DNA sequences (Lusser and Davies, 2013).

In the frame of COST (European Cooperation in Science and Technology) action FA0806: "Plant virus control employing RNA-based vaccines: a novel non-transgenic strategy," research has been carried on using gene silencing approaches to control plant virus spread. This has produced efficient and cost-effective methods for reactive and proactive response to viral diseases of plants, and optimized protocols for production and delivery of suitable resistance inducer molecules.

Epigenetics is currently recognized as one of the most exciting fields of modern science; there is an urgent need for considering a role for the EU research community with respect to this discipline. It is expected that epigenetics will have a positive impact of the advances in agriculture and food biotechnology.

MicroRNAs. Overexpression of OsmiR397, a microRNA from rice in the same species has led to tan increase in grain yield (Zhang et al., 2013). Therefore, it is possible that in the future, other microRNAs could be upregulated in order to improve agronomic and quality traits.

Recently Carol Auer summarized the state-of-the-art of plant biotechnologies with a special focus on new approaches based on small RNAs, RNAi, and production of RNA-mediated traits in plants (Auer, 2011). The potential of RNA-regulated traits in nonfood plants and biofuel producing plants is well acknowledged. Accordingly, new methods for risk analysis are required to perform analyses of off-target effects and persistence of RNAs in the environment.

12.1.7 Accelerated Transformation of Woody Plants and Early Flowering

In wood tree, fruit tree, and stone fruit tree research, the major difficulty is to shorten the long juvenile phase to accelerate the selection of recombinant clones, where phenotypes can be observed only at flowering or fruiting time (Gambino and Gribaudo, 2012). The incorporation of flowering genes may result in reduction of generation time and in juvenile phenotypes. For example, *BpMADS4* has been overexpressed in birch (*Betula pendula* Roth.), *APETALA1* (*AP1*) and *LEAFY* genes have been expressed in citrus. *TERMINAL FLOWER* (*MdTFL1*) gene from apple, an inhibitor of flowering that maintains the identity of inflorescence shoot meristems, was introduced as an RNAi cassette targeting the native pear genes *PcTFL1-1* and *PcTFL1-2* (Freiman et al., 2012), producing a transgenic line of Spadona variety, Early Flowering Spadona (EF-Spa). Pollination of EF-Spa trees generated normal-shaped fruits with viable seeds. F1 seedlings formed shoots and produced flowers 1–33 months after germination and were found devoid of transfer DNA (T-DNA), being recognized to be nontransgenic. *FLOWERING LOCUS T 1* (*MdFT1*) gene from apple was positively used to produce an early flowering phenotype in apple (Kotoda et al., 2010), pear, and poplar without adverse pleiotropic effects such as morphological changes in flowers (Gambino and Gribaudo, 2012).

A cisgenic apple, resistant to apple scab, has been produced by insertion of the *Rvi6* scab resistance gene into the Gala variety (Vanblaere et al., 2014).

Exploiting the properties of modification of tree life cycle, genetically engineered resistance to *Plum pox virus* (causative agent of sharka disease) infection was produced in stone fruit trees (peach, plum, apricot) (Ilardi and Di Nicola-Negri, 2011).

FuturaGene, a Brazilian forestry company, signed recently an agreement with the Donald Danforth Plant Science Center in Saint Louis to exploit their agriculture biotechnology (http://www.futuragene.com/en/default.aspx), already tested in GM eucalyptus and poplar, to boost plant biomass levels, improve crop adaptation to climate changes, and facilitate processing for animal feed in strategic crops. The technology is based on the endo-β-1,4-glucanase *CEL1* gene from *Arabidopsis*, with activity in cell wall metabolism. Expressing the gene produces a relaxation of the crystalline matrix of cells facilitating cell expansion. The regulatory trails are envisaged for eucalyptus (ongoing), the model grass Setaria, and subsequently millet, sorghum, and cassava. The FuturaGene Scientific Advisory Board is chaired by Marc Van Montagu, recent winner of the 2013 World Food Prize.

In Brazil, FuturaGene's GM eucalyptus was recently approved by the Brazilian National Technical Commission on Biosafety, April 2015. FuturaGene merged with Rehovot, Israel, in 2006, and is now owned by Sao Paulo-based Suzano Pulp and Paper company. GM trees produce 20% more wood and are ready for harvest in five and half years instead of seven.

12.2 Non-GMO Methods for Plants Improvement: Mutagenesis

Plant products produced by conventional breeding are more familiar to consumers. In the last twenty years, plants with mutated genes have been introduced in agriculture, since their production confers the ability to eliminate weeds or to grow them in adverse conditions (excessive salinity, adverse climate conditions). Many gene-editing events have been the result of random mutagenesis.

The induction of mutations using various types of mutagenic agents has been used worldwide for crops and trees. At the beginning of 2000, the number of mutated varieties was around 2252 (FAO, 2000), including varieties obtained by crosses with mutated varieties, with cereals summing up to 1172 cases, making the largest part of the total number.

In Italy, 35 varieties obtained by induced mutagenesis have been released to market, among them, 13 varieties of wheat (*Triticum turgidum sbsp. durum*) (Creso variety), 6 for pea, 2 for *T. aestivum*, 3 for eggplant, 1 for olive tree, 1 potato, and 1 for rice (Fulgente variety in 1973, obtained from Maratelli by X-ray mutagenesis).

According to the United Nations Food and Agricultural Organization (FAO), a great number of rice mutated varieties, 434 at the beginning of this century, were present in Italy, Asia, US, Australia, Egypt, and South America.

Considering that the new characters have been obtained through selection and genetic improvement starting from mutant varieties crossed with well-performing ones, the general opinion is that these plants are not transgenic, since any gene, or DNA sequence, from other genomes has been introduced but has passed through an accelerated evolution process.

Therefore, it is at the attention of authorities and regulatory framework that when plants have maintained genome integrity and only one nucleotide substitution is cause of the gene-based phenotype, these plants are assimilated to those obtained by advanced breeding methods. The same equivalence should be granted, or the methods will be assimilated to accelerated evolution and enhanced breeding-like processes when methods for precise editing of genomes, such as CRISPR-cas system, TALENs , and ZFNs will provide more precise intervention protocols without any other change in plant genomes.

12.2.1 Insertional Mutagenesis

An important and direct approach to defining the function of a novel gene is to abolish or activate its function by mutagenesis with physical or chemical mutagens and modern tools such as targeting-induced local lesions in genomes (TILLING) and NGS.

Transposons are genetic mobile elements that occasionally move from one DNA position to another, causing the inactivation of genes, and are based on a transposase and long terminal repeat (LTR) sequences. Insertional mutagenesis, with T-DNA (the transfer DNA on *Agrobacterium* plasmids) or a transposable element, provides opportunities for assigning a function to a particular DNA sequence and isolating the target gene, causing a specific phenotype.

Ds and Spm maize transposable elements have been introduced in other crops to create lines carrying new transposon inserts. The maize Ac/Ds transposon system has been used to generate an insertional mutant population in maize itself, *Arabidopsis*, and rice. An Ac/Ds-based library has several advantages: (1) revertants can be readily obtained and easily identified; and (2) Ds elements prefer to transpose to genetically linked sites (i.e. the same chromosome).

Ac/Ds belong to the hAT super family, with the designation hAT from the *Drosophila* element hobo, maize element Ac, and *Antirrhinum majus* Tam3 element. Ac, Ds, and Spm are introduced by *Agrobacterium*-mediated transformation, and the regenerated rice lines may exhibit somaclonal variation, especially in the first generation, since the variation will tend to be diluted with no further changes in the offspring, once the mutants are crossed with the Ac lines and insert population amplified. New rice hAT elements have been found and are suggested as new candidates to generate insertion mutants in rice. An active 0.6-kb endogenous DNA transposon, nonautonomous DNA-based active rice transposon1 (nDart1), was recently identified to act as a causative fragment. One drawback is the GM nature of the lines obtained using transposons from other species. Though belonging to the same family as Ac/Ds, the use of hAT elements for rice mutagenesis is not subject to the concerns on GM problematics because they are endogenous elements in the rice genome, and their mobilization does not necessitate callus formation. Indeed, the transposition of nDart1 can be triggered by ordinary crossing under natural field conditions. As for other transposons, the remobilization of the element generates a revertant.

12.2.2 Chemical Mutagenesis

Various chemical and physical agents have been applied to induce mutants in plant DNA. Both chemical and ionizing radiation mutagenesis have been routinely used to generate genetic variability in rice varieties.

Among the chemical methods for mutagenesis, ethyl methane sulfonate (EMS) has been widely used in plant studies. In Japan, researchers applied EMS mutagenesis to rice immature embryos immediately after fertilization, and the embryos were allowed to develop to seeds. These seeds were sown to generate M1 plants, in which the majority of EMS mutations were expected to be in the heterozygous state (Fekih et al., 2013; Takagi et al., 2015).

Using this approach, RIKEN leaded researchers have developed a mutant rice variety that is able to grow on salt contaminated fields, possessing a high salt tolerance (Takagi et al., 2015). This will allow reintroduction of rice cultivation in areas touched by the tsunami in the North Kanto area.

MutMap is a whole genome sequencing (WGS)-based method that has been applied to accelerated breeding of a salt-tolerant rice cultivar (Takagi et al., 2015). The gene conferring the improved salt resistance has been identified as *OsDSS1* (Tamiru et al., 2015).

Among herbicides used to control weeds, the family of imidazolinones is widely used for the low induction of tolerance, often used after appearance of resistance to glyphosate, metolachlor, or imazethapyr. The imidazolinones family consists of six active ingredients, such as imazamox; each of which controls a different spectrum of weeds. Imidazolinones are active at low dosage on most weeds, have low toxicity to animals, inhibit plant acetolactate synthetase (ALH), also known as acetohydroxy acid synthetase (AHAS), an enzyme required for the production of essential branched chain amino acids such as valine, isoleucine, and leucine. Several crops have been mutagenized with EMS to obtain imidazolinones-tolerant mutants. The technology presently registered by BASF, Clearfield, exploits these pioneering studies by various universities. After being licensed the intellectual property rights, BASF has launched its portfolio of Clearfield mutant crop varieties. The current Clearfield crops (maize, canola, rapeseed, rice, sunflower, wheat, and lentils) have been developed using enhanced plant breeding methods. The Clearfield Production System is one of only a few herbicide-tolerant systems recognized as nontransgenic by international authorities. This provides farmers with an effective agronomic tool along with global market acceptance. The Clearfield Production

System uses several different herbicides. Each formulation is custom designed to provide exceptional contact and soil activity for broad spectrum, maximum control of the weed species most likely to plague specific crops in each region, while still offering crop rotation flexibility. Elite seed varieties have been developed to tolerate Clearfield herbicides, so the crops thrive as weeds wither and die.

The Clearfield production system for rice

Clearfield rice seed is a nontransgenic, non-GM crop for rice production devoid of weeds, developed with traditional plant breeding techniques. The rice varieties CL 161 and CFX 18, possessing AHAS with an asparagine substitution at position 653, have been crossed with commercial varieties to produce the hybrid. Libero (*Oryza sativa indica*) is a hybrid representing 10% of the rice cropped in Italy. In the US, 300,000 ha are cultivated using Clearfield rice, over 25% of rice cultivated surface in the mid-South (Louisiana, Arkansas, Missouri, Mississippi, Texas), and its use has spread also in Brazil, Colombia, Uruguay, Argentina, and Costa Rica. Rice Clearfield hybrids, cultivated following the Clearfield® technology by BASF, are tolerant to Newpath® herbicide, Clearpath™ herbicide, and Beyond® herbicide.

Oryza sativa (L.) var. *sylvatica* is the main weed contaminating paddy fields. Due to the early and casual seed maturation and dropping into the soil, it is able to scatter around and take the lead at the expense of cultivated rice. To preserve the long-term efficacy of the Clearfield rice technology, certain stewardship practices must be followed, and field evaluation has been assessed (Shan et al., 2007). Clearfield rice producers are asked to help protect and prolong the usefulness of this technology by following specific requirements and recommendations to help prevent weed resistance and gene flow from rice to red rice (Shivrain et al., 2007). When used in a planned sequential program, the Clearfield Production System for rice provides the broadest spectrum control of some of the toughest rice weeds, including red rice, barnyard grass, broadleaf signalgrass, eclipta, hemp sesbania, northern joint-vetch, and many more. Clearfield rice producers are required to help protect and prolong the usefulness of the technology by following specific requirements and recommendations to help prevent weed resistance.

The Clearfield production system for wheat

Winter wheat ranks high in importance as an agricultural crop in the Pacific Northwest states of Washington, Idaho, and Oregon. These states rank 3rd, 7th, and 18th in US winter wheat production with a total estimated value of production of over $633 million in 2000 (National Ag. Statistics Service, 2001). Winter wheat is a winter annual grass that is planted and emerges in the fall, overwinters as a small plant, grows fast and develops tillers in the spring, and is harvested in July and August. Winter annual grass weeds such as jointed goatgrass, downy brome, feral rye, and Italian ryegrass have the same growth cycle as winter wheat and are difficult to control in conventional wheat–fallow rotations. These weeds annually account for millions of dollars of losses in yield and wheat production with reduced quality. There has been moderate success in controlling winter annual grasses in wheat by utilizing multiple-year crop rotations with spring crops and fallow periods, and with chemical control of weeds before and after the wheat crop. Beyond herbicide, the EPA received federal registration for use in Clearfield wheat in December 2001. Spring applications of Beyond can control or suppress summer annual broadleaf weeds, such as common lambsquarters, pigweed, wild buckwheat goatgrass, and feral rye.

12.2.3 Mutagenesis by Physical Agents

Accelerated heavy ion beams, carrying much greater energy than X-rays and gamma rays, have been used in mutation induction (mutagenesis). Researchers at the RIKEN Nishina Center for Accelerator-Based Science, provided with a Cyclotron accelerator, developed protocols to mutate a gene even with only a single particle. The high performance of RIKEN accelerators enables the use of ions of carbon, nitrogen, neon, argon, iron, and other elements, thus offering a greater variation in the types of mutations. Seeds and cuttings from a variety of different plants have been exposed to beams of heavy atomic ions accelerated to half the speed of light, producing mutations and breeding new varieties of flowers, crops, and trees. In contrast to other breeding techniques such as hybridization and gene recombination, the timespan for breeding with heavy ion beams can be shortened to only 2 or 3 years.

Gamma rays. Californian rice Calrose 76 variety has been obtained, treating Calrose rice variety with gamma rays, inducing a single nucleotide mutation in the gene *sd1-c* (*semidwarf*). The mutant gene affects the production of the growth hormones gibberellins, thus producing semidwarf plants adapted to submerged cultivation, giving rise to the rice green revolution (Sasaki et al., 2002).

Other mutations have been obtained using this method. Among these, amino acids substitutions at Ala122, Pro197, Ala205, Trp574, Ser653 in *ALS* gene have been found to induce tolerance to imidazolinones.

12.3 Regulation Harmonization

Regulatory authorities in the EU require to avoid unnecessary DNA segments and to verify the possible effects on the integration of external DNA into plant genomes on gene expression, epigenetics, and other possible effects (disruption of a DNA segment controlling gene expression of possessing regulatory functions, i.e. enhancers, genes producing noncoding RNAs, and others).

Thus, foreign sequences such as selection genes and VBSs are intentionally excluded by performing backcrossing of the first transformed plant. Intragenesis differs from cisgenesis by allowing use of new gene combinations created by in vitro rearrangements of functional genetic elements. Several surveys show higher public acceptance of intragenic/cisgenic crops compared to transgenic crops. Today, several different traits in a variety of crops have currently been modified according to these concepts (Holme et al., 2013).

Thus, while the steps requiring the regulatory approvals based on EU legislation for authorization of release of new GMOs are expensive, the production of NPPs based on consumer-acceptable gene modifications (i.e. plant genomes devoid of bacterial sequences, introduction of genes based on interventions similar to the effects of breeding technologies, producing hybrids devoid of transgenes) could be less costly in terms of time needed by the regulatory authorities to proceed with a preferential road for examination by the evaluation panels.

In the USA, the EPA has opened up a debate on a draft rule that would exempt certain cisgenic organisms from registration as GMOs.

As the scientific community has not yet developed a consensus, national regulatory authorities have put forward their own definitions. The US Department of Agriculture, for example, announced that it does not have the authority to oversee cisgenic plants created without the help of a plant pathogen, whereas the Australian Office of the Gene Technology Regulator considered that certain cisgenic plants might not be regulated. EU member states and the European Commission (EC) are also considering developments in plant breeding and discussing whether new biotechnology-based plant breeding techniques would be captured by, or excluded from, the existing definition of GMOs. In the case of cisgenic plants, the experts have indicated that they fall under GMO legislation, as the definition of a GMO in the EU (http://ec.europa.eu/food/food/biotechnology/index_en.htm) is based on the technique used to produce it.

Various open questions for cisgenic plants illustrate the challenge of harmonizing definitions. These include which genes could be used from primary, secondary, and tertiary gene pools available to conventional breeders, whether the gene(s) to be transferred has been assessed for potential allergenic or toxic effects of the expressed proteins, to what extent the introgressed sequence needs to be identical to the sequence from the donor plant, if small extra sequences are allowed, what should be the maximum size of these, which transformation methods can be used to produce a cisgenic plant, and the level of expression allowed for the inserted gene.

The "breeders' gene pool" is the total of all genes, or genetic information, in any population that can be used by conventional breeders to improve their crops. Breeders might need to overcome barriers to gene transfer, depending on the source of the genes used. The primary gene pool comprises species that interbreed freely with the plant of interest. The secondary gene pool includes species that crossbreed only but with difficulty with the plant of interest, and that produce at least some fertile hybrids. The tertiary gene pool comprises species that are more distantly related with the plant of interest, but such species crossbreed by using advanced techniques such as embryo rescue, induced polyploidy, and bridge crosses.

Recently in the last two years, new biotechnology-based plant breeding techniques and their derived products raise several regulatory challenges, as they do not necessarily fit into known product definitions, regulatory frameworks, and risk assessment approaches for GMPs. Regulators and policymakers will have to decide whether NPPs are actually GMPs as categorized by standard definitions. If they were to be classified as GMPs, it raises the question of whether product definitions should be modified to take into account these new techniques and any future advances in plant breeding methods. It also raises the question of whether the regulatory frameworks and risk assessment approaches implemented for GMPs provide a sustainable and proportionate approach for the regulation and safety assessment of NPPs. In addition, regulators and policymakers have to consider whether the frameworks put in place provide an optimal balance between policy objectives, international harmonization, and equal regulatory oversight for different products that raise similar safety concerns.

Process-based regulatory frameworks. Argentina, Brazil, the EU, and many other countries have put new process-based regulatory systems in place to regulate the use of GMOs, as the techniques used for their production were thought to raise specific safety concerns. In these jurisdictions, a GMO is mainly characterized by the transformation techniques used in its production. The definitions of GMOs used by these countries are often partly or fully based on those put forward by international organizations, such as FAO, and international treaties, such as the Cartagena Protocol.

FAO: GMOs and derived products are produced by using techniques that alter the genetic material of an organism in a way that does not occur naturally by mating and/or natural recombination. Techniques of genetic engineering include but are not limited to: recombinant DNA, cell fusion, micro- and macro-injection, encapsulation, and gene deletion and doubling. GMOs do not include organisms resulting from techniques such as conjugation, transduction, and hybridization (http://www.fao.org/DOCREP/005/Y2772E/y2772e04.htm).

In the Cartagena Protocol, an LMO is defined as any living organism that has a combination of genetic material obtained through the use of modern biotechnology, namely: (1) in vitro nucleic acid techniques, including recombinant DNA and direct injection of nucleic acid into cells or organelles; or (2) fusion of cells beyond the taxonomic family that overcomes natural, physiological reproductive, or recombination barriers, and that are not techniques used in conventional breeding and selection (http://bch.cbd.int/protocol).

Product-based regulatory frameworks. Canada and the USA opted to regulate all plants or products with new traits developed either through genetic engineering or any other plant breeding techniques under the same, yet existing, regulatory system. The transformation techniques were not considered inherently risky. Therefore, the focus of product-based regulatory systems is on the risks of products and new traits or attributes introduced into a plant, rather than the method of production.

The next step is to make sure that regulatory frameworks remain fit for purpose. However, frameworks that use process-based definitions as a trigger for regulatory oversight might not be functional over time. NPPs blur the sharp distinction between GMP and non-GMP, and introduce a new continuum between genetic engineering and conventional breeding.

There is a need to guarantee that regulatory approvals and risk assessment for NPPs will be proportional to the level of risk that the plants may pose to health and environment. Process-based and narrowly defined product-based legislation therefore run the risk of quickly becoming obsolete, given the rate of innovation in the field. Process-based legislation requires not only updates to the lists of new biotechnological plant breeding techniques, but also debate on their classification as GMP or non-GMP. In 2012, the European Food Safety Authority (EFSA) published a scientific opinion to address the safety assessment of plants developed through cisgenesis and intragenesis (EFSA, 2012b). The EFSA Panel on GMOs concluded that cisgenic and conventionally bred plants represent similar hazards, whereas intragenic and transgenic plants could raise new hazards. Whether or not identified hazards would translate into risks to human and animal health and the environment depends on exposure, for instance, the extent to which the plant is cultivated or its derived products are consumed. The Panel's conclusion is consistent with similar reports by the National Academy of Sciences and other risk assessment bodies. However, if cisgenic plants were to be considered GMPs in the EU, the EFSA GMO Panel considered that its existing risk assessment guidelines for plants and products developed through transgenesis would generally apply to cisgenic and intragenic plants, but that it would require less event-specific data, depending on the specific case.

We need to ensure that regulatory frameworks and risk assessment approaches for NPPs remain proportional to the level of risk that these plants might pose to human and animal health and the environment. Despite the fact that genetic engineering techniques have been used to insert recombinant DNA transiently or stably during the development of NPPs, the genomic changes are often similar to those obtained by conventional breeding, such that the end products are indistinguishable from conventionally bred plants. Therefore, from a product-based perspective, the risk profile of certain NPPs resembles conventionally bred plants more closely than GMPs. This questions the proportionality inherent in regulatory frameworks and risk assessment practices.

As many of the changes introduced in NPPs could also be obtained through conventional breeding, it is important to consider whether any unintended changes arising from these techniques are specific to NPPs and differ from those caused by conventional breeding. The in vitro procedures (for example, cell and tissue culture) used to obtain NPPs are also used in conventional plant breeding, so unintended changes owing to somaclonal variation will be similar in both cases. Only unintended changes attributed to the stable or transient presence of recombinant DNA in NPPs are new compared with conventional breeding, and therefore merit further investigation. In the case of NPPs obtained after transient expression of recombinant DNA or after the use of a GM intermediate, it will be essential to verify that the recombinant DNA is no longer present in the genome of the selected plant if such NPPs were to be excluded from legislation for GMPs.

Another challenge is to develop regulatory frameworks and risk assessment practices, not only to prevent harms, but also to stimulate the innovation required to meet other policy objectives such as food security, economic development, and building consumer trust. It can be argued that extensive GMO legislation has actually stimulated the development of some of these new biotechnological plant breeding techniques to circumvent the regulatory burden and improve consumer attitudes. NPPs also provide new opportunities for small companies and research institutes to develop innovative consumer-oriented products that might not be subjected to extensive and expensive regulatory requirements and procedures. However, as NPPs are already being designed, developed, and tested in the field, plant breeders need clarity on the regulatory implications and on whether or not specific NPPs are covered by legislation (Raybould and Poppy, 2012).

To ensure that NPPs and their development are accepted by citizens and consumers, it will be essential to clarify and accommodate factual and normative premises about the governance of new biotechnology-based plant breeding techniques. This will help to increase accountability and improve inclusivity through the involvement of diverse stakeholders. A proactive engagement involving stakeholders and the public might help to create a shared responsibility for the governance of new biotechnological plant breeding techniques (Kuzma and Kokotovich, 2011). This would also reduce the risk of market failures and mistrust among citizens. Although there are significant challenges in addressing the wide range of societal concerns, European policymakers have had a public hearing on cisgenesis (http://ecrgroup.eu/?p=4521). In addition, the EC has commissioned several reports on specific aspects of new biotechnological plant breeding techniques. In the USA, the EPA has opened up a debate on a draft rule that would exempt certain cisgenic organisms from registration as GMOs (Reardon, 2011).

The development of regulatory frameworks for NPPs should allow international harmonization. So far, there is a lack of consensus on detailed product definitions for NPPs. At an international workshop organized by the EC's Joint Research Center, it became clear that the definition of a GMO differs between jurisdictions and that this determines whether or not NPPs are classified as GMO. Products generated through targeted mutation, for example, will probably be considered as non-GMO in many countries, although they might be defined as a GMO in other jurisdictions (Lusser and Davies, 2013).

As NPPs might be defined and regulated differently in different jurisdictions, the existing problem of market approval time differences could be further amplified. The outcome of continuing discussions will have far-reaching consequences for the development of future NPPs, international trade and requirements for labeling and detection. Differences in regulatory systems, and the fact that many jurisdictions still have to establish a consistent regulatory framework for existing GMPs, suggest this is not a straightforward process (Podevin et al., 2012).

Finally, here is a need to avoid disparities in risk assessment practices between products with equal potential to cause harm. Several authors have argued that process-based regulatory approaches lack consistency because conventionally bred products can raise safety concerns similar to those for their transgenic counterparts (Podevin et al., 2012). According to these authors, there are no convincing arguments for applying more stringent regulatory requirements for one particular technique if another one might result in similar adverse impacts. The European Policy Evaluation Consortium, commissioned by the EC to evaluate the EU GMO legislation, came to a similar conclusion. Only the Canadian legislative approach enables consistent evaluation of plants with similar new traits, irrespective of the techniques used. To ensure consistent and proportionate risk assessment practices, it is therefore important that regulatory requirements for NPPs are based on risk assessments associated with the plant species, traits, receiving environments and intended uses, and the combination of these characteristics, rather than the production method itself.

In addition, to build and maintain public and consumer trust in NPPs, new regulations must be built on factual and normative premises for governance. Importantly, the regulatory environment needs to support innovation, wider policy objectives, and consumer acceptance of NPPs. Therefore, regulatory systems for NPPs need to be dynamically scalable to the rate of innovation and advances in the field, remain proportionate to the level of risk that the use of the technology might pose, prevent harm without jeopardizing other policy objectives, support international harmonization, and avoid disparities in risk assessment practices.

12.4 Conclusion

In conclusion, the NPPs being developed raise various regulatory challenges that need to be addressed urgently. It requires broad international consensus on whether, and how, NPPs will be regulated to avoid potential adverse effects on human and animal health and the environment, while stimulating innovation to meet other policy objectives. The recent opening in a speech by president Obama in 2015 on plant biotechnologies and food security in the next years has put attention on the need to bypass FDA constraints to pave the way for protocols less expensive in the to generation of regulatory dossiers. Similarly, a new awareness on the same topic and needs by the EU scientific community may compel the public authorities to sustain plant science and agriculture innovation, at least for industrial-oriented NPPs.

References

Alatwi, H.E., Downs, J.A., 2015. Removal of H2A.Z by INO80 promotes homologous recombination. EMBO Reports 16 (8), 986–994.

Auer, C., 2011. Small RNAs for crop improvement. In: Erdmann, V.A., Barciszewski, J. (Eds.), Non Coding RNAs in Plants. Springer-Verlag, Berlin, pp. 461–484.

Bajaj, S., Puthigae, S., Templeton, K., Bryant, C., Gill, G., Lomba, P., Zhang, H., Altpeter, F., Hanley, Z., 2008. Towards engineering drought tolerance in perennial ryegrass using its own genome. 6th Canadian plant genomics workshop, 62 Abstract.

Bajaj, S., Puthigae, S., Bryant, C., Whittaker, D., Elborough, K., Hanley, Z., 2010. Ryegrass genes- a potential source for improving rice and grass species. AsPac J. Mol. Biol. Biotechnol. 8, 71–73.

Belhaj, K., Chaparro-Garcia, A., Kamoun, S., Nekrasov, V., 2013. Plant genome editing made easy: targeted mutagenesis in model and crop plants using the CRISPR/Cas system. Plant Methods 9, 39.

Breyer, D., Herman, P., Brandenburger, A., Gheysen, G., Remult, E., Soumillon, P., et al., 2007. Genetic modification trough oligonulceotide-madiated mutagenesis. A GMO regulatory challenge? Environmental Biosafety Research 8, 57–64.

Chen, K., Gao, C.X., 2014. Targeted genome modification technologies and their applications in crop improvements. Plant Cell Reports 33 (4), 575–583.

Chen, Z., Jaafar, L., Agyekum, D.G., Xiao, H., Wade, M.F., Kumaran, R.I., Spector, D.L., Bao, G., Porteus, M.H., Dynan, W.S., Meiler, S.E., 2013. Receptor-mediated delivery of engineered nucleases for genome modification. Nucleic Acids Research 41 (19), e182.

Chu, V.T., Weber, T., Wefers, B., Wurst, W., Sander, S., Rajewsky, K., Kuhn, R., 2015. Increasing the efficiency of homology-directed repair for CRISPR/Cas9-induced precise gene editing in mammalian cells. Nature Biotechnology 33 (5), 543–548.

Deng, D., Yan, C., Pan, X., Mahfouz, M., Wang, J., Zhu, J.K., Shi, Y., Yan, N., 2012. Structural basis for sequence-specific recognition of DNA by TAL effectors. Science 335, 720–723.

Dow, L.E., Fisher, J., O'Rourke, K.P., Muley, A., Kastenhuber, E.R., Livshits, G., Tschaharganeh, D.F.I., Socci, N.D., Lowe, S.W., 2015. Inducible in vivo genome editing with CRISPR-Cas9. Nature Biotechnology 33 (4), 390–394.

EFSA Panel on Genetically Modified Organisms (GMO), 2012a. Scientific opinion addressing the safety assessment of plants developed using zinc finger nucleases and other site-directed nucleases with similar function. EFSA Journal 10, 2943.

EFSA Panel on Genetically Modified Organisms (GMO), 2012b. Scientific opinion addressing the safety assessment of plants developed through cisgenesis and intragenesis. EFSA Journal 10, 2561.

Endo, M., Toki, S., 2014. Toward establishing an efficient and versatile gene targeting system in higher plants. Biocatalysis and Agricultural Biotechnology 3 (1), 2–6.

Fekih, R., Takagi, H., Tamiru, M., Abe, A., Natsume, S., Yaegashi, H., Sharma, S., et al., 2013. MutMap+: genetic mapping and mutant identification without crossing in rice. PLoS One 8 (7), e68529.

Freiman, A., Shlizerman, L., Golobovitch, S., Yablovitz, Z., Korchinsky, R., Cohen, Y., Chevreau, E., Le Roux, P.-M., Patocchi, A., Flaishman, M.A., 2012. Development of a transgenic early flowering pear (*Pyrus communis* L.) genotype by RNAi silencing of *PcTFL1-1* and *PcTFL1-2*. Planta 235, 1239–1251.

Gaj, T., Gersbach, C.A., Barbas, C.F., 2013. ZFN, TALEN, and CRISPR/Cas-based methods for genome engineering. Trends in Biotechnology 31, 397–405.

Gambino, G., Gribaudo, I., 2012. Genetic transformation of fruit trees: current status and remaining challenges. Transgenic Research 21, 1163–1181.

Holme, I.B., Wendt, T., Holm, P.B., 2013. Intragenesis and cisgenesis as alternatives to transgenic crop development. Plant Biotechnology Journal 11, 395–407.

Hsu, P.D., Scott, D.A., Weinstein, J.A., Ran, F.A., Konermann, S., Agarwala, V., Li, Y., Fine, E.J., Wu, X., Shalem, O., Cradick, T.J., Marraffini, L.A., Bao, G., Zhang, F., 2013. DNA targeting specificity of RNA-guided Cas9 nucleases. Nature Biotechnology 31, 827–832.

Ilardi, V., Di Nicola-Negri, E., 2011. Genetically engineered resistance to *Plum pox virus* infection in herbaceous and stone fruit hosts. GM Crops 2, 24–33.

Jiang, W., Zhou, H., Bi, H., Fromm, M., Yang, B., Weeks, D.P., 2013. Demonstration of CRISPR/Cas9/sgRNA-mediated targeted gene modification in Arabidopsis, tobacco, sorghum and rice. Nucleic Acids Research 41 (20), e188.

Konermann, S., Brigham, M.D., Trevino, A.E., Joung, J., Abudayyeh, O.O., Barcena, C., Hsu, P.D., Habib, N., Gootenberg, J.S., Nishimasu, H., Nureki, O., Zhang, F., 2015. Genome-scale transcriptional activation by an engineered CRISPR-Cas9 complex. Nature 517 (7536), 583–588.

Kotoda, N., Hayashi, H., Suzuki, M., Igarashi, M., Hatsuyama, Y., Kidou, S., Igasaki, T., et al., 2010. Molecular characterisation of FLOWERING LOCUS T –like genes of apple (*Malus* × *domestica* Borkh.). Plant and Cell Physiology 51, 561–575.

Kuzma, J., Kokotovich, A., 2011. Renegotiating GM crop regulation. Targeted gene-modification technology raises new issues for the oversight of genetically modified crops. EMBO Reports 12, 883–888.

Lusser, M., Parisi, C., Plan, D., Rodríguez-Cerezo, E., 2011. New Plant Breeding Techniques: State-of-the-art and Prospects for Commercial Development. Publications Office of the European Union, Luxembourg, Luxembourg. http://ftp.jrc.es/EURdoc/JRC63971.pdf.EUR.24760.

Lusser, M., Parisi, C., Plan, D., Rodríguez-Cerezo, E., 2012. Deployment of new biotechnologies in plant breeding. Nature Biotechnology 30, 231–239.

Lusser, M., Davies, H.V., 2013. Comparative regulatory approaches for groups of new plant breeding techniques. New Biotechnology 30, 437–446.

Mahiny, A., Dewerth, A., Mays, L.E., Alkhaled, M., Mothes, B., Malaeksefat, E., Loretz, B., et al., 2015. In vivo genome editing using nuclease-encoding mRNA corrects SP-B deficiency. Nature Biotechnology 33 (5), 584–586.

Mak, A.N., Bradley, P., Bogdanove, A.J., Stoddard, B.L., 2013. TAL effectors: function, structure, engineering and applications. Current Opinion in Structural Biology 23, 93–99.

Maruyama, T., Dougan, S.K., Truttmann, M.C., Bilate, A.M., Ingram, J.R., 2015. Increasing the efficiency of precise genome editing with CRISPR-Cas9 by inhibition of nonhomologous end joining. Nature Biotechnology 33 (5), 538–542.

Mussolino, C., Cathomen, T., 2012. TALE nucleases: tailored genome engineering made easy. Current Opinion in Biotechnology 23, 644–650.

Nekrasov, V., Staskawicz, B., Weigel, D., Jones, J.D.G., Kamoun, S., 2013. Targeted mutagenesis in the model plant *Nicotiana benthamiana* using Cas9 RNA-guided endonuclease. Nature Biotechnology 31, 691–693.

New Plant Breeding Techniques. Report of a Workshop Hosted by Food Standards Australia New Zealand, 2012. http://www.foodstandards.gov.au/publications/Documents/New%20Plant%20Breeding%20Techniques%20Workshop%20Report.pdf.

Officially released mutant varieties - the FAO/IAEA Database, Dec 2000. Mutation breeding review; (no.12). 84 p; IAEA; Vienna (Austria) https://inis.iaea.org/search/search.aspx?orig_q=RN:32006534; http://www.iaea.org/inis/collection/NCLCollectionStore/_Public/32/006/32006534.pdf.

Pauwels, K., Podevin, N., Breyer, D., Carroll, D., Herman, P., 2014. Engineering nucleases for gene targeting: safety and regulatory considerations. New Biotechnology 31, 18–27.

Podevin, N., Devos, Y., Davies, H.V., Nielsen, K.M., 2012. New biotechnology-based plant breeding techniques and the regulatory landscape. EMBO Reports 13, 1057–1061.

Podevin, N., Davies, H.V., Hartung, F., Nogué, F., Casacuberta, J.M., 2013. Site-directed nucleases: a paradigm shift in predictable, knowledge-based plant breeding. Trends in Biotechnology 31, 375–383.

Raybould, A., Poppy, G.M., 2012. Commercializing genetically modified crops under EU regulations: objectives and barriers. GM Crops Food 3, 9–20.

Reardon, S., 2011. EPA proposal would exempt some GMOs from registry. Science 332, 652.

Regulation (EC) No. 258/97 of the European Parliament and of the Council concerning novel foods and novel food ingredients. CELEX-EUR Official J. L 43, 1997, pp. 1–7. http://faolex.fao.org/docs/texts/eur18638.doc.

Sasaki, A., Ashikari, M., Ueguchi-Tanaka, M., Itoh, H., Nishimura, A., Swapan, D., Ishiyama, K., et al., 2002. Green revolution: a mutant gibberellin-synthesis gene in rice. Nature 416, 701–702.

Scholze, H., Boch, J., 2011. TAL effectors are remote controls for gene activation. Current Opinions in Microbiology 14, 47–53.

Sha, X.Y., Linscombe, S.D., Groth, D.E., 2007. Field evaluation of imidazolinone tolerant Clearfield rice (*Oryza sativa* L.) at nine Louisiana locations. Crop Science 47, 1177–1185.

Shan, Q., Wang, Y., Li, J., Zhang, Y., Chen, K., Liang, Z., Zhang, K., Liu, J., Xi, J.J., Qiu, J.-L., Gao, C., 2013. Targeted genome modification of crop plants using a CRISPR-Cas system. Nature Biotechnology 31, 686–688.

Shivrain, V.K., Burgos, N.R., Anders, M.M., Rajguru, S.N., Moore, J., Sales, M.A., 2007. Gene flow between Clearfield™ rice and red rice. Crop Protection 26 (3), 349–356.

Takagi, H., Tamiru, M., Abe, A., Yoshida, K., Uemura, A., Yaegashi, H., Obara, T., et al., 2015. MutMap accelerates breeding of a salt-tolerant rice cultivar. Nature Biotechnology 33, 445–449.

Tamiru, M., Undan, J.R., Takagi, H., Abe, A., Yoshida, K., Undan, J.Q., Natsume, S., et al., 2015. A cytochrome P450, OsDSS1, is involved in growth and drought stress responses in rice (*Oryza sativa* L.). Plant Molecular Biology 88 (1–2), 85–99.

Tsuji, S., Futaki, S., Imanishi, M., 2013. Creating a TALE protein with unbiased 5′-T binding. Biochemical and Biophysical Research Communications 441, 262–265.

Upadhyay, S.K., Kumar, J., Alok, A., Tuli, R., 2013. RNA-Guided genome editing for target gene mutations in wheat. G3 (Genes, Genomes, Genetics) 3, 2233–2238.

Vanblaere, T., Flachowsky, H., Gessler, C., Broggini, G.A.L., 2014. Molecular characterization of cisgenic lines of apple 'Gala' carrying the *Rvi6* scab resistance gene. Plant Biotechnology Journal 12, 2–9.

Waltz, E., 2011. GM grass eludes outmoded USDA oversight. Nature Biotechnology 29, 772–773.

Waltz, E., 2015a. USDA approves next-generation GM potato. Nature Biotechnology 33 (1), 12–13.

Waltz, E., 2015b. Nonbrowning GM apple cleared for market. Nature Biotechnology 33 (4), 326–327.

Waltz, E., 2015c. EC approves loing overdue GM plants. Nature Biotechnology 33 (6), 578.

Wei, F.-J., Droc, G., Guiderdoni, E., Hsing, Y.C., 2013a. International consortium of rice mutagenesis: resources and beyond. Rice 6 (1), 39.

Wei, C., Liu, J., Yu, Z., Zhang, B., Gao, G., Jiao, R., 2013b. TALEN or Cas9-Rapid, efficient and specific choices for genome modifications. Journal of Genetics and Genomics 40, 271–279.

Wheat Situation and Outlook Yearbook, March 2001. Market and Trade Economics Division, Economic Research Service. U.S. Department of Agriculture, WHS-2001. http://usda.mannlib.cornell.edu/usda/ers/WHS-yearbook/2000s/2001/WHS-yearbook-03-27-2001.pdf.

Xie, K., Yang, Y., 2013. RNA-guided genome editing in plants using a CRISPR-Cas system. Molecular Plant 6, 1975–1983.

Xu, R.-F., Li, H., Qin, R.-Y., Li, J., Qiu, C.-H., Yang, Y.-C., Ma, H., Li, L., Wei, P.-C., Yang, J.-B., 2015. Generation of inheritable and "transgene clean" targeted genome-modified rice in later generations using the CRISPR/Cas9 system. Scientific Reports 5, 11491.

Zhang, Y.-C., Yu, Y., Wang, C.-Y., Li, Z.-Y., Liu, Q., Liao, J.-Y., Wang, X.-J., et al., 2013. Overexpression of microRNA OsmiR397 improves rice yield by increasing grain size and promoting panicle branching. Nature Biotechnology 31, 848–852.

Zhao, D., Song, G.-Q., 2014. Rootstock-to-scion transfer of transgene-derived small interfering RNAs and their effect on virus resistance in nontransgenic sweet cherry. Plant Biotechnology Journal 12 (6), 1319–1328.

Index

'*Note:* Page numbers followed by "f" indicate figures, "t" indicate tables.'

A

A4A. *See* Airlines for America (A4A)
Abiotic stress, 314
ACC. *See* Aminocyclopropane (ACC)
Accelerated transformation of woody plants and early flowering, 328–329
Acetoclastic methanogens, 93
Acetogenesis, 93
Acetogens, 93
Acetohydroxy acid synthetase (AHAS). *See* Acetolactate synthetase (ALH)
Acetolactate synthetase (ALH), 319, 331–332
Acibenzolar-S-methyl (ASM), 167
Acid hydrolysis, 137–138
Acidic HG (HGA), 223–224
Acidogenesis, 92–93
Active solvents, 259–261
AD. *See* Anaerobic digestion (AD)
Adsorption, 101, 103, 263, 274–276
AFEX. *See* Ammonia fiber explosion (AFEX)
Agricultural facilities, 75–76
Agricultural mulch films, 50
Agricultural residues, 6
Agricultural waste, 91, 125–126
  polyhydroxybutyrate, enzymes, and succinic acid from, 298
Agrobacterium-mediated transformation, 163–164
Agroindustrial by-products transformation, 201
Agro–industrial fruit wastes and potential as biomass, 41–42
Agronomic traits, 55, 317
Aircraft fuel, bioeconomics for, 178–181
Airlift bioreactor, 194–195, 195f
Airlines for America (A4A), 175
Algae, 59
Alginate, 281
ALH. *See* Acetolactate synthetase (ALH)
Alkaline hydrolysis, 138–139
Alkaline pretreatment, 147–148
Alkyl-poly-pentosides (APPs), 65
1-Allyl-3-methylimidazonium chloride (AMIMCl), 145
ALS. *See* Acetolactate synthetase (ALH
AMIMCl. *See* 1-Allyl-3-methylimidazonium chloride (AMIMCl)
Aminocyclopropane (ACC), 320
Ammonia (NH_3), 104
  pretreatment, 142–143
Ammonia fiber explosion (AFEX), 142–143
Ammonia recycle percolation (ARP), 142–143
Anaerobic digestion (AD), 71, 90–121
  biogas treatment, 100–105
  biogas valorization, 105–106
  for electricity production, fertilizer production and effect of different feeds, 302
  energy recovery and recycling by, 299–302
  input, 90–91
  methods, 95–100
  posttreatment steps, 100
  pretreatment steps, 91–92
  principle, 90
  process, 92–95, 92f
    factors affecting digestion process, 93–95
    microbial communities, 92–93
  SWOT analysis, 106–116
  within TRANSBIO Project, 116–121
    long-term evaluation of biogas production, 118–121
    short-term evaluation of biogas potential, 116–118
Ananas comosus. *See* Pineapple (*Ananas comosus*)
Antioxidant activity, 34–35, 35t
APHIS. *See* Department of Agriculture Animal and Plant Health Inspection Service (APHIS)
Apple, 57
Apple pomace, 10–11
APPs. *See* Alkyl-poly-pentosides (APPs)
ARP. *See* Ammonia recycle percolation (ARP)
Ascorbic acid, 32
ASM. *See* Acibenzolar-S-methyl (ASM)
Asparagine synthetase-1 gene (*Asn*1), 317
Asparagus waste, 3f
Aspergillus niger (*A. niger*), 203
Aspergillus oryzae (*A. oryzae*), 210–211
Aspergillus terreus (*A. terreus*), 242
ASTM D1655, 176
ASTM D7566, 176–177

ATJ, 170
Authorization, 61
Automotive and building industry, 50, 56
Aviation Climate Change Commitments, 173
Azelaic acid, 243

B

Bacillus thuringiensis (Bt), 310–311
Bacteria, 200, 210–211
Bamboo, 59
Banana
  biological activities of fruits from Costa Rica and Central America region, 39
  Costa Rica, 24–25
  wastes, 58
Batch digestion, SWOT analysis of, 114–115
Batch stirred tank reactor, 191–192, 191f
Baumuster composting categories, 83, 84t
BDO. *See* 1,4-Butanediol (BDO)
Beet necrotic yellow vein virus (BNYVV), 315
BEFS. *See* Bioenergy and Food Security project (BEFS)
Biobased biochemicals, 57–60
Biobased chemicals, drivers for, 237–239
Biobased dicarboxylic acids. *See also* Dicarboxylic acids
  downstream processing
    effective designing, 263–264
    factors influencing, 257–258
    reactive extraction and back extraction system, 260f
    state of the art and new trends, 258–263
    steps in, 255–257, 256t
  production
    linking strain physiology, 252–254
    requirements for scale-up toward commercial operation, 254–255
    from strain isolation to initial physiological studies, 251–252
Biobased products, 237–238
Biocell reactors, 87
Biochemical conversion routes, 227
Biodegradable plastics, 64
Biodiesel, energy recovery by conversion to, 302
Bioeconomy, 237–238
  bioeconomics for aircraft fuel, 178–181
  of biofuels, 167–173
    production of biofuels and chemicals, 172f
    prospects of transformation of MSWs, 171f
Bioenergy, 228, 230
  crops, 61–62
Bioenergy and Food Security project (BEFS), 227
Biofuel production
  bioeconomy, 167–173
  polysaccharides and lignin
    cell wall recalcitrance, 227–228
    hemicellulose and pectin modification, 229–230
    lignin modification to improving cellulose accessibility, 229
    modification of cell wall structure, 228
    plants cell wall as source of energy, 226–227
Biogas
  production, 97, 100
  treatment, 100–105
    CO_2 removal, 103–104
    H_2S removal, 101–103
    removal of other components, 104–105
    water removal, 101
  valorization, 105–106
    electricity production, 105
    heat production, 105
    natural gas substitute production, 106
Biogas production potential (BPP), 116
Biokerosene deployment and sustainability issues, 173–178
Biological activities of fruits from Costa Rica and Central America region, 39–41. *See also* Physicochemical composition of fruits
  banana, 39
  mango, 39–40
  pineapple, 40
  watermelon, 41
Biological H_2S removal
  in AD reactor, 101–102
  outside AD reactor, 102
Biological pretreatment of biomass, 145–147
  enzymatic pretreatment, 147
  with fungal species, 146
Biomass, 8, 198–199
  agro–industrial fruit wastes and potential as, 41–42
  hubs, 63–64
  logistic requirements of, 6–7
  pretreatment strategies, 125–126
    biological pretreatment, 145–147
    chemical pretreatments, 137–141
    combined methods, 147–148
    industrial implementation of lignocellulosic biomass pretreatments, 148–149
    physical and chemical characteristic of lignocellulosic biomass, 126–131
    physical pretreatments, 131–136
    physicochemical pretreatments, 141–145
    pretreatments, 131
  SSF, 208–210
  temporal availability and compositions of, 5–6
  valorization methods, 288
Biomolecule production
  SmF, 201–202
  SSF, 213
Bioplastics, novamont case study on, 64–65
Bioreactors
  SmF, 191–196
    airlift, 194–195, 195f
    batch stirred tank, 191–192, 191f

bubble column, 193–194, 194f
continuous stirred tank, 192–193, 192f
fed-batch culture, 193, 193f
fluidized bed, 195, 196f
trickle bed, 196, 197f
SSF, 202
characteristics, 203t
continuous screw reactors, 205–206, 207f
packed bed column reactor, 203, 204f
rotary drum reactor, 204–205, 206f
tray bioreactors, 203–204, 205f
Bioresins for plastic materials, 50
Biosolids, 75
Biosurfactants, 65
Biotechnology, 309
Biotic stress, 314
selection of tobacco varieties, 166–167
Biowaste Directive, 72–73
Black shank, 166
BMIMCl. *See* 1-Butyl-3-methylimidazonium chloride (BMIMCl)
BNYVV. *See* Beet necrotic yellow vein virus (BNYVV)
Boeing Airlines, 175–176
BPP. *See* Biogas production potential (BPP)
Breeders gene pool, 334
°Brix/acid ratio, 30
Broccoli, 11
Bromelain, 40
Bt. *See Bacillus thuringiensis* (Bt)
Bubble column reactor, 193–194, 194f
Bulk agricultural raw materials, 1
1,4-Butanediol (BDO), 239
1-Butyl-3-methylimidazonium chloride (BMIMCl), 145

C

C/N ratio. *See* Carbon/nitrogen ratio (C/N ratio)
C4H. *See* Cinnamate 4-hydroxylase (C4H)
Caffeine, 35
Caffeoyl CoA, 222–223
CAFOs. *See* Concentrated Animal Feeding Operations (CAFOs)
Carbon dioxide explosion pretreatment (CDE pretreatment), 143–144
"Carbon neutral" growth, 173
Carbon sources, 257–258
Carbon/nitrogen ratio (C/N ratio), 79
Carboxymethyl cellulose, 136
Cargill starches, 63
Cartagena Protocol, 335
Cas9. *See* CRISPR-associated protein 9 (Cas9)
Cases for electronics products or household appliances, 50
Catabolic response, 200–201
Cauliflower, 11
Cavendish group bananas, 25
ccr. *See* Cinnamoyl-CoA-reductase (ccr)
CDE pretreatment. *See* Carbon dioxide explosion pretreatment (CDE pretreatment)
Cell wall modification/deconstruction, 223
enzymes
cellulose, 224–225
hemicellulose, 224
pectins, 223–224
learning from microbes, 225–226
Cell wall recalcitrance, 227–228
hemicellulose and pectin modification, 229–230
Cell wall structure modification, 228
Cellobiose, 127
Cellulases, 223
Cellulose, 6, 49, 55–56, 126–128, 127f, 219–221, 224–225
fiber, 59–60
microfibrils, 127
polymers, 125
pulp fibers, 50, 56
Cellulose synthase catalytic subunit (CESA), 220–221
Cellulosic biomass, 49
Cellulosic material, 174
Central America, 21–22
biological activities of fruits from, 39–41
fruit industrialization in region, 28–30
fruit production in region, 25–27
Central American countries, 25–27
CESA. *See* Cellulose synthase catalytic subunit (CESA)
CFR. *See* Code of Federal Regulations (CFR)
cgRNA. *See* Chimeric guide RNAs (cgRNA)
Chemical cross-linking method, 276
Chemical H_2S removal
in AD reactor, 102
outside AD reactor, 102–103
Chemical mutagenesis, 331–333
Chemical pretreatments of biomass, 137–141. *See also* Biological pretreatment of biomass; Physical pretreatments of biomass
acid hydrolysis, 137–138
alkaline hydrolysis, 138–139
Organosolv pretreatment, 139–140
ozonolysis, 140–141
peroxide pretreatment, 140
peroxyformic acid pretreatment, 141
CHEMREC Bio-DME project, 169
Chimeric guide RNAs (cgRNA), 323–324
Chipping, 131–132
CHP. *See* Combined heat and power (CHP)
Cinnamate 4-hydroxylase (C4H), 164–165, 222–223
Cinnamoyl-CoA-reductase (ccr), 164–165, 222–223
Cisgenesis, 324–325
Citraconic acid, 242–243
Citrulline, 41
Citrus fruits, 10–11
Citrus paradisi. See Grapefruit (*Citrus paradisi*)
Clean Water Act, 75–76
Clearfield Production System, 331–332
for rice, 332
for wheat, 332–333
Clustered regularly interspaced short palindromic repeats (CRISPR), 309, 323–324
CO_2 emissions, 174f

CO_2 removal, 103–104. *See also* H_2S removal
adsorption, 103
cryogenic separation, 104
membrane separation, 104
PSA, 103–104
VPSA, 103–104
Code of Federal Regulations (CFR), 74
Codigestion, 96
SWOT analysis, 114
Coffee, 35–37, 36t, 58
Coffee silver skin (CS skin), 35
Combinatorial biological pretreatment, 148
Combined heat and power (CHP), 107, 169
Combined methods, 147–148
alkaline and mechanical pretreatment, 147–148
alkaline pretreatment
and gamma irradiation, 147
and wet oxidative pretreatment, 148
combinatorial biological pretreatment, 148
Competitiveness Index, 21–22
Complexity, 96
Composites, 50
Compost facility management, 76
Composting, 78
energy recovery and recycling by anaerobic digestion with, 299–302
recycling by, 298–299
Concentrated Animal Feeding Operations (CAFOs), 75–76
Concentrated strong acids, 137
Condensation, 101
Coniferyl-alcohol, 130, 130f
Construction materials, 50
Continuous screw reactors, 205–206, 207f
Continuous stirred tank reactor, 192–193, 192f
Continuously fed, completely mixed digestion
SWOT analysis, 115
Continuously fed, plug flow digestion, SWOT analysis of, 116
Cosmetics, 272–273
COST. *See* European Cooperation in Science and Technology (COST)
Cost barriers leveling playing field, 61–62
Costa Rica, 21
biological activities of fruits from, 39–41
fruit industrialization in region, 27–28
fruit production in region, 22–25
banana, 24–25
pineapple, 23–24
physicochemical composition of fruits of, 30–35
Cotton, 58–59
Covalent bonds, 274–276
Crabtree effect, 252–253
CRISPR. *See* Clustered regularly interspaced short palindromic repeats (CRISPR)
CRISPR-associated protein 9 (Cas9), 323–324
CRISPR/Cas9 nuclease system, gene modification using, 322–324
Cross-linking, 276
Crotonic acid, 243
Cryogenic separation, 104
CS skin. *See* Coffee silver skin (CS skin)

D

D-type cyclin (*CycD*2), 165
Degree of polymerization (DP), 128
Delta endotoxin, 316
Department of Agriculture Animal and Plant Health Inspection Service (APHIS), 309–310
Dicarboxylic acids. *See also* Biobased dicarboxylic acids
challenges for sustainable microbial production, 264–265
current uses and production
biotechnological routes, 242–243
fumaric acid, 240–241
glutaric acid, 241
itaconic acid, 241–242
succinic acid, 239–240
yeast as platform organisms for production, 244
Diethyl maleate, 239
Digestate, 97, 100
Dilute acid pretreatment, 137–138
Dish applications, 283–284
Disk milling, 131–132
Dissolving pulp, 58
DME, 169
Double-strand breaks (DSBs), 322
Downstream processing (DSP), 255
biobased dicarboxylic acids
effective designing, 263–264
factors influencing, 257–258
reactive extraction and back extraction system, 260f
state of the art and new trends, 258–263
steps in, 255–257, 256t
DP. *See* Degree of polymerization (DP)
Drum reactors, 87, 87f
Dry digestion, 95, 97
SWOT analysis of, 111–112
Dry feed production, 297
DSBs. *See* Double-strand breaks (DSBs)
DSP. *See* Downstream processing (DSP)

E

E-LCA. *See* Environmental LCA (E-LCA)
Early flowering accelerated transformation, 328–329
Early Flowering Spadona (EF-Spa), 328
EBFPI. *See* European Biofuels Flight Path Initiative (EBFPI)
EBTP. *See* European Biofuels Technology Platform (EBTP)
EC. *See* European Commission (EC)
ECHA. *See* European Chemical Agency (ECHA)
Eco-indicator 99 impact assessment method, 300–301

Economic sustainability assessment, 294–295
EcoTILLING, 320
EDTA. *See* Ethylenediaminetetraacetic acid (EDTA)
EF-Spa. *See* Early Flowering Spadona (EF-Spa)
Effector triggered immunity (ETI), 315
EFSA. *See* European Food Safety Authority (EFSA)
"Egg-box" structures, 223–224
Eggplant, 316
eIF5A, 317
Electric field strength, 136
Electricity production, 105
Electrodialysis, 261–262, 262f
ELISA. *See* Enzyme-linked immunosorbent assays (ELISA)
EMPYRO pyrolysis oil plant, 169
EMS. *See* Ethyl methane sulfonate (EMS)
Encapsulation, 276–282
endo-PGs. *See* Polygalacturonases (endo-PGs)
Endo-β-1,4-glucanases, 224
Endocellulases, 225–226
Endoplasmic reticulum (ER), 317
Endpoint methods, 294
Energy
  crops, 91
  demand, 132
  recovery, 288–289
    by anaerobic digestion, 299–302
    by conversion to biodiesel, 302
    by incineration, 303–304
  and soil improver, 106
Energy Tobacco Variety Solaris, 178–181
  content in fatty acids, 180t
  sustainability of production and exploitation, 181–185
  variety comparison, 182f
Enerkem, 171
5-Enolpyruvylshikimate-3-phosphate synthase (EPSPS), 317–318
Entrapment, 276
Environmental LCA (E-LCA), 292
Environmental Protection Agency (EPA), 74, 310
Environmental sustainability assessment, 292–294
Enzymatic hydrolysis, 139
Enzymatic pretreatment, 147
Enzyme-linked immunosorbent assays (ELISA), 273–274
Enzyme(s), 271
  agricultural waste from, 298
  enzyme-assisted extraction, 14
  immobilization, 274, 275f
  in industry, 273–274
    encapsulation, 276–282
    fluid bed agglomeration, 279
    fluid bed coating, 279
    gel entrapment, 280–281
    granulation, 278
    immobilization, 274–276
    liposomes, 282
    nanogels, 282
    spray drying, 277–278, 278f
    spray drying and fluid bed agglomeration combination, 279
  in technical applications, 272–273
  uses of immobilized enzymes, 282–284
EPA. *See* Environmental Protection Agency (EPA)
EPA Resource Conservation programs, 74
Epigenetics, 328
EPSPS. *See* 5-Enolpyruvyl-shikimate-3-phosphate synthase (EPSPS)
ER. *See* Endoplasmic reticulum (ER)
Escherichia coli (*E. coli*), 239–240
Ethyl methane sulfonate (EMS), 331
Ethylenediaminetetraacetic acid (EDTA), 283
ETI. *See* Effector triggered immunity (ETI)
Eucalyptus, 55
Europe, 70–73
  Biowaste Directive, 72–73
  EU business analysis of case studies, 60–62
  EU textile industries, 57
  Landfill Directive, 71
  Renewables Directive, 71–72
  Waste Framework Directive, 70–71
European Biofuels Flight Path Initiative (EBFPI), 175
European Biofuels Technology Platform (EBTP), 175
European Chemical Agency (ECHA), 177–178
European Commission (EC), 334
European Cooperation in Science and Technology (COST), 328
European Food Safety Authority (EFSA), 311, 325, 336
European hemp fibers, 62
European Policy Evaluation Consortium, 338
Exo-PGs, 223–224
Exocellulases, 225–226
Exotic fruits industry processing, 11
Expansins, 223–224
Extraction, 259–261
Extrusion pretreatment, 133

F

FAA. *See* Federal Aviation Administration (FAA)
Fabrics, 50, 57, 59
FAO. *See* Food and Agriculture Organization (FAO)
FDA. *See* Food and Drug Administration (FDA)
Fed-batch culture reactor, 193, 193f
Federal Attorney for Environmental Protection (PROFEPA), 77
Federal Aviation Administration (FAA), 175
Feedstock biomasses, 55, 163–164, 164f
Fermentable sugars, 128
Fermentation industry, feedstock for, 14–15
Fermentative production methods
  SmF, 189–202, 214t
  SSF, 202–213, 214t
Fermenter, 97
Fiber applications, 57–60

"First-generation" biofuels, 168
"First-generation" feedstocks oils, 163
Fluid bed agglomeration, 279
Fluid bed coating, 279
Fluidized bed bioreactor, 195, 196f
Food
  by-products, 9
  industry, 282–283
  processors, 50
Food and Agriculture Organization (FAO), 58, 329
Food and Drug Administration (FDA), 310
Food Standards Australia New Zealand (FSANZ), 320
Food waste(s), 10
  biodiesel from, 302
  conversion to feed, composting, 296–297
Fourier transform infrared spectroscopy (FTIR), 162
Free trade agreements (FTAs), 22
Freeze pretreatment, 135
Fresh compost, 80–83
Fruit and shoot borer (FSB), 316
Fruit industrialization in region, 27–30
  Central America, 28–30
  Costa Rica, 27–28
Fruit processing
  agro–industrial fruit wastes and potential as biomass, 41–42
  biological activities of fruits, 39–41
  fruit industrialization in region, 27–30
  fruit production in region, 22–27
  physicochemical composition and biological activity, 30–39
Fruit production in region, 22–27
  Central American countries, 25–27
  Costa Rica, 22–25
  Mexico, 27
Frumat Company, 57
FSANZ. *See* Food Standards Australia New Zealand (FSANZ)
FSB. *See* Fruit and shoot borer (FSB)
FSK. *See* Fully synthetic kerosene (FSK)
FTAs. *See* Free trade agreements (FTAs)
FTIR. *See* Fourier transform infrared spectroscopy (FTIR)
Fully automated turning machine, 85, 86f
Fully synthetic kerosene (FSK), 169–170
Fumaric acid, 240–241
Fungi, 210–211
Fungus resistant plants, 315

G

G2M flow. *See* Genetics to market flow (G2M flow)
GA 20-oxidase overexpressing plants, 165
β-Galactosidases, 224–226
Gamma irradiation, 135–136, 147
Gamma-butyrolactone, 239
Garlic residues, 3f
GAs. *See* Gibberellins (GAs)
Gasification, 227
GAX. *See* Heteroxylans (GAX)
Gel entrapment, 280–281
Gene editing, 323, 329
Gene modification, 322–324
Gene overexpression technologies, 327
Gene silencing, 327
Generally Regarded As Safe (GRAS), 244
Genetic engineering, 315
Genetically modified crops
  in European Union, 312t
  around world, 311t
Genetically modified organisms (GMOs), 310–311
Genetically modified plants (GM plants), 309
  accelerated transformation of woody plants and early flowering, 328–329
  agronomic traits, 317
  cisgenesis, 324–325
  containment, 314–315
  eggplant, 316
  EU Commission, 311
  FDA, 310
  fungus resistant plants, 315
  gene modification, 322–324
  genetically modified crops
    in European Union, 312t
    around world, 311t
  genetically modified rootstock grafting, 325–327
  intragenesis, 324–325
  novel plant products, 317–322
  plant improvement techniques, 327–328
  product quality traits, 316–317
  virus-resistant, 315–316
Genetically modified rootstock grafting, 325–327
Genetics to market flow (G2M flow), 183f
GHG emissions. *See* Greenhouse gas emissions (GHG emissions)
GHs. *See* Glycosyl hydrolases (GHs)
Giant recombinant (GR), 166
Gibberellins (GAs), 165
β-Glucosidases, 224–226
Glutaric acid, 241
Glycosyl hydrolases (GHs), 223
Glycosyltransferase 8 (GT8), 222
Glyoxylate cycle oxidative pathway, 247–248
Glyoxylic acid, 239
GM plants. *See* Genetically modified plants (GM plants)
GMOs. *See* Genetically modified organisms (GMOs)
GR. *See* Giant recombinant (GR)
Graft union, 325–326
Grafting, 325
Granulation, 278
Grapefruit (*Citrus paradisi*), 38–39
GRAS. *See* Generally Regarded As Safe (GRAS)
Green bananas, 41
Green bean by-products, 4f
Greenhouse gas emissions (GHG emissions), 168, 226
Grinding, 131–132

Growth kinetics
SmF, 198–199, 198f
SSF, 208–210
GT8. *See* Glycosyltransferase 8 (GT8)

H

H-oxygen radical absorbance capacity assay (H-ORAC assay), 34–35
H_2S removal, 101–103. *See also* CO_2 removal
biological H_2S removal
in AD reactor, 101–102
outside AD reactor, 102
chemical H_2S removal
in AD reactor, 102
outside AD reactor, 102–103
HCHL. *See* Hydroxycinnamoyl-CoA hydratase-lyase (HCHL)
HDI. *See* Human Development Index (HDI)
HDR. *See* Homology-directed repair (HDR)
Heat generation, 198
Heat production, 105
Heat transfer
SmF, 198
SSF, 208
Heavy Fraction FT Liquids (HFTL), 170
HEFA. *See* Hydrotreated esters and fatty acids (HEFA)
Hemicellulose, 6, 49, 126–130, 129f, 221, 224
modification to overcoming cell wall recalcitrance, 229–230
Hemp, 59
HempFlax Company, 60
Herbicide tolerance (HT), 314
Heterogeneous waste products, 106–107
Heteroxylans (GAX), 221
HFTL. *See* Heavy Fraction FT Liquids (HFTL)
HGA. *See* Acidic HG (HGA)
High oleic–lower linoleic varieties, 181
High shear granulation, 278
High-rate decomposition phase, 80–81
HMF. *See* 5-Hydroxymethylfurfural (HMF)
Homogeneity of feedstock, 96
Homology recombination (HR), 322–323
Homology-directed repair (HDR), 322
Horse chestnut extracts, 11–12
HR. *See* Homology recombination (HR)
HT. *See* Herbicide tolerance (HT)
5-HT. *See* 5-Hydroxytryptamine (5-HT)
HT and insect resistance (HT-IR), 315
HT-IR. *See* HT and insect resistance (HT-IR)
Human Development Index (HDI), 21
Hydraulic mixing, 98
Hydrochloric acid, 137
Hydrogen
bonding, 127, 128f
peroxide, 140
Hydrogenotrophic methanogens, 93
Hydrolysis, 92
Hydrolytic lipases, 273
Hydrotreated esters and fatty acids (HEFA), 170
Hydroxycinnamoyl-CoA hydratase-lyase (HCHL), 229
5-Hydroxymethylfurfural (HMF), 126, 133
5-Hydroxytryptamine (5-HT), 39
Hygiene products, 273–274, 283
Hypersensitive reaction, 315

I

IEA. *See* International Energy Agency (IEA)
ILCD. *See* International Life Cycle Data System (ILCD)
ILs pretreatment. *See* Ionic liquids pretreatment (ILs pretreatment)
ILUC. *See* Indirect land use change (ILUC)
Imidazolinones, 331–332
Immobilization, 274–276
Immobilized enzymes, 282–284
Incineration, energy recovery by, 303–304
indel. *See* Insertion/deletion (indel)
India food industry wastes, 50–53
barriers, 50–53
opportunities, 53
potential of mango waste components, 51t–52t
processed mango, 50
Indirect land use change (ILUC), 174
Industrial biotechnology processes, 49
Industrial composting, 78–90
input, 79–80
C/N ratio, 79
moisture content, 79–80
structure, 80
methods, 83–87
drum reactors, 87, 87f
in piles, 83–87
tunnel composting, 87
organic matter transformation, 78f
posttreatment steps, 88
pretreatment steps, 80
principle, 78
process, 80–83, 81f
SWOT analysis, 88–89
temperature and acidity profile, 82f
within TRANSBIO project, 89–90
Industrial implementation of lignocellulosic biomass pretreatments, 148–149
Industrial waste, 91
Industrial-scale submerged fermentation processes, 197
INE. *See* National Ecology Institute (INE)
Infinite Fuels GmbH, 169
Inositol oxygenase (inox), 323–324
inox. *See* Inositol oxygenase (inox)
Insect resistance (IR), 315–316
Insertion/deletion (indel), 323–324
Insertional mutagenesis, 330–331

Integration-based plant breeding techniques, 319
International Energy Agency (IEA), 149
International Life Cycle Data System (ILCD), 293
InteSusAl, 169
Intragenesis, 324–325
Ionic liquids pretreatment (ILs pretreatment), 145
IR. *See* Insect resistance (IR)
Irradiation, 135–136
Issatchenkia orientalis (*I. orientalis*), 250–251
Istituto Italiano di Tecnologia in Genova, 58
Itaconic acid, 241–242

K

Ketone–alcohol oil (KA oil), 242–243
KIC InnoEnergy Germany, 169
Kjeldahl nitrogen content (Kj–N content), 116

L

Laccases, 222–223, 225–226
Landfill Directive, 71
Laundry applications, 283–284
LCA. *See* Life cycle analysis (LCA)
LCC. *See* Life cycle costing (LCC)
LCI analysis. *See* Life cycle inventory analysis (LCI analysis)
LCIA. *See* Life cycle impact assessment phase (LCIA)
LFTL. *See* Light Fraction FT Liquids (LFTL)
LHW. *See* Liquid hot water (LHW)
Libero (*Oryza sativa indica*), 332
Life cycle analysis (LCA), 56, 168, 292
  phases and interactions, 292f
Life cycle assessment. *See* Life cycle analysis (LCA)
Life cycle costing (LCC), 294
Life cycle impact assessment phase (LCIA), 292
Life cycle inventory analysis (LCI analysis), 292
Light Fraction FT Liquids (LFTL), 170
Lightweight structural composites, 56
Lignin, 6, 49, 125, 130–131, 222–223
  modification to improving cellulose accessibility, 229
  optimizing biofuel production
    cell wall recalcitrance, 227–228
    hemicellulose and pectin modification, 229–230
    modification of cell wall structure, 228
    plants cell wall as source of energy, 226–227
Lignin-specific peroxidase (prx), 164–165
Lignocellulosic biomass, 55, 125
  physical and chemical characteristic of, 126–131
    cellulose, 127–128
    hemicellulose, 128–130, 129f
    lignin, 130–131
  effect of pretreatments methods on composition, 132t
Lignocellulosic residues, 169
Lime pretreatment, 139, 148
Linkage mapping (LM), 165
Liposomes, 282
Liquid hot water (LHW), 126
  pretreatment, 133–134
Living modified organisms (LMOs), 310–311
LM. *See* Linkage mapping (LM)
LMOs. *See* Living modified organisms (LMOs)
Loading rate, 99–100
Logistic requirements of biomass, 6–7
Long terminal repeat sequences (LTR sequences), 330
Low shear granulation, 278
LTR sequences. *See* Long terminal repeat sequences (LTR sequences)
Lyocell process, 59–60

M

M-QTL. *See* Main quantitative trait loci (M-QTL)
MAE. *See* Microwave-assisted extraction (MAE)
Main quantitative trait loci (M-QTL), 165
Maize, 91
Male sterility (MS), 314
Maleic acid, 239, 243
Mangiferin, 40
Mango, 39–40
Mango by-products, 53
Mannans, 224
Manure, 91
Markets and applications
  dicarboxylic acids current uses and production, 239–243
  drivers for biobased chemicals, 237–239
MASBI. *See* Midwest Aviation Sustainable Biofuels Initiative (MASBI)
Material recovery, 289
Maturation phase. *See* Slow-rate decomposition phase
Mature compost, 83
MCC. *See* Microcrystalline cellulose (MCC)
Mechanical comminution, 131–132
Mechanical pretreatment, 147–148
Mechanistic models, 209
Medical and pharmaceutical applications, 283
Medium Fraction FT Liquids (MFTL), 170
Membrane separation, 104
Mesophilic digestion, 95
  SWOT analysis of, 109–110
Mesophilic incubation phase, 80–81
Metabolic engineering strategies
  for biotechnological production of SA
    I. orientalis, 250–251
    Saccharomyces cerevisiae, 248–250
    Y. lipolytica, 250
  improving succinate production, 249t

Methanogenesis, 93
Mexico, 27
  physicochemical composition of fruits production, 35–39
MFC. *See* Microfibrillar cellulose (MFC)
MFTL. *See* Medium Fraction FT Liquids (MFTL)
Microbial communities, 92–93
Microbial enzymes, 271–272
  markets for, 272
Microcrystalline cellulose (MCC), 283–284
Microfibrillar cellulose (MFC), 55–56
Microorganisms
  producing succinic acid, 245
  SmF, 200
  SSF, 210–211
Microwave irradiation, 136
Microwave-assisted extraction (MAE), 12, 14
Midpoint methods, 293
Midwest Aviation Sustainable Biofuels Initiative (MASBI), 175
Mild pyrolysis, 134
Military certification constraints, 178
Milling, 131–132
Milox process, 141
miRNA sponges, 327
Miscanthus, 55
Mixed-linkage glucans (MLG), 221
Mixing methods, 97–99
MLG. *See* Mixed-linkage glucans (MLG)
Moisture content, 79–80, 95–96
Mono-digestion, SWOT analysis of, 113
Monolignols, 222–223
MS. *See* Male sterility (MS)
MSWs. *See* Municipal solid wastes (MSWs)
Multistage digestion, SWOT analysis of, 112–113
Multistage drying, 279
Municipal organic waste, 91
Municipal solid wastes (MSWs), 170, 298–299
  feedstock, 172
  prospects of transformation of, 171f
Municipal wastewater treatment plant facilities, 75
Mutagenesis, 329
  chemical mutagenesis, 331–333
  insertional mutagenesis, 330–331
  by physical agents, 333
MutMap, 331

N

Nanogels, 282
Nanoparticles, 282
Naphtha, 170
National Ecology Institute (INE), 77
National Pollutant Discharge Elimination System (NPDES), 75–76
Natural gas substitute production, 106
NBGT. *See* Nuclease-based gene targeting (NBGT)
nDart1. *See* nonautonomous DNA-based active rice transposon1 (nDart1)
Nematode resistance, selection of tobacco varieties for, 166–167
Nettle, 59
New plant products (NPPs), 318–319
Next-generation sequencing (NGS), 320
NGS. *See* Next-generation sequencing (NGS)
Nicotiana genus, 161
Nitric acid, 137
Nitrogen (N_2), 104, 134
NMPs. *See* Nutrient Management Plans (NMPs)
non-GM plants. *See* Nongenetically modified plants (non-GM plants)
Non-GMO methods for plants improvement, 329
nonautonomous DNA-based active rice transposon1 (nDart1), 330–331
Nongenetically modified plants (non-GM plants), 309. *See also* Genetically modified plants (GM plants)
Nonrenewable fuels, biodiesel from, 302
Nonrenewable resources, 291
Nonrenewable sources, heat and electricity from, 303–304
North American Free Trade Agreement, 76–77
Novel plant products (NPPs), 317–318
  recombinant DNA
    stable integration, 319–322
    stable introduction, 318–319
    transient introduction, 318
NPDES. *See* National Pollutant Discharge Elimination System (NPDES)
NPPs. *See* New plant products (NPPs); Novel plant products (NPPs)
Nuclease-based gene targeting (NBGT), 322
Nutrient Management Plans (NMPs), 75–76

O

O-Methyltransferase, 222–223
ODM. *See* Oligonucleotidedirected mutagenesis (ODM)
Odor control, 107
Oil cake, 162, 183
Oil producing plants, 163
Oligonucleotidedirected mutagenesis (ODM), 309
Olive mill pomace, 10
Operating mode, 96–97
Operating temperature, 95
OR. *See* Origins of replication (OR)
Orange, 57
Orange Fiber Company, 57
Organic fractions, 79
Organic loading rate, 99
Organic recycling of agricultural waste today, 69–70
  AD, 90–121
  Europe, 70–73

Organic recycling of agricultural waste today (*Continued*)
industrial composting, 78–90
South, Central, and North America, 73–78
Organic waste, 287
sustainability, 290–296
waste hierarchy, 288–289, 289f
Organic waste valorization pathways, 288–289
environmental sustainability, 296
energy recovery and recycling, 299–302
energy recovery by conversion to biodiesel, 302
energy recovery by incineration, 303–304
recycling by composting, 298–299
recycling by conversion to higher value chemicals, 297–298
reuse as feed, 296–297
Organosolv pretreatment, 139–140
Origins of replication (OR), 317
Oryza sativa indica. *See* Libero (*Oryza sativa indica*)
Oxidation, 283
Oxidative pathway of glyoxylate cycle, 247–248
Oxygen (O_2), 104, 199
Ozonolysis, 140–141

P

p-coumaryl-alcohol, 130, 130f
Packed bed column reactor, 203, 204f
Paddle mixers, 98
PAMP. *See* Pathogen associated molecular pattern (PAMP)
Pathogen, selection of tobacco varieties of, 166–167
Pathogen associated molecular pattern (PAMP), 315
PBS. *See* Polybutylene succinate (PBS)
Pdc–. *See* Pyruvate decarboxylase-negative (Pdc–)
pds. *See* Phytoene desaturase (pds)
Pectin methylesterases (PMEs), 222
inhibitors, 223–224
Pectinases, 223, 225–226
Pectins, 9, 222–224
modification to overcoming cell wall recalcitrance, 229–230
PEF pretreatment. *See* Pulsed electric field pretreatment (PEF pretreatment)
PEG. *See* Polyethylene glycol (PEG)
Pentoses, 65
Percolation, 99
Peroxidases, 222–223, 225–226
Peroxide pretreatment, 140
Peroxyformic acid pretreatment, 141
PH. *See* Plant height (PH)
pH value, SSF, 210
PHB. *See* Polyhydroxybutyrate (PHB)
Phenolic compounds content, 33–34, 34t
Phenylalanine ammonia lyase, 222–223
Phosphoric acid, 137
Php gene, 166
Physical agents, mutagenesis by, 333
Physical cross-linking method, 276
Physical pretreatments of biomass, 131–136. *See also* Biological pretreatment of biomass; Chemical pretreatments of biomass
extrusion pretreatment, 133
freeze pretreatment, 135
irradiation, 135–136
LHW pretreatment, 133–134
mechanical comminution, 131–132
PEF pretreatment, 136
pyrolysis, 134–135
torrefaction, 134–135
USP, 136
Physicochemical composition of fruits. *See also* Biological activities of fruits from Costa Rica and Central America region
of Costa Rica, 30–35, 31t
antioxidant activity, 34–35, 35t
phenolic compounds content, 33–34, 34t
vitamin C, 32–33, 32t
produced in Mexico, 35–39
coffee, 35–37, 36t
Grapefruit (*Citrus paradisi*), 38–39
pomegranate (*Punica granatum*), 37–38, 38t
Physicochemical pretreatments of biomass, 141–145
AFEX, 142–143
ammonia pretreatment, 142–143
CDE pretreatment, 143–144
ILs pretreatment, 145
steam explosion, 141–142
WOP, 144
Phytoene desaturase (pds), 323–324
Phytophthora nicotianae (Pn). *See* Black shank
Pineapple (*Ananas comosus*), 23–24, 32
biological activities of fruits, 40
PLA. *See* Polylactic acid (PLA)
Plant biomass as feedstock, 3–15
by-product calendar of subproducts, 5f
current applications, 8–9
feedstock for fermentation industry, 14–15
logistic requirements, 6–7
temporal availability and compositions, 5–6
value-added compounds, resource, 9–14, 12t
vegetable waste amount in 2012, 4t
Plant cell walls, 219. *See also* Cell wall modification/deconstruction
dynamic interface, 219–220
organization and biosynthesis of wall components
cellulose, 220–221
hemicellulose, 221
lignin, 222–223
pectins, 222
primary cell walls, 220
Plant height (PH), 165

Plant(s)
- biomass, 227
- cell wall as source of energy, 226–227
- hormones, 165
- improvement techniques, 327–328
- market making in bioeconomy of plants by-products, 53–65
  - bioplastics, novamont case study on, 64–65
  - Cargill starches, 63
  - cellulose, 55–56
  - EU business analysis of case studies, 60–62
  - EU textile industries and vegetal fibers, 57
  - feedstock biomasses, 55
  - innovative companies, 57–60
  - Tomatoes Plant Waste Recycling Industrial Group, 63–64
  - Wheatoleo, 65
- residues recycling, 57–60

PMEs. *See* Pectin methylesterases (PMEs)
Pneumatic mixing, 98–99
PNRSV-hpRNA. *See Prunus necrotic ringspot virus*-hpRNA (PNRSV-hpRNA)
Polybutylene succinate (PBS), 239
Polyethylene glycol (PEG), 281
Polygalacturonases (endo-PGs), 223–224
Polyhydroxybutyrate (PHB), 297
- agricultural waste from, 298

Polylactic acid (PLA), 297
Polyphenol oxidase (PPO), 316–317
Polysaccharides optimizing biofuel production
- cell wall recalcitrance, 227–228
- hemicellulose and pectin modification, 229–230
- lignin modification to improving cellulose accessibility, 229
- modification of cell wall structure, 228
- plants cell wall as source of energy, 226–227

Pomegranate (*Punica granatum*), 37–38, 38t
Poplar, 55
Potato peels, 10
PPO. *See* Polyphenol oxidase (PPO)
PQ. *See* Product quality (PQ)
Precipitation, 258–259
Preliminary phase. *See* High-rate decomposition phase
Pressure swing adsorption (PSA), 103–104
Pressurized solvent extraction (PSE), 12–13
Process control
- SmF, 196–199
  - biomass and growth kinetics/time and yield, 198–199, 198f
  - temperature and heat transfer, 198
- SSF, 206–207
  - biomass and growth kinetics, 208–210
  - pH value, 210
  - substrate moisture and water activity, 207–208
  - temperature and heat transfer, 208

Process-based regulatory frameworks, 335
Product quality (PQ), 314
- traits, 316–317

Product-based regulatory frameworks, 335
PROFEPA. *See* Federal Attorney for Environmental Protection (PROFEPA)
Propeller mixers, 98
Proven technology, 107
Prunus necrotic ringspot virus-hpRNA (PNRSV-hpRNA), 327
prx. *See* Lignin-specific peroxidase (prx)
PSA. *See* Pressure swing adsorption (PSA)
PSE. *See* Pressurized solvent extraction (PSE)
Pulsed electric field extraction, 14
Pulsed electric field pretreatment (PEF pretreatment), 136
Punica granatum. *See* Pomegranate (*Punica granatum*)
Pyrolysis, 134–135
Pyruvate decarboxylase-negative (Pdc–), 248–249

Q

QTL qPH-12, 165
Quantitative trait loci (QTL), 165

R

Raimbault column. *See* Packed bed column reactor
Rapid Trait Development System (RTDS), 319
RB. *See* Reverse breeding (RB)
RCRA. *See* Resource Conservation and Recovery Act (RCRA)
RdDM. *See* RNA-dependent DNA methylation (RdDM)
RDF. *See* Refuse-derived fuel (RDF)
Recirculation, 88
Recombinant DNA
- stable integration, 319–322
- stable introduction, 318–319
- transient introduction, 318

RED. *See* Renewable Energy Directive (RED)
REDD. *See* Reducing emissions from deforestation and forest degradation (REDD)
Reducing emissions from deforestation and forest degradation (REDD), 173–174
Reductive branch of tricarboxylic acid cycle, 247
Refining, 88
Reformed standard NOM-083-SEMARNAT-2003, 77
Refuse-derived fuel (RDF), 100
Regulation harmonization, 333–338
Renewable Energy Directive (RED), 173, 175
Renewable fuels, biodiesel from, 302
Renewable raw materials, 238
Renewable resources, 290–291

Renewable sources, heat and electricity from, 303–304
Renewables Directive, 71–72
Renewcell, 58
Residence time, 99
Resource Conservation and Recovery Act (RCRA), 74–75
Resource Technology Strength, 175
Reuse as feed, 296–297
Reverse breeding (RB), 309
RG-I. *See* Rhamnogalacturonan I (RG-I)
RG-II. *See* Rhamnogalacturonan II (RG-II)
Rhamnogalacturonan I (RG-I), 222
Rhamnogalacturonan II (RG-II), 222
Rice, Clearfield Production System for, 332
Rice straw, 57
Ripe banana, 41
Ripe mangoes, 50
RNA interference (RNAi), 180–181, 316–317
RNA-dependent DNA methylation (RdDM), 318
RNAi. *See* RNA interference (RNAi)
Rootstock, 325
Rotary drum reactor, 204–205, 206f
Rottegrad, 81
Roundtable and sustainable materials (RSBs), 184
Row composting, 87
RSBs. *See* Roundtable and sustainable materials (RSBs)
RTDS. *See* Rapid Trait Development System (RTDS)

S

S-LCA. *See* Social life cycle analysis (S-LCA)
SA biotechnological production
  metabolic engineering strategies
    I. orientalis, 250–251
    S. cerevisiae, 248–250
    Y. lipolytica, 250
SAA. *See* Soaking aqueous ammonia (SAA)
Saccharification, 227–228
  modification of cell wall structure, 228
Saccharomyces cerevisiae (*S. cerevisiae*), 248–250
  succinate pathways in, 246f
SAGARPA. *See Secretaría de Agricultura, Ganadería, Desarrollo Rural, Pesca y Alimentación* (SAGARPA)
SAR. *See* Systemic acquired resistance (SAR)
SAS. *See* South African Airways (SAS)
SCAR. *See* Standing Committee on Agricultural Research (SCAR)
SCG. *See* Spent coffee grounds (SCG)
SDNs. *See* Site-directed nucleases (SDNs)
se4h. *See* Cinnamate 4-hydroxylase (C4H)
SeaCell, 59
Second-generation cellulosic ethanol, 163
"Second-generation" feedstocks lignocellulose, 163
Secretaría de Agricultura, Ganadería, Desarrollo Rural, Pesca y Alimentación (SAGARPA), 27
Seed cake, 183
Seed production technology (SPT), 309
Senesco proprietary technology, 317
SFE. *See* Supercritical fluid extraction (SFE)
sgRNA. *See* Single-guide RNA (sgRNA)
Sieving, 88
Siloxanes, 104–105
Sinapyl alcohol, 130, 130f, 222–223
Single-guide RNA (sgRNA), 323–324
Single-stage digestion, SWOT analysis of, 112
SIP. *See* Synthesized isoparaffin (SIP)
siRNAs. *See* Small interfering RNAs (siRNAs)
Site-directed nucleases (SDNs), 320
SKA. *See* Synthetic paraffinic kerosene with aromatics (SKA)
SkyNRG, 182
Slow-rate decomposition phase, 80–81
Sludge, 90
Small interfering RNAs (siRNAs), 327
SmF. *See* Submerged fermentation (SmF)
Soaking aqueous ammonia (SAA), 142–143
Social and economic aspects, 310–311
Social life cycle analysis (S-LCA), 292
Social sustainability assessment, 296
SOD. *See* Superoxide dismutase (SOD)
Solar energy, 126–127
Solaris, 178, 181–183
Solaris plants
  from G2M flow, 183f
  TSO from, 184f
Solaris-based jet fuel, 184
Solid-state fermentation (SSF), 189, 202. *See also* Submerged fermentation (SmF)
  biomolecule production, 213
  bioreactor types, 202–206
  biotechnological potential, 202
  general considerations, 213–215, 213–214
  microorganisms, 210–211
  process control, 206–210
  substrates, 211–212
Solid–liquid extraction, 13
Sonication, 136
South, Central, and North America, 73–78
  agricultural facilities, 75–76
  integration
    with existing compost facilities, 76
    with landfills or refuse-derived fuel facilities, 76

municipal wastewater treatment plant facilities, 75
stand-alone facilities, 76–78
South African Airways (SAA). *See* South African Airways (SAS)
South African Airways (SAS), 176, 182–183
Soxhlet extraction, 13
Soya, 59
Spent coffee grounds (SCG), 35
SPK, 170
Spray drying, 277–278, 278f
agglomerating process, 280f
and fluid bed, 280f
and fluid bed agglomeration combination, 279
SPT. *See* Seed production technology (SPT)
SSF. *See* Solid-state fermentation (SSF)
SSN-1 techniques, 319
SSN-2 techniques, 319
Stabilization phase. *See* Slow-rate decomposition phase
Stable compost, 83
Stand-alone facilities, 76–78
Standardization, 61
Standing Committee on Agricultural Research (SCAR), 54
Steady-state production, 190
Steam explosion, 141–142, 149
Sterols, 40
Stoichiometric models, 209–210
Stomoxys calcitrans fly, 29
Strategic Working Group "Sustainable Bio-Resources for Growing Bioeconomy" (SWG SBGB), 54
Strengths, Weaknesses, Opportunities, Threats analysis (SWOT analysis), 88–89
for anaerobic digestion, 106–109
of batch digestion, 114–115
of codigestion, 114
of continuously fed, completely mixed digestion, 115
of continuously fed, plug flow digestion, 116
of dry digestion, 111–112
industrial composting, 88–89
air emissions during process, 89
difficulties in marketing, 89
low investment cost, 88
no energy production, 89
price of compost, 89
production of compost, 88
reduction of biowaste going to landfills, 89
replacement by end-of-life systems with energy recuperation, 89
simplicity, 88
of mesophilic digestion, 109–110
of mono-digestion, 113
of multistage digestion, 112–113
of single-stage digestion, 112
of thermophilic digestion, 110
of wet digestion, 110–111
Submerged fermentation (SmF), 189–202. *See also* Solid-state fermentation (SSF)
advantages and application, 201–202
biomolecules production, 201–202
bioreactor types, 191–196
biotechnological potential, 190–191
general considerations, 213–215, 213–214
microorganisms, 200
process control, 196–199
substrates, 200–201
Substrates
moisture of SSF, 207–208
SmF, 200
agroindustrial by-products transformation, 201
catabolic response, 200–201
SSF, 211
agroindustrial by-products transformation, 212
catabolite repression, 212
Succinate formation pathways, 245
oxidative pathway of glyoxylate cycle, 247–248
oxidative tricarboxylic acid cycle, 245–247
pathways in *Saccharomyces cerevisiae*, 246f
reductive branch of tricarboxylic acid cycle, 247
Succinic acid, 239–240, 254
agricultural waste from, 298
microorganisms producing, 245
Succinic anhydride, 239, 242
Sulfuric acid, 137
Supercritical fluid extraction (SFE), 12–13
Superoxide dismutase (SOD), 273
Sustainability
assessment, 291
economic sustainability assessment, 294–295
environmental sustainability assessment, 292–294
impact assessment methods, 293t–294t
life cycle phases of product, 291f
social sustainability assessment, 296
natural resource depletion, 290–291
sustainable development, 290
Sustainable biomass feedstock, 163
Sustainable development, 290
Swedish Biofuels AB, 169
SWG SBGB. *See* Strategic Working Group "Sustainable Bio-Resources for Growing Bioeconomy"(SWG SBGB)
Switchgrass, 55
SWOT analysis. *See* Strengths, Weaknesses, Opportunities, Threats analysis (SWOT analysis)
Synthesis gas, 227
Synthesized isoparaffin (SIP), 170
Synthetic fuels, 227
Synthetic paraffinic kerosene with aromatics (SKA), 170
Systemic acquired resistance (SAR), 167

T

T-DNA. *See* Transfer DNA (T-DNA)
Table composting, 85
Tackling regulatory barriers, 61

TALEN, gene modification using, 322–324
TALENs. *See* Transcription activator-like effector nucleases (TALENs)
Targeting-Induced Local Lesions in Genomes (TILLING), 320, 330
TCA cycle. *See* Tricarboxylic acid cycle (TCA cycle)
TCN. *See* Tobacco cyst nematode (TCN)
Technical enzymes, 272
Temperature
 SmF, 198
 SSF, 208
Terminator technology, 314
Tetrahydrofuran (THF), 239
Textifood exhibition, 58
Textiles, 57–60, 271–272
ThermoChem Recovery International, 172
Thermochemical conversion routes, 227
Thermophilic digestion, SWOT analysis of, 110
Thermophilic phase, 80–81
THF. *See* Tetrahydrofuran (THF)
TILLING. *See* Targeting-Induced Local Lesions in Genomes (TILLING)
Tobacco, 161
Tobacco cyst nematode (TCN), 167
Tobacco mutagenesis and genetic studies, 164–166
Tobacco seed oil (TSO), 162
 bioeconomics for aircraft fuel, 178–181
 bioeconomy of biofuels, 167–173
  production of biofuels and chemicals, 172f
  prospects of transformation of MSWs, 171f
 biokerosene deployment and sustainability issues, 173–178
 characteristics, 181t
 feedstock biomasses, 163–164, 164f
 selection of tobacco varieties, 166–167
 from Solaris plants, 184f
 tobacco mutagenesis and genetic studies, 164–166
Tomatoes, 11
Tomatoes Plant Waste Recycling Industrial Group, 63–64
Torrefaction, 134–135
Total phenolic content (TPC), 33
Toxic Substances Control Act, 310
TPC. *See* Total phenolic content (TPC)
TRANSBIO project, 89–90, 298
 AD within, 116–121
 industrial composting within, 89–90
Transcription activator-like effector nucleases (TALENs), 309
Transfer DNA (T-DNA), 328
Transgenic cherry rootstocks (TRs), 327
Tray bioreactors, 203–204, 205f
Tricarboxylic acid cycle (TCA cycle), 245
 reductive branch, 247
Trickle bed reactor, 196, 197f
Tropical fruits, 22
TRs. *See* Transgenic cherry rootstocks (TRs)
TSO. *See* Tobacco seed oil (TSO)
Tunnel composting, 87

U

UA. *See* United Airlines (UA)
UDP-glucuronic acid decarboxylase (uxs), 164–165
UE. *See* Ultrasound-assisted extraction (UE)
Ultrasonic energy, 136
Ultrasonic pretreatment (USP), 136
Ultrasound-assisted extraction (UE), 12–13
United Airlines (UA), 173
United Fruit Company, 24–25
US Department of Agriculture (USDA), 314, 334
US National Bioeconomy Blueprint, 175
USDA. *See* US Department of Agriculture (USDA)
USP. *See* Ultrasonic pretreatment (USP)
uxs. *See* UDP-glucuronic acid decarboxylase (uxs)

V

Vacuum pressure swing adsorption (VPSA), 103–104
Value-added compounds, resource for, 9–14
VBSs. *See* Vector backbone sequences (VBSs)
Vector backbone sequences (VBSs), 317
Vegetable, garden, and fruit waste (VGF waste), 91
Vegetable oils, 162
Vegetal cashmere, 59
Vegetal fibers, 57
Vegetal leather, 50, 57
Verdezyne, 242–243
VGF waste. *See* Vegetable, garden, and fruit waste (VGF waste)
Vibratory ball milling, 131–132
Viral diseases, 315
Virus resistance (VR), 315
Vitamin C, 32–33, 32t
VPSA. *See* Vacuum pressure swing adsorption (VPSA)
VR. *See* Virus resistance (VR)

W

Waste Framework Directive, 70–71
Waste hierarchy, 288–289, 289f
Waste processing facilities, 75
Wastewater, 90
Wastewater treatment plants (WWTP), 75
Water activity of SSF, 207–208
Water removal, 101
Watermelon, 41
Wet digestion, SWOT analysis of, 110–111
Wet oxidation pretreatment (WOP), 144, 148
WGS-based method. *See* Whole genome sequencing-based method (WGS-based method)

Wheat, Clearfield Production System for, 332–333
Wheatoleo (French company), 65
Whole genome sequencing-based method (WGS-based method), 331
Willow, 55
Windrow and static pile systems, 81
Windrow composting, 83–85, 85f
Winter wheat, 332–333
Wood trees, 55
Woody plants accelerated transformation, 328–329
WOP. *See* Wet oxidation pretreatment (WOP)
WWTP. *See* Wastewater treatment plants (WWTP)

X

XyG transglycosylase/hydrolase (XTH), 224
Xylans, 222
Xyloglucanases, 223
Xyloglucans (XyG), 221–222
α-Xylosidases, 224

Y

Yarrowia lipolytica (*Y. lipolytica*), 250
Yeasts, 210–211
  as platform organisms for dicarboxylic acid production, 244
  strains selection and improvement, 243
  metabolic engineering strategies for biotechnological production of SA, 248–251
  microorganisms producing succinic acid, 245
  succinate formation pathways, 245–248
Yield of SmF, 198–199

Z

Zinc-finger nucleases (ZFNs), 309
gene modification using, 322–324

FSC
www.fsc.org
MIX
Papier | Fördert gute Waldnutzung
FSC® C083411

Kolibri 360 GmbH
Ferdinand-Jühlke-Straße 7
99095 Erfurt, Deutschland
produktsicherheit@kolibri360.de